High Energy e^+e^- Interactions

(Vanderbilt, 1980)

AIP Conference Proceedings
Series Editor: Hugh C. Wolfe
Number 62
Particles and Fields Subseries No. 20

High Energy e^+e^- Interactions

(Vanderbilt, 1980)

Editors
R.S. Panvini and S.E. Csorna
Vanderbilt University

American Institute of Physics
New York 1980

L.C. Catalog Card No. 80-53377
ISBN 0-88318-161-4
DOE CONF- 800588

FOREWORD

This was the fourth in a series of international high energy physics conferences held at Vanderbilt University since 1973. Unlike the previous ones which covered a wide spectrum of topics in high energy physics, the latest meeting dealt only with the topic of high energy e^+e^- interactions. This conference was open to anyone who wished to attend, as were the first three.

The title of this conference was the International Symposium on High Energy e^+e^- Interactions, Vanderbilt (1980). To cover the field, we had reports on the latest data from Cornell, DESY, and SLAC. In addition, six talks were devoted to various aspects of theory, three were devoted to plans for new accelerators, and one was an overview of instrumentation for e^+e^- experiments.

It would not have been possible to plan the meeting without the expert advice of several individuals. K. Berkelman (Cornell), D. Coyne (Princeton), E. Lohrmann (DESY), gave invaluable advice and assistance in organizing the variety of talks on experiments. E. Berger (Argonne) together with J. D. Bjorken (Fermilab) advised us on the theory talks and A. Odian (SLAC) helped us decide on what might be included in terms of instrumentation.

The major credit for the hotel arrangements, the coffee breaks, registration, banquet, etc. goes to my wife, Doria, who has an extraordinary talent for organization combined with a real enthusiasm for helping people and for making things work.

Financial support for this 1980 Vanderbilt Symposium was provided by the National Science Foundation, the Department of Energy, and by the Vanderbilt University Research Council.

R. S. Panvini
Conference Chairman
July, 1980

TABLE OF CONTENTS

<u>Instrumentation</u>

QED TESTS AND NEUTRAL-CURRENT STRUCTURE

J. J. Sakurai
University of California, Los Angeles, Ca. 90024

ABSTRACT

Recent PETRA experiments to test QED are shown to provide non-trivial information on the structure of weak neutral-current interactions.

I would like to present a short report on the relevance of QED tests to the structure of neutral currents. Some of the new things I am going to say are based on work done in collaboration with a student of mine, Nigel Wright.[1,2]

Since the early days of QED there have been many attempts to test its validity, usually by comparing data with some ad hoc modifications of QED. Already in the '40s, proposals were made to modify the photon propagator by introducing a convergence factor as follows:[3]

$$\frac{1}{s} \to \frac{1}{s}\left[\Lambda_+^2/(\Lambda_+^2 - s)\right] = \frac{1}{s} - \frac{1}{s-\Lambda_+^2} . \tag{1}$$

In this kind of approach, convergence--i.e. finite QED--is achieved at the expense of negative metric; put in another way, we are introducing a ghost of mass Λ_+ coupled with a coupling constant whose square is negative. Of course, as $\Lambda_+ \to \infty$, the conventional QED is recovered; so limits on Λ_+ quantitatively characterize the validity of QED. For example, we may attempt to "measure" Λ_+, or set a lower limit to Λ_+, in electron-positron annihilation into muon pairs

$$e^+ + e^- \to \mu^+ + \mu^- , \tag{2}$$

which can be done by comparing the observed cross section to the QED prediction:

$$\sigma_{obs}/\sigma_{QED} = [\Lambda_+^2/(\Lambda_+^2 - s)]^2 . \tag{3}$$

It is also possible to give a somewhat different interpretation to the "cut-off" parameter Λ_+ appearing in (3). Let us suppose, for definiteness, that in reaction (2) the electron and the virtual photon are normal, i.e. pointlike, while the muon turned out to have "structure." The factor $\Lambda_+^2/(\Lambda_+^2 - s)$ appearing in (3) can then be viewed as the Dirac form factor for the muon; the muon charge radius would then be given by

ISSN:0094-243X/80/620001-13$1.50

$$\langle r^2 \rangle_{muon} = 6/\Lambda_+^2 . \tag{4}$$

In contrast to (1), it has also been tried to parametrize possible QED violation by considering

$$\frac{1}{s} \to \frac{1}{s} + \frac{1}{s-\Lambda_-^2} . \tag{5}$$

Notice that this does not help convergence. There is, of course, no deep reason for choosing either sign for the residue of the heavy particle pole. I'll turn to the particle interpretation of the second term of (5) in a moment.

The first serious attempt to deduce limits on $\Lambda_\pm$ from experimental data was made by Drell in 1958.[4] The fact that the calculated value of the Lamb shift agreed with observation within 0.2 MHz was used to infer that $1/\Lambda_\pm$ cannot be larger than 0.5×10^{-13} cm, or

$$\Lambda_\pm > 0.4 \text{ GeV} . \tag{6}$$

Some of the senior members of the audience may recall that the major reason for building e^-e^- and e^+e^- colliding rings was to test QED at energies considered to be astronomically high, especially when translated into equivalent laboratory energies, e.g. $\sqrt{s} \simeq 1$ GeV $\Longrightarrow E_{lab} \simeq 1$ TeV. In the late '60s Møller scattering studied at the Stanford e^-e^- colliding ring with $\sqrt{s} \lesssim 1.1$ GeV gave limits[5]

$$\Lambda_- > 2.4 \text{ GeV} , \quad \Lambda_+ > 4.0 \text{ GeV} . \tag{7}$$

Around that time similar limits were quoted for Bhabha scattering at Orsay ACO.[6] The early '70s were the glorious days of Frascati[7] (Adone) and CEA (Bypass);[8] typical limits obtained were

$$\Lambda_+ > 10.6 \text{ GeV} , \quad \Lambda_- > 10.0 \text{ GeV} \tag{8}$$

for Bhabha scattering at $\sqrt{s} \simeq 5$ GeV. Then SPEAR came into operation, which is responsible for the best pre-PETRA limits[9]

$$\Lambda_+ > 38 \text{ GeV} , \quad \Lambda_- > 39 \text{ GeV} . \tag{9}$$

With PETRA breaking through the $s = 1000$ GeV^2 barrier dramatic advances have been made in the continuing race to obtain the best limits on $\Lambda_\pm$. Using electron-positron annihilation into muon pairs, the Mark J Collaboration has reported[10]

$$\Lambda_- > 160 \text{ GeV} , \quad \Lambda_+ > 120 \text{ GeV} , \tag{10}$$

while the PLUTO Collaboration,[11] using Bhabha scattering, has established

$$\Lambda_- > 230 \text{ GeV} , \quad \Lambda_+ > 80 \text{ GeV} . \tag{11}$$

At this Symposium similar limits are also being reported by the JADE[12] and the TASSO Collaboration.[13]

This looks great! A cut-off parameter limit of 200 GeV would mean that QED has been tested down to a distance of

$$\hbar c/200 \text{ GeV} \simeq 10^{-16} \text{ cm} , \tag{12}$$

which represents an improvement by a factor of ~500 since Drell's 1958 paper. But before we start celebrating, let us note that this kind of analysis presupposes that the electron and the muon have no interactions other than pure QED. In reality the electron and the muon are known to have weak neutral-current interactions. Currently popular gauge models of electroweak interactions all have one (or more) neutral Z boson of mass $m_Z \sim 100$ GeV with dimensional coupling strength $g \sim e$. The Λ_- parametrization (5) would correspond to, in addition to conventional photon exchange, the exchange of a neutral Z boson of mass and coupling strength

$$\begin{aligned} m_Z &= \Lambda_- , \\ g_{Z\bar{e}e} &= g_{Z\bar{\mu}\mu} = e . \end{aligned} \tag{13}$$

From this point of view, the so-called "QED tests" are actually tests of <u>electroweak</u> models at unprecedentedly high values of q^2 or s. Conversely, the fact that pure QED still holds at PETRA energies is seen to impose stringent constraints on electroweak models.[14]

In this talk I would like to concentrate on two "QED reactions" --annihilation into muon pairs and Bhabha scattering. With μe universality, the most general V, A four-fermion interaction appropriate for these reactions is given by

$$\begin{aligned} L = -(G/\sqrt{2})\,[h_{VV}(\bar{e}\gamma_\lambda e + \bar{\mu}\gamma_\lambda\mu)(\bar{e}\gamma_\lambda e + \bar{\mu}\gamma_\lambda\mu) \\ + 2h_{VA}(\bar{e}\gamma_\lambda e + \bar{\mu}\gamma_\lambda\mu)(\bar{e}\gamma_\lambda\gamma_5 e + \bar{\mu}\gamma_\lambda\gamma_5\mu) \\ + h_{AA}(\bar{e}\gamma_\lambda\gamma_5 e + \bar{\mu}\gamma_\lambda\gamma_5\mu)(\bar{e}\gamma_\lambda\gamma_5 e + \bar{\mu}\gamma_\lambda\gamma_5\mu)] . \end{aligned} \tag{14}$$

The standard SU(2) ⊗ U(1) model[15] predicts

$$h_{AA} = \frac{1}{4} , \quad h_{VV} = \frac{1}{4}\,(1-4\sin^2\theta_W)^2 , \quad h_{VA} = \frac{1}{4}\,(1-4\sin^2\theta_W) . \tag{15}$$

In any single Z model we must have the factorization relation

$$h_{VA}^2 = h_{VV}h_{AA} \ . \tag{16}$$

Let us start with annihilation into muon pairs. If forward-backward asymmetry in the angular distribution is not studied, only the h_{VV} term contributes to order G, and we can readily obtain for the fractional deviation from the pure QED prediction as follows:[16]

$$\begin{aligned}\Delta\sigma/\sigma_{QED} &\equiv (\sigma_{exp}-\sigma_{QED})/\sigma_{QED} \\ &= -(G/\sqrt{2}\,\pi\alpha)\, h_{VV}s \ .\end{aligned} \tag{17}$$

Comparing this with the fractional deviation based on an expansion of (1) and (5)

$$\Delta\sigma/\sigma_{QED} \simeq \pm 2s/\Lambda_{\pm}^2 \tag{18}$$

valid for $\Lambda_{\pm}^2 >> s$, we see that limits on Λ_- imply limits on h_{VV}, or within the context of the standard SU(2) ⊗ U(1) model, limits on $\sin^2\theta_W$. When we include finite Z mass corrections--practically negligible at $\sqrt{s} \simeq 30$ GeV if $m_Z \simeq 90$ GeV--and also terms quadratic in G, the formula becomes

$$\begin{aligned}\Delta\sigma/\sigma_{QED} &= -4(s/4\pi\alpha)[G(s,m_Z^2)/\sqrt{2}] \\ &\times \{h_{VV} - (s/4\pi\alpha)[G(s,m_Z^2)/\sqrt{2}](h_{VV}+h_{AA})^2\}\end{aligned} \tag{19}$$

where

$$G(s,m_Z^2) \equiv G/(1-s/m_Z^2) \xrightarrow{s<<m_Z^2} G \ . \tag{20}$$

In Fig. 1 we present the allowed and the forbidden region in a h_{VV}-h_{AA} coupling constant plane within the framework of a single Z boson model assuming that the QED violation parameter Λ_- is determined to be larger than 160 GeV at s = 1000 GeV^2.[1,17] We see that the standard model prediction with $\sin^2\theta_W = 0.23$ is comfortably within the allowed region. However, some once-popular models appear to be in difficulty. For example, the V-A SU(2) symmetric model[18] (whose predictions coincide with those of the standard SU(2) ⊗ U(1)

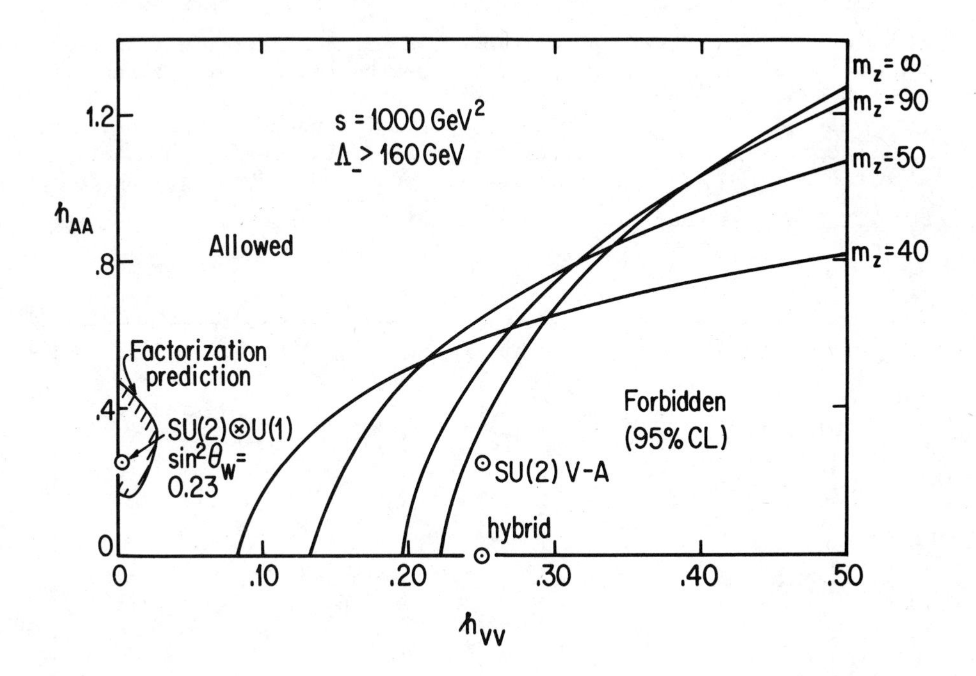

Fig. 1. Allowed region for h_{VV} and h_{AA}.

model with $\sin^2\theta_W = 0$) and the "hybrid model,"[19] a "non-standard" SU(2) ⊗ U(1) model with

$$\begin{pmatrix} \nu_e \\ e^- \end{pmatrix}_L \quad \text{and} \quad \begin{pmatrix} N^o \\ e^- \end{pmatrix}_R \tag{21}$$

to make the electron current purely vectorial, are both ruled out if $\Lambda_- > 160$ GeV at $s \simeq 1000$ GeV^2. As is well known, both models were already killed by other neutral-current experiments. However, it is sobering to contemplate that had the PETRA QED tests been available prior to the 1978 SLAC Massacre (SLAC-Yale ... asymmetry measurements in inelastic electron-deuteron scattering[20]) they would have made a very significant impact on our understanding of neutral-current structure.

In single Z models, studies of the $\nu q \to \nu q$, $\nu e \to \nu e$ and $eq \to eq$ interactions with small spacelike values of q^2, when supplemented by the assumption of μe universality, completely determine the neutral-current couplings involved in annihilation into muon pairs and Bhabha scattering for $s \ll m_Z^2$. This is because the Z couplings to the electron can be inferred by writing down various factorization relations.[21] The allowed range predicted on the basis of factorization and other neutral-current data is shown also in Fig. 1.

Instead of looking at fixed s, we may compare, as a function of s, the cross section predicted by the QED violation parametrization with the SU(2) ⊗ U(1) standard model prediction. This is done in Fig. 2.[1] We see that $\sin^2\theta_W$ as large as 0.5, or as small as 0, runs

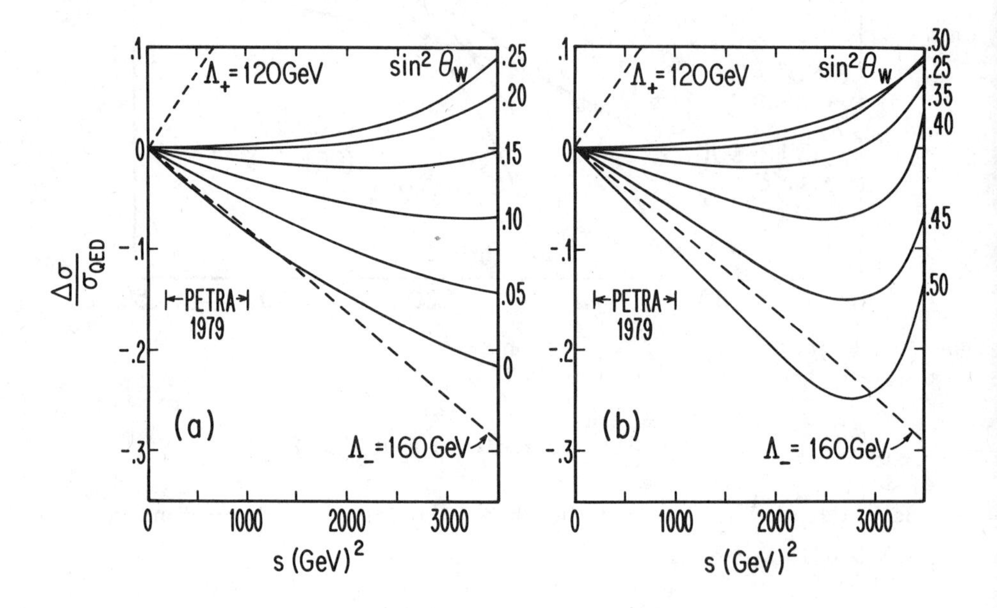

Fig. 2. Cross section for annihilation into muon pairs in the standard SU(2) ⊗ U(1) model.

into difficulty with $\Lambda_- > 160$ GeV at $s \simeq 1000$ GeV^2.

The existing limits on $\Lambda_\pm$ deduced from annihilation into muon pairs hardly constrain h_{AA}. This is not surprising because the $\Lambda_\pm$ determinations were based on the integrated cross section to which only h_{VV} contributes to order G; as is well known, the best way to isolate the axial-vector coupling parameter h_{AA} is to study forward-backward asymmetry:[16]

$$\langle A\rangle \equiv \left[\int_{\theta=0}^{\theta=\pi/2} (d\sigma/d\Omega)\,d\Omega - \int_{\theta=\pi/2}^{\theta=\pi} (d\sigma/d\Omega)\,d\Omega\right] \Bigg/ \int_{\theta=0}^{\theta=\pi} (d\sigma/d\Omega)\,d\Omega \; . \tag{22}$$

The standard model prediction for this quantity is -8% at s = 1000 GeV^2; if you can establish $|A| < 10\%$ at s = 1000 GeV^2, then you can say $h_{AA} < 0.3$ provided $m_Z \simeq 90$ GeV, to be compared with the standard model prediction $h_{AA} = 1/4$.

At this Symposium I became aware of various attempts to measure <A> at PETRA. Unfortunately the errors quoted are still quite large, typically ±10%, due to poor statistics. For example, if we use the JADE determination we obtain[12,17]

$$h_{AA} = 0.26 \pm 0.43 \tag{23}$$

with $s \ll m_Z^2$ assumed.

We may also take the point of view that the standard model predictions at <u>low</u> energies are now firmly established from other experiments and that the data should be analyzed within the context of single Z models (not necessarily gauge models[22]) that give the same low energy predictions; m_Z then is the only adjustable parameter. Fig. 3 shows the asymmetry predictions of such single Z models for

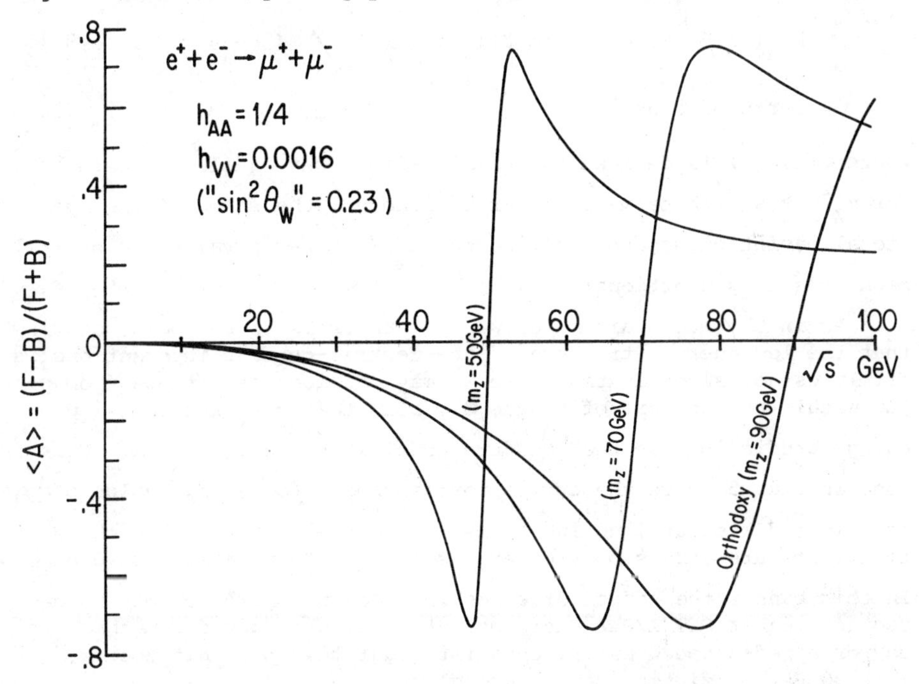

Fig. 3. Forward-backward asymmetry in annihilation into muon pairs in single Z models that give the same predictions at low s as the standard model.

various values of m_Z; the curves are drawn in such a way that the low energy predictions $h_{AA} = 1/4$ and $h_{VV} = 0.0016$ (corresponding to $\sin^2\theta_W = 0.23$) are not upset. To give an example, if future experiments establish $|A| < 10\%$ at $s = 1000\ \mathrm{GeV}^2$, then m_Z can be concluded to be greater than 55.5 GeV. Ultimately the energy dependence of A must be studied; in this way, a finite Z mass effect may be detected even before we reach $\sqrt{s} \simeq m_Z$.[23]

We now consider Bhabha scattering where γ and Z appear in both the t channel and the s channel. The fractional deviation from the pure QED prediction in the angular distribution for Bhabha scattering can be written as[24]

$$\Delta(d\sigma/d\Omega)/(d\sigma/d\Omega)_{QED}$$

$$= (s/\sqrt{2}\pi\alpha)\Big\{h_{VV}[(3+\cos^2\theta)/(1-\cos\theta)]\Big[\frac{1}{2}(3+\cos\theta)\,G(s,m_Z^2)$$

$$+\cos\theta\,G(t,m_Z^2)\Big] - h_{AA}\Big[\frac{1}{2}(7+4\cos\theta+\cos^2\theta)\,G(s,m_Z^2)\Big]$$

$$- h_{AA}[(1+3\cos^2\theta)/(1-\cos\theta)]\,G(t,m_Z^2)\Big\}\Big/[(3+\cos^2\theta)/(1-\cos\theta)]^2$$

$$+ \text{terms of order } G^2 \qquad (24)$$

where $G(t,m_Z^2)$ is defined as in (20) with t set equal to $-s\sin^2\theta/2$. For $m_Z^2 \gg s$ with terms of order G^2 ignored, the shape of the fractional deviation depends only on the h_{VV}/h_{AA} ratio while the overall magnitude is proportional to $(h_{VV}+h_{AA})s$. This is shown in Fig. 4.

As in the muon pair case, we may subscribe to the point of view that the low energy structure of the neutral-current interactions is firmly established; in that case we may consider deviations from pure QED within the context of single Z models that give the same low energy predictions as the standard model with $\sin^2\theta_W = 0.23$. This is done in Fig. 5 where the curves corresponding to various value of m_Z are plotted without ignoring terms of order G^2. Notice how the deviation from pure QED is particularly sensitive to m_Z at backward angles. In this manner the effect of a possible low mass Z boson may be detected even at PETRA/PEP energies. At least, it gives something to do for a red-blooded experimentalist eager to prove that the $m_Z \simeq 90$ GeV prediction may be wrong!

Also shown in Fig. 5 are two curves based on the $\Lambda_\pm$ parametrization. We note that there is no resemblance whatsoever between these curves and the electroweak curves. With limits on $\Lambda_\pm$ now in the 100-250 GeV range, it is no longer meaningful to use the $\Lambda_\pm$ parametriza-

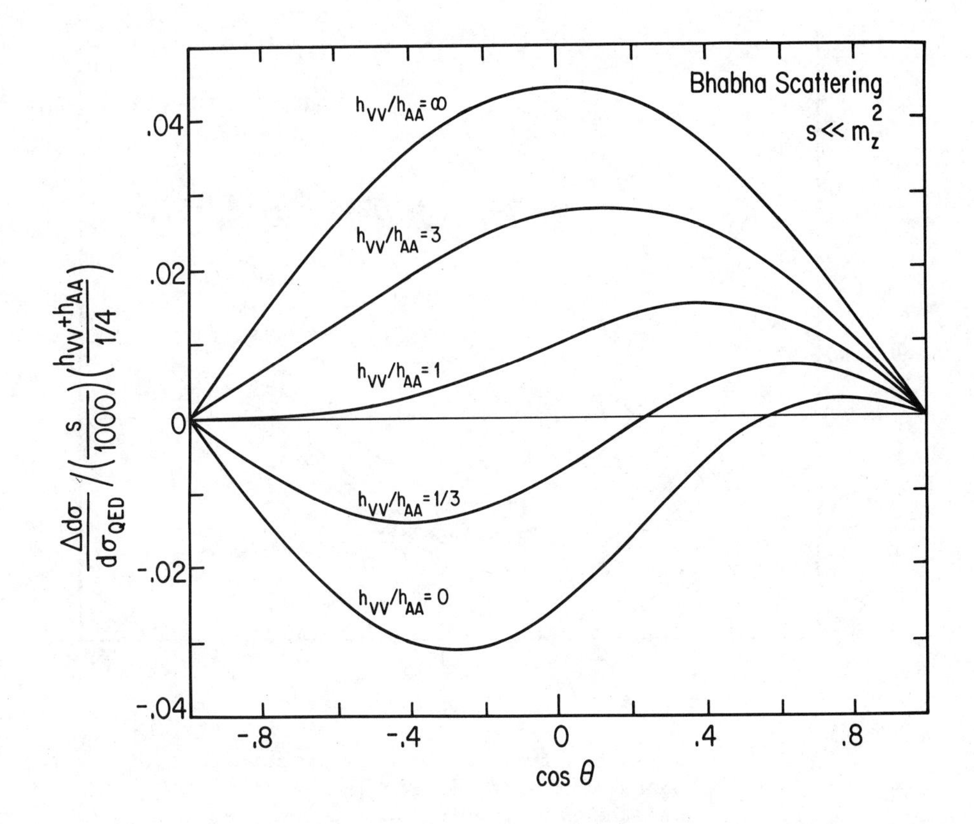

Fig. 4. Deviations from the pure QED predictions in Bhabha scattering for $s \ll m_Z^2$.

tion.

Finally I wish to make a few comments on two-Z boson models. The most popular two-Z boson model is a right-left symmetric model based on $SU(2)_L \otimes SU(2)_R \otimes U(1)$[25] where the right-left symmetry is broken by a Higgs sector arranged in such a way to give

$$m_{W_R}^2 \gg m_{W_L}^2 , \tag{25}$$

say, by a factor of 10 or more. Actually there are several variants along this line, and in the version that survived the SLAC Massacre, the predictions of the model are essentially identical to those of the standard $SU(2) \otimes U(1)$ model until we reach LEP energies. Neutral-current physics at PEP/PETRA energies is pretty much controlled by the first Z boson predicted to be around 90 GeV, just as in the standard model.[26]

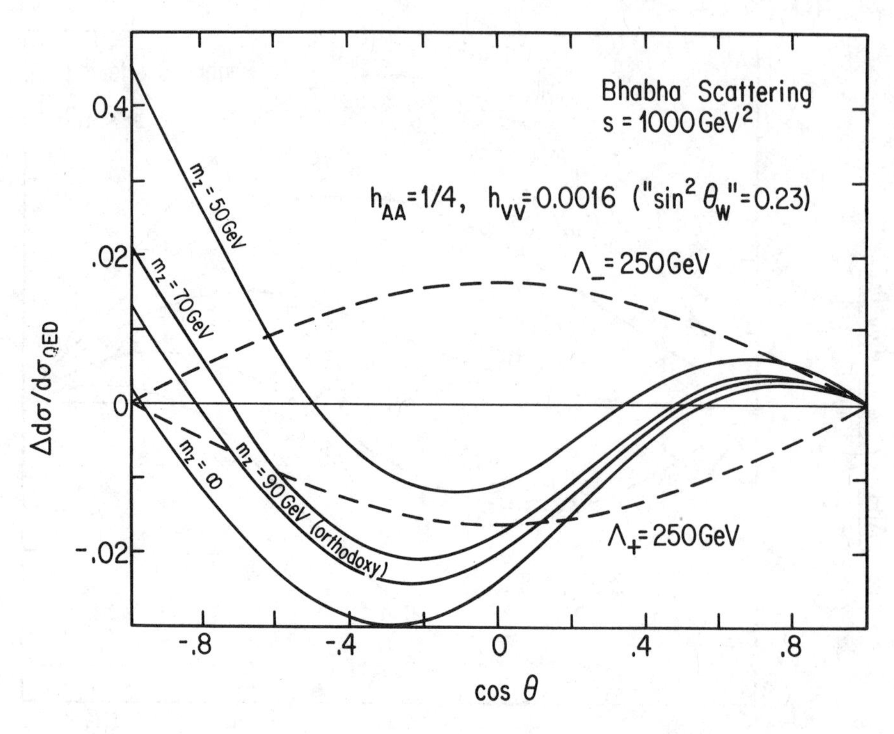

Fig. 5. Deviations from the pure QED predictions in Bhabha scattering in single Z models that give the same predictions at low s as the standard model.

There is yet another class of two Z boson models which gained some popularity in the past year--a model proposed by De Groot, Gounaris and Schildknecht[27] and a subsequent variant by Barger, Keung and Ma.[28] Here the basic group is either SU(2) ⊗ U(1) ⊗ U(1) or SU(2) ⊗ SU(2) ⊗ U(1) where in the second case both SU(2) refer to the left-handed weak isospin. Two salient features common to both versions are: (a) There are two Z's such that[29]

$$m_{Z_1} < m_Z < m_{Z_2} \tag{26}$$

where m_Z is the mass predicted by Weinberg's SU(2) ⊗ U(1) mass formula. (b) In the low q^2 limit the effective neutral-current interactions are given by

$$L = (4G/\sqrt{2})[(J_\lambda{}^3 - \sin^2\theta_W J_\lambda{}^{e.m.})^2 + c(J_\lambda{}^{e.m.})^2] \,. \tag{27}$$

Notice that (27) differs from the standard model prediction only by the presence of the c term. Because the neutrino does not have electric charge, the model makes exactly the same predictions as the standard model for neutrino-induced reactions. Furthermore, because the extra term is purely parity-conserving, it gives no contribution to the SLAC (-Yale ...) electron-deuteron asymmetry. In fact, electron-positron annihilation into muon pairs and Bhabha scattering are the best place to test the presence of the c term, far better than the anomalous magnetic moment of the muon known to 8 significant figures.

Phenomenologically we can set a limit on the c term by relating it to h_{VV}:

$$h_{VV} = 4c + \frac{1}{4}\,(1-4\,\sin^2\theta_W)^2 \quad . \tag{28}$$

Because $\sin^2\theta_W$ is known to be close to 1/4, c is essentially the same as $h_{VV}/4$, so it is unlikely that c is larger than 0.06.[30] In this type of model it is possible to relate the lower limit on Λ_- to the lower limit of m_{Z_1} for an assumed value of $\Gamma(Z_1)$ and also to establish allowed regions in an m_{Z_1}-m_{Z_2} plane. See Fig. 6 taken from a

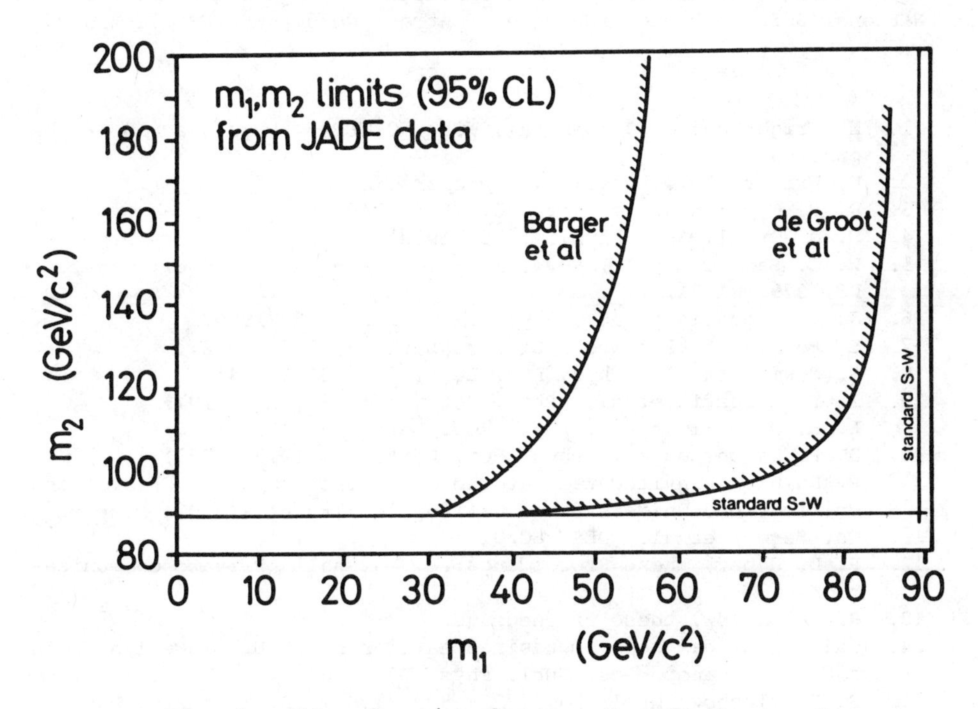

Fig. 6. Allowed region for m_1 and m_2 in the two-Z boson models of Refs. 27 and 28.

contribution presented at this Symposium by the JADE Collaboration.[12]

To summarize:

(a) Weak effects should not be ignored in interpreting "QED tests" at PETRA/PEP energies.

(b) The so-called QED tests are actually "electroweak tests" at unprecedentedly high values of q^2.

(c) If Λ_- is really bigger than 160 GeV, once popular models such as the hybrid model with purely vectorial $\bar{e}e$, $\bar{\mu}\mu$ currents are ruled out by the PETRA data alone. In the context of the standard SU(2) $\otimes$ U(1) model, $\sin^2\theta_W$ as large as 0.5, or as small as 0, appears to be in difficulty.

(d) More precise data on $\mu^+\mu^-$ asymmetry will be most informative.

(e) The QED violation parametrization based on $\Lambda_\pm$ should be discouraged particularly for Bhabha scattering.

(f) Instead, in Bhabha scattering it is proposed to relate the shape of $\Delta(d\sigma/d\Omega)/(d\sigma/d\Omega)_{QED}$ to the V-to-A ratio, the Z mass etc.

(g) Interesting constraints on some two-Z boson models have been obtained.

I am indebted to Dr. R. Marshall of the JADE Collaboration for an interesting communication. This research is supported in part by the National Science Foundation under Contract Number NSF/PHY 78-21502.

REFERENCES

1. N. Wright and J. J. Sakurai, UCLA/80/TEP/3, Phys. Rev. D (to be published).
2. N. Wright, Ph.D. Thesis (in preparation).
3. R. P. Feynman, Phys. Rev. 76, 769 (1949).
4. S. D. Drell, Ann. Phys. 4, 75 (1958).
5. W. C. Barber et al., Phys. Rev. Lett. 16, 1127 (1966); Phys. Rev. D3, 2796 (1971).
6. J. E. Augustin et al., Phys. Lett. 31B, 673 (1970).
7. E. Borgia et al., Nuovo Cimento Lett. 3, 115 (1972).
8. H. Newman et al., Phys. Rev. Lett. 32, 483 (1974).
9. J. E. Augustin et al., Phys. Rev. Lett. 34, 233 (1975); L. H. O'Neill et al., Phys. Rev. Lett. 37, 395 (1976).
10. D. P. Barber et al., Phys. Rev. Lett. 43, 1915 (1979); P. Duinker, invited talk at the Conference on Color, Flavor and Unification, University of California, Irvine (1979).
11. Ch. Berger et al., DESY 80/01.
12. R. D. Heuer, these Proceedings; R. Marshall (Private communication).
13. R. J. Barlow, these Proceedings.
14. This point has been emphasized earlier by C. H. Llewellyn Smith and D. V. Nanopoulos, Nucl. Phys. B78, 205 (1974).
15. S. L. Glashow, Nucl. Phys. 22, 579 (1961); S. Weinberg, Phys. Rev. Lett. 19, 1264 (1967); A. Salam, Elementary Particle Theory, ed. N. Svartholm (Almquist and Wiksell, Stockholm, 1968), p. 367.

16. Weak interaction effects in reaction (2) have been discussed by many authors. Some of the earliest papers may include: N. Cabibbo and R. Gatto, Phys. Rev. 124, 1577 (1961); T. Kinoshita et al., Phys. Rev. D2, 910 (1970); J. Godine and A. Hankey, Phys. Rev. D6, 3301 (1972); R. Budny, Phys. Lett. 41, 449 (1973). For reviews see, e.g., L. Wolfenstein, AIP Proceedings 23, 84 (1974); J. J. Sakurai, AIP Proceedings 51, 138 (1979).
17. A similar plot has been prepared by the JADE Collaboration.
18. S. A. Bludman, Nuovo Cimento 9, 443 (1958).
19. P. Fayet, Nucl. Phys. B78, 14 (1974); T. P. Cheng and L.-F. Li, Phys. Rev. Lett. 38, 381 (1977).
20. C. Prescott et al., Phys. Lett. 77B, 347 (1978); 84B, 524 (1979).
21. P. Q. Hung and J. J. Sakurai, 69B, 323 (1977); 88B, 91 (1979).
22. J. D. Bjorken, Phys. Rev. D19, 335 (1979); P. Q. Hung and J. J. Sakurai, Nucl. Phys. B143, 81 (1978).
23. Realistic counting estimates along this line have recently been made by J. H. Field, Internal Report DESY F 36-80/01.
24. Eq. (24) is a model independent way of writing down a formula derived by R. Budny and A. McDonald, Phys. Rev. D10, 3107 (1974); R. Budny, Phys. Lett. 55B, 227 (1975).
25. J. Pati and A. Salam, Phys. Rev. D10, 275 (1974); H. Fritzch and P. Minkowski, Nucl. Phys. B103, 61 (1976); M. A. Bég et al., Phys. Rev. Lett. 38, 1252 (1977); R. N. Mohapatra and D. P. Sidhu, Phys. Rev. Lett. 38, 667 (1977).
26. For a recent model calculation see, e.g., T. G. Rizzo, Phys. Rev. D21, 1214 (1980).
27. E. H. De Groot, D. Schildknecht, and G. J. Gounaris, Phys. Lett. 85B, 399 (1979).
28. V. Barger, W. Y. Keung, and E. Ma, Phys. Rev. Lett. 44, 1169 (1980).
29. For a more general discussion of the inequality (26) based on $SU(2) \otimes U(1) \otimes G$ see H. Georgi and S. Weinberg, Phys. Rev. D17, 275 (1978).
30. A similar limit on c was obtained by the JADE Collaboration. See Ref. 12.

JETS, GLUONS, QCD

T. F. Walsh
Deutsches Elektronen-Synchrotron DESY, Hamburg

Vanderbilt Symposium on e^+e^- Interactions

Vanderbilt University
Nashville, Tennessee
May 1-3, 1980

ABSTRACT

1. The evidence from e^+e^- collisions for QCD's gluons is considered. The criterion for a convincing experimental test is that plausible alternative scenarios fail to describe data.

2. Future experiments to establish the gauge nature of QCD are discussed.

I. INTRODUCTION

If quantum chromodynamics is really the theory of the strong interactions, we should be able to produce convincing experimental evidence that its elementary colored quarks and gluons exist. (In fact, we are obliged to do so.) This is not easy if color is confined. But it is possible - convincing evidence for quarks comes from the $q\bar{q}$ and qqq, $\overline{qqq}$ spectra, and from quark jets. Asymptotic freedom ensures perturbatively calculable rates for quanta made at short distances [1]. It is an assumption that energetic colored quanta made at short distances appear as narrow jets of colorless hadrons [2]. Data and popular myth support this. Someday it will appear as a logical consequence of the theory. But we can use this now, to produce evidence for the gluons of QCD and of the local gauge nature of the theory.

There are two aims in this talk. First, we look at the evidence

ISSN:0094-243X/80/620014-16$1.50

for gluons (gluon jets) and ask: how convincing is it? Scenarios are our tool. Evidence is convincing if no trivial experimental adjustments can reproduce data without QCD. Theoretical scenarios check whether data is really sensitive to the gluons' properties. They have spin one, color (and a self-interaction), but no flavor. Our second aim is to look for future experiments which will produce evidence that QCD is a local gauge theory. This means looking for the predicted three-gluon vertex in short distance reactions. (There is also a 4G vertex, which will be harder to prove real.) Strong gluon self-coupling at large distances will produce bound glue hadrons. We remark briefly on this.

II. SCENARIOS CONTRA QCD

a. Υ(9.46) DECAY

Fig. 1 recapitulates the evidence that $\Upsilon \to 3G \to$ hadrons is the Υ decay mechanism.[3] Υ thrust distributions agree with the 3G decay and not with a two jet or phase space model (the dashed lines on Figs. 1a and 1b.) P_{out} distributions (perpendicular to the event plane) also agree. So does the polar angle dependence of the thrust axis (solid line in Fig. 1c.) Everything looks satisfactory. But how convincing is it?

EXPERIMENTAL SCENARIO

One can imagine a novel "phase space" model which reproduces the T distribution in Fig. 1a. Perhaps one can come close to 1b, too. (The phase space angular distribution on Fig. 1c should, however, be flat.) Since no one has tried this, it would be unreasonable to claim that it is impossible. (The experimentalists do tell us that trivial changes, such as including meson resonances in the model, are not enough.) This scenario can be destroyed two ways

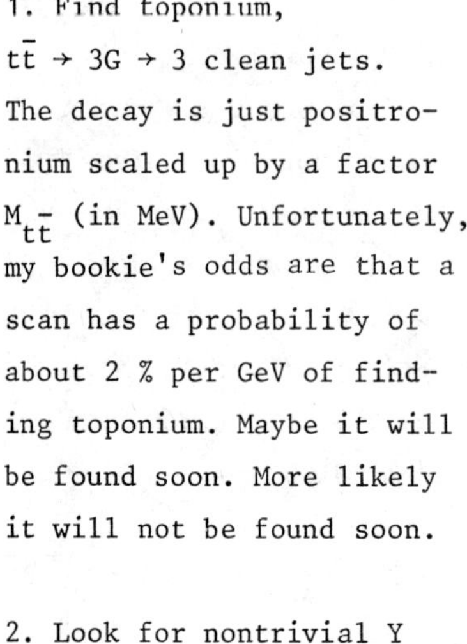

1. Find toponium,
$t\bar{t} \rightarrow 3G \rightarrow 3$ clean jets.
The decay is just positronium scaled up by a factor $M_{t\bar{t}}$ (in MeV). Unfortunately, my bookie's odds are that a scan has a probability of about 2 % per GeV of finding toponium. Maybe it will be found soon. More likely it will not be found soon.

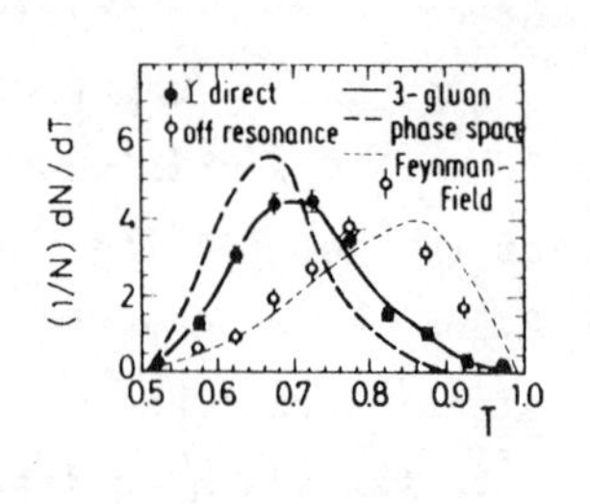

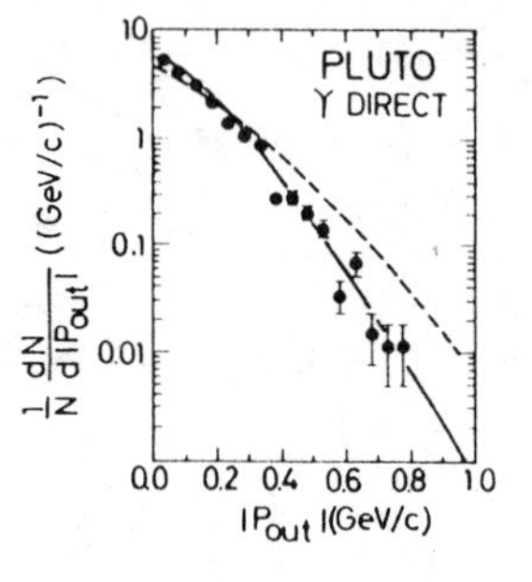

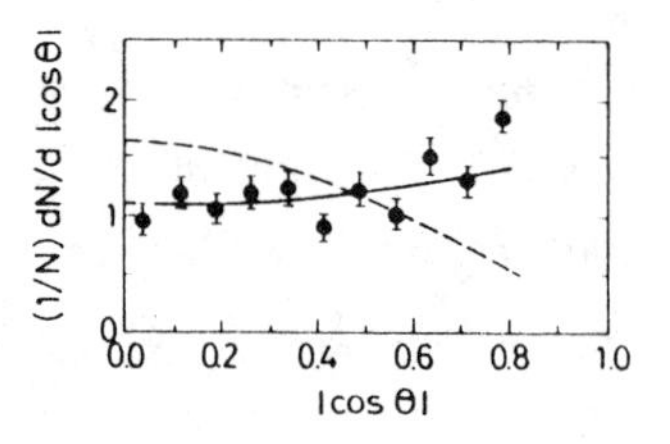

Fig. 1

Distributions in Υ decay

2. Look for nontrivial Υ decay angular distributions. The thrust axis distribution is of the form [4]

$$\frac{d\sigma}{d\cos\theta_T} \propto 1+\alpha(T)\cos^2\theta_T \qquad (1)$$

where θ_T is the angle of the T axis to the e^+e^- beams. We also have the decay $\Upsilon \rightarrow \gamma GG$. The γ angular distribution is [5]

$$\frac{d\sigma}{d\cos\theta_\gamma} \propto 1+\alpha(x_\gamma)\cos^2\theta_\gamma \qquad (2)$$

where θ_γ is the angle between the photon and the e^+e^- beams and $x_\gamma = E_\gamma/E_B$ (E_B is the e^+ or e^- beam energy). These are shown in Fig. 2. (The average $<\alpha(T)> = .39$[4] was used in Fig. 1c.) Angular distributions are important. Remember that the angular distribution of $e^+e^- \rightarrow q\bar{q} \rightarrow 2$ jets being $\propto 1+\cos^2\theta$ settled the question of whether or not 2 jet events were being seen at SPEAR, as well as establishing that the jets were from spin 1/2 partons.

TWO THEORY SCENARIOS

One may wonder how much Υ or toponium decay distributions depend on the essential elements of QCD. So we change them to see [6].

First, what would happen if gluons had no color? Then the decay

$$Q\bar{Q} \to g \to q\bar{q} \to 2 \text{ jets} \qquad (3)$$

would dominate over $Q\bar{Q}\to 3g$. This is because (3) is allowed for colorless g but not for QCD's colored G. The two jet decay (3) is excluded by data for Υ. So is the colorless gluon scenario.

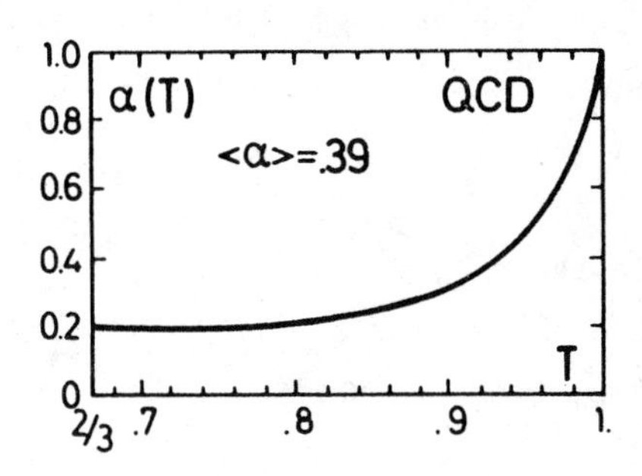

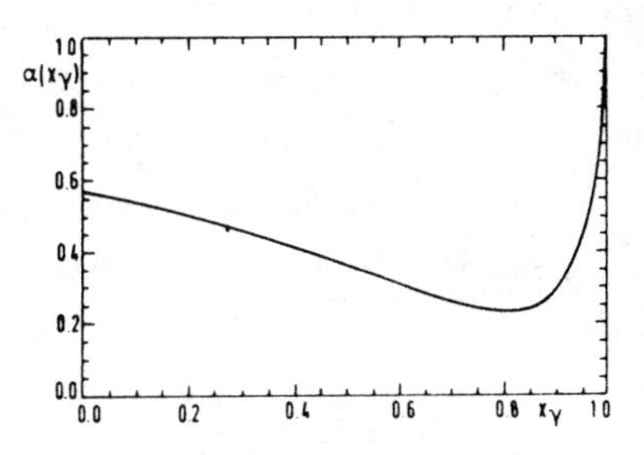

Fig. 2

$\alpha(T)$ versus T and $\alpha(x_\gamma)$ vs x_γ

Second, we ask what would happen if gluons were colored but had no spin. $Q\bar{Q} \to 3S$ (S a scalar gluon) is needed for planarity. The matrix element for $Q\bar{Q} \to 3S$ has zeros at the center of the Dalitz plot $E_1 = E_2 = E_3 = M_\Upsilon/3$ and at the midpoints of the sides $E_1 = M_\Upsilon/2$, $E_2 = E_3$ (and permutations). This is because of the symmetry of the final state. It is <u>not</u> possible to trivially reproduce the approximately constant squared matrix element of QCD. For the scalar case symmetry alone - no dynamics - enforces a kinematic configuration where one gluon is softer than the other two. Koller and Krasemann noticed that in the scalar case, there is no recoil factor suppressing soft scalar gluon emission. So the decay $Q\bar{Q}\to$hard S + hard S + soft S dominates. The average $\alpha(T)$ from helicity is then [7]

$$\langle\alpha_{SCALAR}(T)\rangle \simeq -1 \tag{4}$$

This is shown as a dashed line on Fig. 1c. It clearly disagrees with data, so the scalar scenario is ruled out.

From these scenarios, it's clear that $Q\bar{Q}$ decays are sensitively dependent on QCD being a theory with colored vector gluons. The observations (3) and (4) are fairly insensitive to conjectural higher order corrections to the colorless or scalar cases. Agreement of QCD with data even for Y(9.46) is clearly nontrivial.

A remaining question is whether higher-order QCD corrections seriously modify the lowest-order predictions we have discussed. At a fixed renormalization point some of these corrections are indeed large [8]. However things may not be so bad as they seem at this first glance [9].

b. $e^+e^- \to q\bar{q} + q\bar{q}G \to$ 2 JETS + 3 JETS

The theoretical prediction for the α_s/π QCD correction to $\sigma(e^+e^- \to$ hadrons) is one of the cleanest tests of the theory. It should also be possible to describe the final state at PETRA and PEP energies as due to 2 and 3 jet events [10]. This is theoretically somewhat less clean. But it may offer us convincing evidence that gluons are real.

Two jet events $e^+e^- \to q\bar{q}$ have parton thrust $T_{PARTON} = 1$. Three jet events $e^+e^- \to q\bar{q}G$ have $2/3 \leq T_{PARTON} \leq T_o$. (The cutoff T_o is used to distinguish three jet events - which are perturbatively small in rate at $O(\alpha_s/\pi)$ - from two jet events.) Another way of putting this is in terms of the invariant mass of a parton or pair of partons,

$$\begin{aligned} q\bar{q}:\ p^2 &= Q^2(1-T_{PARTON}) = 0 \\ q\bar{q}G:\ p^2 &= Q^2(1-T_{PARTON}) \geq p_o^2. \end{aligned} \tag{5}$$

In an appropriate gauge p^2 is the virtual $(\text{mass})^2$ of a fragmenting parton. The $q\bar{q}G$ process was chosen to be for parton masses $p^2 \geq p_o^2$, or for parton rest frame distances $\lesssim O(1/\sqrt{p_o^2})$. A lower limit for p_o^2 is p_{NP}^2, the range of fluctuation in parton masses which can arise from the formation of two (confinement or nonperturbative) jets in $e^+e^- \to q\bar{q}$,

$$p_{NP}^2 \approx \langle M_{JET}^2 \rangle = \langle (\sum_i E_i)^2 - (\sum_i \vec{p}_i)^2 \rangle \cong \left[Q^2 \langle p_\perp^2 \rangle \right]^{1/2}, \tag{6}$$

where this sum is over the hadrons in the jet. (Essentially the same result comes from the requirement that the virtual parton not travel further than $\Lambda^{-1} \approx 1/500$ MeV in the lab before decaying to qG.)

A separation into 2 and 3 jet events is not obviously meaningful below p_{NP}^2. Higher order QCD corrections will appear. So will long-range confining forces. It also no longer makes sense to add rates rather than amplitudes. (At some very high energy, higher order effects will appear at some p_o^2 larger than the p_{NP}^2, where confinement is important. These effects will probably be hard to distinguish at PETRA and PEP.) We choose $p_o^2 \cong p_{NP}^2 \cong 20\text{-}50\ \text{GeV}^2$. At $Q^2 = 50\ \text{GeV}^2$ $e^+e^- \to q\bar{q} \to 2$ jets is barely resolvable. So we expect that $q(p) \to qG \to 2$ jets is also resolvable. One can argue about whether or not the cutoff should be placed at $p_o^2 = 50\ \text{GeV}^2$ or $20\ \text{GeV}^2$ (i.e. $T_o = .95$ or $.98$ at $E_{CM} = 30$ GeV).

It is important to realize that one can calculate observables at infinite energy ignoring such a cut [11]. It is a measure of our ignorance about finite energy effects. If an observable depends sensitively on T_o at finite energy, it is a poor test of QCD.

Take the mean $\langle 1-T \rangle$ in the final state as an example. To $O(\alpha_s/\pi)$ this has the form

$$\langle 1-T_{PARTONS} \rangle = \frac{8\alpha_s}{3\pi} \int_{2/3}^{T_o} \frac{dT}{T} \ln \frac{2T-1}{1-T} + \dots . \tag{7}$$

The integrand diverges as $T \to 1$, although $<1-T_{PARTONS}>$ is finite. (7) therefore depends on T_o. This is a problem with many linear observables. The mean value $<p_\perp^n>$ of hadrons relative to a jet axis depends hardly at all on T_o for n = 2, and does depend on T_o for n = 1.

If linear observables [11] depend on nonperturbative effects at present energies, quadratic variables have their own deficiency. They depend on quark and gluon fragmentation functions. See the Table

variable	linear (e.g. $<p_\perp>$)	quadratic (e.g. $<p_\perp^2>$)
$D^h(z,Q^2)$ dependent?	no	yes
"infrared safe"?	yes	yes
T_o or p_{np}^2 dependent at low energy?	yes	no

(Notice that "infrared safe" merely means that an observable is finite in perturbation theory.)

Since quadratic variables can be calculated more reliably at low energy, we [12] have done so. The dependence on T_o and on quark and gluon fragmentation functions is indeed weak. Two things which can be calculated analytically are the overall $<p_\perp^2>$ relative to a jet axis and $<p_\perp^2(z)>$, the mean $<p_\perp^2>$ of a hadron with fractional momentum $z = p/E_{BEAM}$. We also produced a Monte Carlo $e^+e^- \to q\bar{q} + q\bar{q}G$ model [12]. Fig. 3 shows the analytic and Monte Carlo results. Clearly the Monte Carlo is quantitative. It can be used to test QCD. Fig. 4 shows data [13]. The agreement is not a result of fiddling with parameters. The calculations and Monte Carlo preceeded the data by some months.

Quantitative agreement with such global QCD predictions is important. More important is the presence of three jet events [14] with the expected rates and distributions. This evidence is convincing

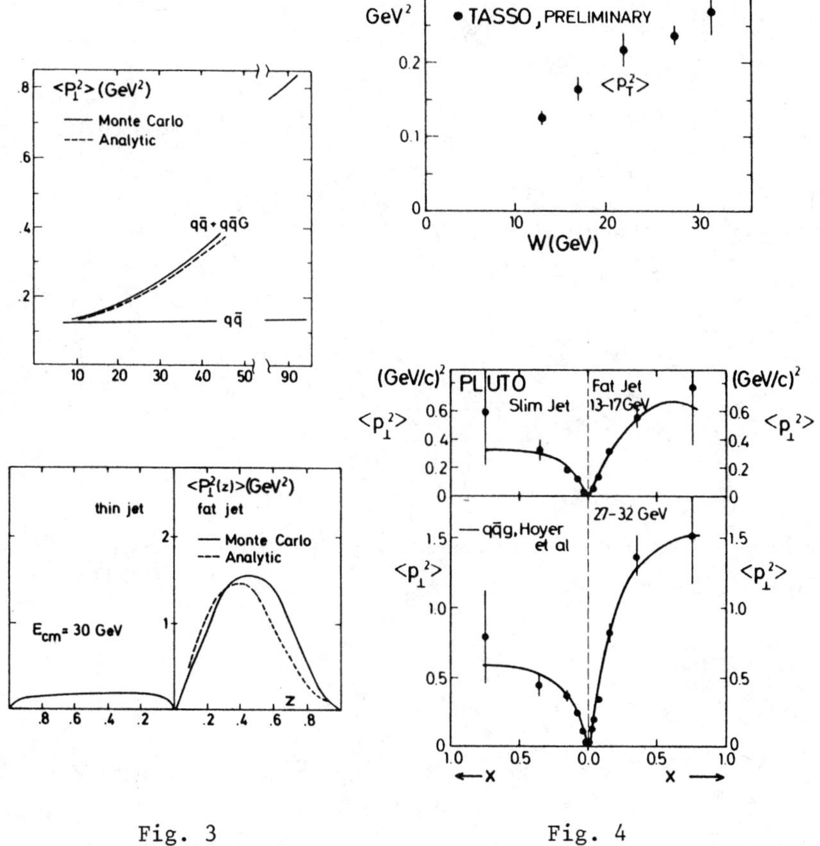

Fig. 3

Analytic Calculations and Monte Carlo Results.

Fig. 4

TASSO data on $\langle p_{\perp}^{2} \rangle$ and PLUTO data on $\langle p_{\perp}^{2}(z) \rangle$.

enough that we need not consider experimental scenarios as for Υ(9.46).

ALTERNATIVE THEORY SCENARIOS

1. It isn't QCD. Maybe jet broadening and three jet events are real, but not due to QCD. We invent a scenario. But there must be ground rules. These are that a model must have plausible dimensional parameters and must be successful. Take the example in Fig. 5, [15]

where a high $p_\perp$ meson (or cluster) is radiated:

In QCD, this would be a "semi-perturbative" process. It is not so complicated as jet formation but far more model dependent than $e^+e^- \to q\bar{q}G$. The mean $\langle p_\perp^2 \rangle$ of the cluster or recoil q or $\bar{q}$ has the form

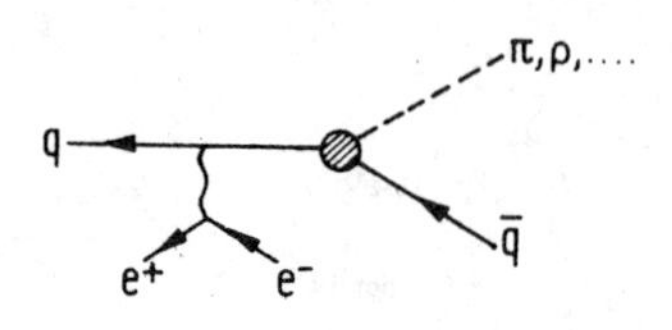

Fig. 5

A Simple Model for 3 jets

$$\langle p_\perp^2 \rangle_{\substack{\text{CLUSTER} \\ \text{or } q,\bar{q}}} \approx \text{const} \int_0^{Q^2/12} \frac{dp_\perp^2}{(p_\perp^2+\mu_o^2)^n} \, p_\perp^2 \tag{8}$$

and, choosing $n = 3$ as an example (your favorite model may have something different, but no matter),

$$\langle p_\perp^2 \rangle_{\text{meson}} = \langle p_\perp^2 \rangle_{NP} + \langle p_\perp^2 \rangle_{SP}\left(1 - \frac{18\mu_o^2}{Q^2} + \ldots\right) \tag{9}$$

The dimensional parameters are $\langle p_\perp^2 \rangle_{SP}$ (the overall additive $\langle p_\perp^2 \rangle$ of this process at $Q^2 \to \infty$) and the scale μ_o^2. Naively, $\langle p_\perp^2 \rangle_{SP}$ ought to be much less than $\langle p_\perp^2 \rangle_{NP}$ (the FF value with a Gaussian $p_\perp$ cutoff). Also, we expect the scale $\mu_o^2 \lesssim 1\ \text{GeV}^2$ as for a typical hadron mass. Even if we arbitrarily set $\langle p_\perp^2 \rangle_{SP} = \langle p_\perp^2 \rangle_{NP}$, the Q^2 dependence of (9) cannot reproduce the data. (9) only increases by $< 10\ \%$ over the PETRA range $Q^2 = 200\text{-}1000\ \text{GeV}^2$ for $\mu_o^2 = 1\ \text{GeV}^2$. The data shows a factor ~ 2.5 increase of $\langle p_\perp^2 \rangle$.

Another flaw of the model of Fig. 5 is that it does not naturally lead to the large observed jet multiplicities.

We conclude that there is no simple and plausible alternative scenario.

Remember that there is such a "higher twist" scenario for scaling violations in deep inelastic scattering [16]. That scenario depends on the fact that higher twist effects (and QCD) lead to no scaling violations (and very tiny scaling violations) at large Q^2. Here the QCD effects <u>increase</u> with energy, e.g. $\langle p_\perp^2\rangle \sim (\alpha_s/\pi) Q^2$. Semiperturbative effects saturate. There is a qualitative distinction.

Of course, this is not to say that such semiperturbative effects are absent in data. Only that they seem small compared to $q\bar{q}G$ (and are possibly smaller than other model dependent effects such as the precise shape of jet distributions).

2. <u>Gluons are Colorless.</u> Through this order, $e^+e^- \to q\bar{q}G$ is like $e^+e^- \to 1\gamma \to \mu^+\mu^-\gamma$. The color of gluons only appears in the normalization of $q\bar{q}G$ relative to $\mu^+\mu^-\gamma$. But this is only a factor and cannot be seen in present data. $q\bar{q}G$ is not sensitive to this feature of QCD.

3. <u>G is Spinless.</u> To exclude this and find nontrivial agreement with QCD, one has to prove that the $q\bar{q}G$ matrix element is as predicted. Another way is <u>via</u> angular distributions (Fig. 6). The angular distributions of the normal and thrust axis must be of the form

$$d\sigma/d\cos\theta_N \propto 1+\alpha_N \cos^2\theta_N;$$
$$d\sigma/d\cos\theta_T \propto 1+\alpha(T)\cos^2\theta_T \qquad (10)$$

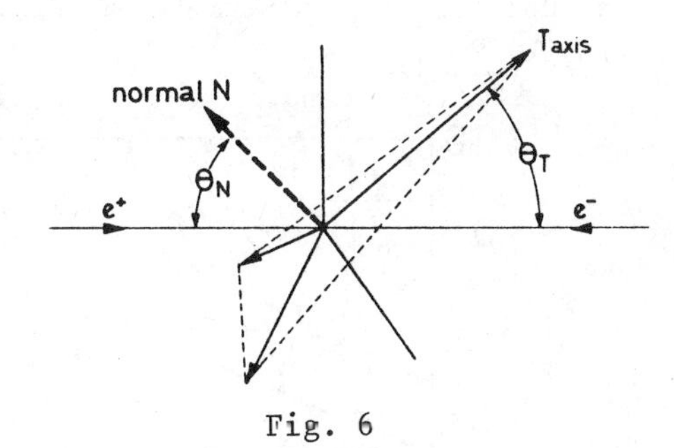

Fig. 6

The variables θ_N and θ_T

These are shown on Fig. 7 [17]. In the table α_N for two jet events is for the case when a plane is "accidentally" fit due to fluctuations in the event.

Measurement of α_N clearly proves nothing. That is why scenarios can be useful. Also, clear evidence for vector gluons from $\alpha(T)$ (or

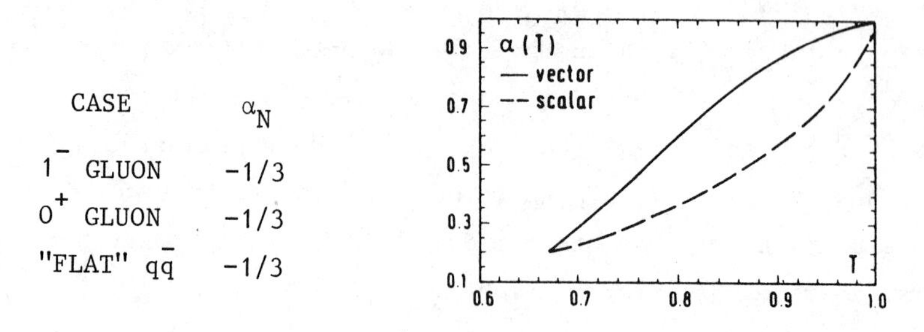

CASE	α_N
1^- GLUON	-1/3
0^+ GLUON	-1/3
"FLAT" $q\bar{q}$	-1/3

Fig. 7

α_N and $\alpha(T)$ versus thrust

other angular distributions) will need lots of data.

4. <u>The third jet is not neutral.</u> Rather than invent a scenario, we just show on Fig. 8 the charge correlation of the two most energetic jets [12],

$$C_{12}(T) = \langle \sum_{i \in 1} e_i \sum_{j \in 2} e_j \rangle \tag{11}$$

as a function of T. The model of Fig. 5 would not give $C_{12}(2/3)/C_{12}(1)=1/3$.

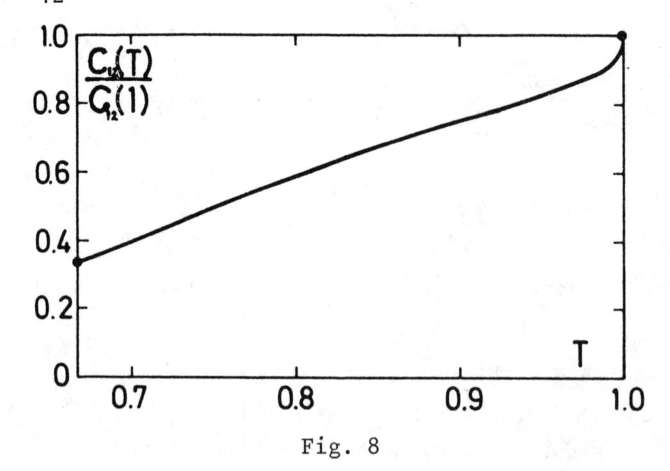

Fig. 8

The charge correlation

This check will also require a lot of data.

III. THE GAUGE NATURE OF QCD

1. TOPONIUM DECAYS

One nice way to find evidence for the 3G vertex of QCD, and therefore its gauge nature, is analogous to $e^+e^- \to q\bar{q} + q\bar{q}G$. Namely the $C = +Q\bar{Q}$ decay

$$Q\bar{Q} \to GG + GGG \tag{12}$$

(plus $Gq\bar{q}$, which is tiny). Figure 9 shows $d\sigma/dT$ for the 3G decay of the ^3P_o $Q\bar{Q}$ state [18].

Because of the 3G vertex the 3 jet rate should be about 9/4 times that in $e^+e^- \to q\bar{q}+q\bar{q}G$ at the same energy [19]. (This can be large.) The ^3P_J states have to be reached via the radiative decay of a ^3S state produced by e^+e^-. The decay

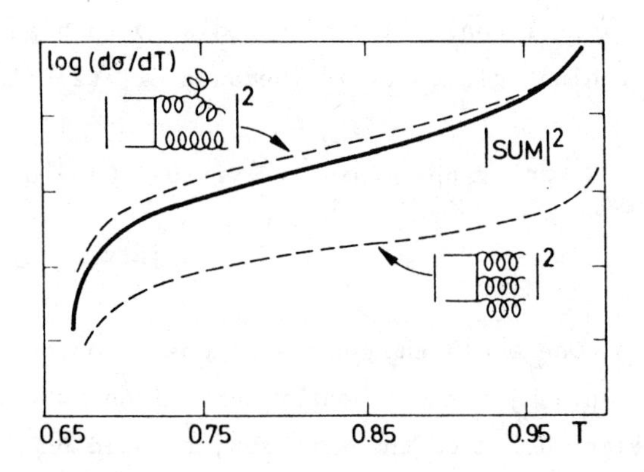

Fig. 9

Thrust distribution for $Q\bar{Q} \to 3G$

$$1^3S_1\ Q\bar{Q} \to GGG + GGGG \tag{13}$$

(plus a small $GGq\bar{q}$ rate) will be easier to see but harder to analyze. (There is also a $Q\bar{Q} \to 1\gamma \to q\bar{q} + q\bar{q}G$ background, absent in (12), and at really large $Q\bar{Q}$ mass the weak decay background of the heavy bound quarks, $Q \to qq\bar{q}$ and $\bar{Q} \to \bar{q}\bar{q}q$).

2. Z^o DECAYS

There is another chance to see the 3G vertex in decays

$$Z^o \to q\bar{q}GG \to 4 \text{ jets} \tag{14}$$

where the 3G vertex appears (although it is not leading for large M_Z). This will require some angular gymnastics. There is also a "background" from $Z^o \to t\bar{t} \to 3q+3\bar{q} \to 6$ jets if M_t is large.

Angular asymmetries for $e^+e^- \to q\bar{q}GG$ have been calculated [20] and appear large enough to measure. (14) would then be an easy related experiment for the SLAC single-pass collider.

3. G JET SPLITTING

A virtual gluon produced at very high $p_\perp$ can radiate a bremsstrahlung gluon <u>itself</u> (because of its color charge). Namely

$$\begin{array}{l} p\bar{p} \text{ or } pp \to (\text{high } p_\perp \text{ virtual G}) + \text{other stuff} \\ \qquad\qquad\qquad\quad \searrow \\ \qquad\qquad\qquad\quad GG \to 2 \text{ jets} \end{array} \tag{15}$$

One would then see a gluon jet often "resolved" into a hard gluon subjet and a nearby soft gluon subjet. The energy and angle distribution of the soft subjet would be $\propto dE/E\ d\theta/\theta$ - just as for the photon in $e \to e\gamma$ or the gluon in $q \to qG$. However, (15) should happen 9/4 times as often as for $q \to qG$. This is about the most elegant way to see that QCD is a local gauge theory. The problem is that jets in $p\bar{p}$ do not come labelled as to whether they are from gluon or quark ancestors at short distance.

Monte Carlo calculations of jet evolution indicate that the subjet structure in (15) will be easily visible for jets with $p_{\perp,JET} \sim 300$ GeV at the FNAL 2 TeV $p\bar{p}$ collider, if it reaches a luminosity of $10^{30}\ cm^{-2}\ sec^{-1}$. [21] The same applies to Isabelle. Demonstrating $G \to GG$ breakup at the CERN $p\bar{p}$ collider is harder because $p_{\perp,JET}$ is much less.

4. LOOKING FOR BOUND GLUE

QCD should have hadrons made of bound glue (glueballs, gluonia or glue bacteria depending on individual fantasy). These mesonic glue states will be made in gluon jets, but the rates are surely quite model dependent. (Υ(9.46) decays will be a good place to look.)[22]

A hopefully model independent source of bound glue is

$$Q\bar{Q} \to \gamma + \text{bound glue state}$$

(Fig. 10). $Q\bar{Q} \to \gamma GG$ is a short-distance source of colorless GG pairs. At long distances, GG → glue bound glue or $GG \to q\bar{q}$ meson. The latter should be Zweig suppressed. Naive duality suggests that the former produce oscillations about the Born cross section for γGG (Fig. 11)[5].

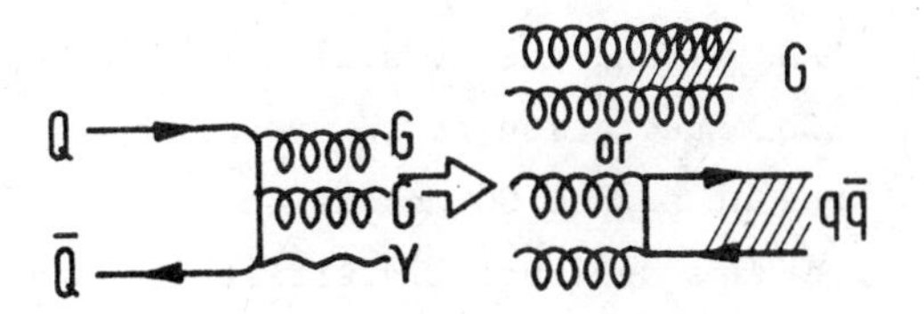

Fig. 10
$Q\bar{Q}$ Radiative Decay

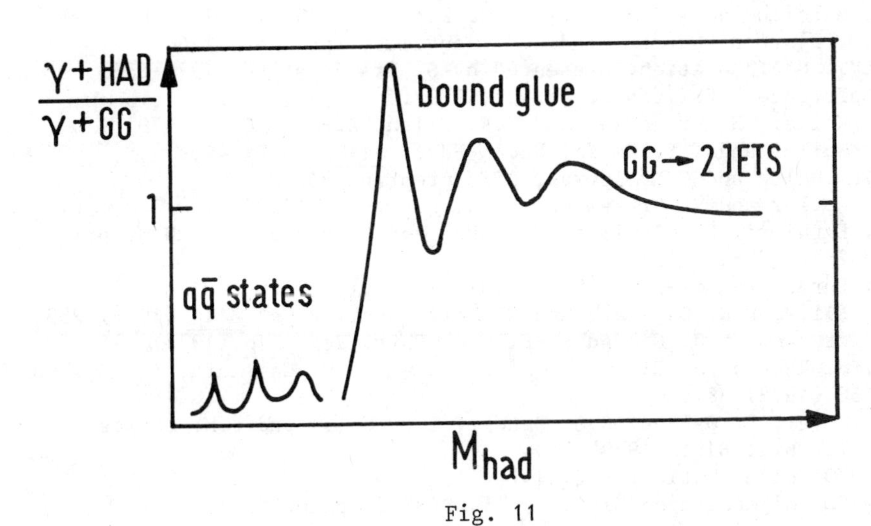

Fig. 11

Naive duality gives rates. Approximate the Born γ spectrum by $(1/\Gamma_{\gamma GG})d\Gamma/dx_\gamma \cong 2x_\gamma$. Then the single γ branching ratio is the same as the fractional rate of γGG in a GG mass slice of width $\Delta m_{\mathcal{G}}$,

$$\frac{\Gamma(\gamma\mathcal{G})}{\Gamma_{\gamma GG}} \cong \frac{4m_{\mathcal{G}}\,\Delta m_{\mathcal{G}}}{M^2}\left(1 - \frac{m_{\mathcal{G}}^2}{M^2}\right) \tag{16}$$

or, for J/ψ, $B(\gamma\mathcal{G}) \sim 2\ \%\ \Delta m_{\mathcal{G}}/(1/2\ \mathrm{GeV})$. Zweig's rule would then imply a $\gamma f^o(1260)$ rate perhaps $\sim 10^{-1}$ this or $\sim 0.2\ \%$. This is roughly correct. (Of course, (16) could easily be wrong by a factor 2.)

It is interesting that activity is seen in $J/\Psi \to \gamma$ + hadrons by MARK II and the Crystal Ball [23]. Rates fall below the lowest order QCD expectation at low recoil mass. Then a resonance (E(1420)?) appears and at higher GG masses the rates are large. This resembles Fig. 11 - perhaps accidentally. If it is no accident, then some of the activity seen at higher masses might be due to overlapping resonances, not resolved in the Crystal Ball γ spectrum. The E(1420) might also have a nonstandard J^P, perhaps 0^-.

REFERENCES

1. cf. H. Politzer, Phys. Rep. 14C (1974) 129.
   W. Marciano and H. Pagels, Phys. Rep. 36C (1978) 137.
2. J.D. Bjorken, Current Induced Reactions, Springer 1976.
3. PLUTO Collaboration, presented by S. Brandt at the 1979 EPS Conference, DESY report 79/42 (1979).
4. K. Koller, H. Krasemann and T.F. Walsh, Z.Phys. C.1 (1979) 71.
5. K. Koller and T.F. Walsh, Nucl. Phys. B140 (1978) 449.
6. T.F. Walsh and P.M. Zerwas, DESY preprint 80/20
7. K. Koller and H. Krasemann, Phys. Lett. 88B (1979) 119.
8. R. Barbieri, E. d'Emilio, G. Curci and E. Remiddi, CERN preprint TH 2622 (Jan. 1979).
9. A. Buras, Fermilab-Pub-80/43 THY (1980).
10. J. Ellis, M.K. Gaillard and G. Ross, Nucl. Phys. B111 (1976) 253.
    T. deGrand, Y.J. Ng and S.-H. Tye, Phys. Rev. D16 (1977) 3251.
11. A. de Rújula, J. Ellis, E. Floratos and M.K. Gaillard, Nucl. Phys. B138 (1978) 387.
12. P. Hoyer, P. Osland, H.G. Sander, T.F. Walsh and P.M. Zerwas, Nucl. Phys. B161 (1979) 349.
13. PLUTO Collaboration, ref. 14
    TASSO Collaboration; in G. Wolf, DESY preprint 80/13

14. TASSO Collaboration, Phys. Rev. 86B (1979) 243.
MARK J Collaboration, Phys. Lett. 43 (1979) 830.
PLUTO Collaboration, Phys. Lett. 86B (1979) 418.
JADE Collaboration, DESY preprint 79/80.
15. T. deGrand et al., ref. 10.
16. L. Abbott, W. Atwood and R.M. Barnett, SLAC-PUB 2400 (1979).
17. G. Kramer, G. Schierholz and J. Willrodt, Phys. Lett. 79B (1978) 249 (E: 80B (1979) 433).
G. Schierholz, DESY preprint 79/71 (1979).
K. Koller, H. Sander, T.F. Walsh and P.M. Zerwas, DESY preprint 79/87 (1979).
Another independent test is due to:
J. Ellis and I. Karliner, Nucl. Phys. B148 (1979) 141.
18. K. Koller et al. (unpublished).
Another 3G vertex test is in K. Koller, T.F. Walsh and P.M. Zerwas, Phys. Lett. 82B (1979) 263.
19. K. Shizuya and S.-H. Tye, Phys. Rev. Lett. 41 (1978) 787; (E) 1195.
M.B. Einhorn and B.G. Weeks, Nucl. Phys. B146 (1978) 445.
20. J. Körner, G. Schierholz and J. Willrodt, private communication.
21. C.-H. Lai, J.L. Peterson and T.F. Walsh, NBI preprint NBI-HE-80-8.
See also: K. Kajantie and E. Pietarinen, DESY preprint 80/19 (1980) and S. Wolfram, XV Rencontre de Moriond, Caltech preprint CALT-68-778.
22. P. Roy and T.F. Walsh, Phys. Lett. 78B (1978) 62.
23. See the presentations by the MARK II and Crystal Ball collaboration at this conference.

LOW ENERGY SIGNALS FOR HYPERCOLOR*

E. Eichten†
Lyman Laboratory of Physics
Harvard University
Cambridge, Massachusetts 02138

ABSTRACT

The low energy phenomenological implications of theories with dynamical breaking of the weak interaction symmetries are reviewed. The expected signals include: (1) a rich spectrum of spinless pseudo-Goldstone bosons, (2) flavor-changing neutral-current interactions mediating rare decays, and (3) an electric dipole moment for the neutron of approximately 10^{-24} e cm.

1. MOTIVATION FOR DYNAMICAL SCHEMES OF ELECTROWEAK SYMMETRY BREAKING

The fundamental scale of the weak interactions is determined by the familiar Fermi coupling G_F to be $\Lambda_W \equiv (\sqrt{2}\, G_F)^{-1/2} \approx 250$ GeV. Within the Weinberg-Salam (W-S) SU(2) × U(1) model[1] the masses of the gauge bosons $W^{\pm}$ and Z^0 are given in terms of this scale: $M_W = e\Lambda_W/(2 \sin\theta_W)$ and $M_Z = M_W/\cos\theta_W$. Using the weighted average of the present experimental measurements of the parameter $\sin^2\theta_W = 0.218 \pm 0.020$,[2] the masses of the $W^{\pm}$ and Z^0 are expected to be approximately 80 GeV and 90 GeV respectively. Undoubtedly, these massive carriers of the electroweak force will be observed experimentally in the not too distant future.

The requirement of a renormalizable gauge theory of electroweak interactions necessitates that the masses of the $W^{\pm}$ and Z^0 arise spontaneously. This means that the polarization tensor for the $W^{\pm}$ and Z^0 propagators must acquire a pole at $k^2 = 0$. This pole can arise in two ways:

(1) The usual Higgs mechanism with elementary scalar fields. Here the minimum of the scalar potential occurs for some nonzero vacuum expectation value, v, of the scalar fields. Upon shifting the fields by their vacuum expectation value, some fields acquire a positive mass and appear as physical scalar particles, while the others are massless Goldstone bosons associated with the symmetries broken by $v \neq 0$. Since these Goldstone bosons couple to the gauge currents, each will produce a pole in the polarization tensor of the associated gauge boson, and become the longitudinal component of the resulting massive vector particle.

(2) The symmetry breaking is dynamical. In this case the Goldstone bosons are bound states produced by a new interaction which becomes strong at an energy scale Λ_W. The W-S model with scalar multiplets should be interpreted as an effective low-energy

*Talk presented at the International Symposium on High Energy e^+e^- Interactions, Vanderbilt University, Nashville, TN (May 1980).

†Research supported in part by the National Science Foundation under Grant No. PHY77-22864, and the Alfred P. Sloan Foundation.

ISSN:0094-243X/80/620030-18$1.50

($<\Lambda_W$) Lagrangian. The scalar fields are actually composite-bound states of a fermion and an antifermion.

For electroweak symmetry breaking of the first type, the W-S model with one complex doublet of elementary scalars provides a simple and highly successful theory. In this standard model a single scalar field acquires mass (the neutral Higgs meson) while the other three members of the multiplet remain massless and couple to the charged and neutral weak currents to give masses to the $W^{\pm}$ and Z^0. The value of v is determined from the known strength of the charged weak interaction at low energies; $v = \Lambda_W/2 = 123$ GeV. In addition to the consistency of the value of $\sin^2\theta_W$ measured in many different physical processes, the value[2] of $\rho \equiv (M_W/(M_Z \cos\theta_W))^2 = 0.985 \pm 0.023$ is in excellent agreement with the theoretical expectation $\rho = 1 + O(\alpha)$.

However, this theory appears to be incomplete because the couplings of the scalar potential which determine the weak scale $\Lambda_W = 2v$ and the Higgs particle's mass, and the Yukawa couplings which determine the current algebra quark masses and lepton masses are arbitrary parameters. A more fundamental problem with this theory is that the small values needed for Yukawa couplings ($\Gamma_u \approx \frac{1}{2}\Gamma_d \approx 10\Gamma_e \approx 4\times10^{-5}$) must be considered accidental. This is because there is no continuous symmetry associated with the limit of zero Yukawa couplings; and therefore the small values of these couplings cannot be associated with a weakly broken symmetry - they are unnatural.[3]

This unpleasant feature of the Yukawa couplings is not cured by embedding the theory in a grand unified theory at superhigh energies. Also, in a grand unified theory the number of independent parameters may be reduced but a new problem of naturalness arises. For the electroweak scale $\Lambda_W = 246$ GeV to appear in a theory with a fundamental scale of $\sim 10^{15}$ GeV some parameters in the scalar potential must be adjusted to one part in 10^{26}.[4]

To avoid both these problems with naturalness, the breaking of weak interaction symmetries must be accomplished without introducing elementary scalar fields. This alternative - a dynamical theory for weak symmetry breaking - requires a new strong interaction (Section 2) and has numerous implications for physics below ~1 TeV. These low energy signals of the dynamics at the electroweak scale include light spinless particles which can be produced in e^+e^- as well as hadronic processes (Section 3), new very weak flavor-changing neutral currents which may mediate such rare processes as $K_L \to \mu e$, $K^+ \to \pi^+\mu e$, $\mu \to e^+\gamma$, etc., and possibly CP symmetry violating interactions which produce an electric dipole moment for the neutron larger than previously expected (Section 4).

2. NEW STRONG INTERACTIONS

To obtain the masses of the $W^{\pm}$ and Z^0 dynamically, a new strong interaction is required. Assume there are no elementary scalar fields in the Lagrangian - only fermions and spin one fields interacting through gauge interactions. The basic Lagrangian must be invariant under the gauged $SU(2) \times U(1)$ electroweak symmetry and thus no bare fermion mass terms are allowed (such terms explicitly break the $SU_L(2)$ symmetry). Since the electroweak symmetries are spontaneously broken there must be Goldstone bosons, i.e. massless spinless particles. By assumption there are no elementary scalars, thus there must be bound-state spinless particles. This requires an interaction strong enough to produce massless bound states. None of the known interactions, color SU(3) or electroweak, are sufficiently strong at the energy scale Λ_W to produce the required bound states.

The general formalism of dynamically broken gauge theories was worked out long ago[5,6] and applied to the weak interactions by Weinberg.[7] The new strong interaction is postulated to be a non-Abelian vector gauge interaction, G_H, called here hypercolor,[8] but also called technicolor,[9] or the extra-strong interaction.[7] Two properties of G_H are postulated in analogy with QCD:

Confinement. The vector gauge interaction itself is unbroken and therefore confining as in QCD. This property is suggested by the principle of the maximally attractive channel.[10] The scale of the interaction Λ_H is the scale at which the running coupling constant becomes strong - $\alpha_H(\Lambda_H^2) \sim 1$. Fermions which have hypercolor will be denoted hyperfermions. The confinement of hypercolor explains how a new very strong interaction could escape detection at energies $\ll \Lambda_H$. It is simply that the associated physical states have masses of the order of Λ_H. (The exceptions are discussed in Section 3.)

(2) Chiral symmetry breaking. In addition to the gauged symmetries there may be global (flavor) symmetries of the fermion fields. Ignoring the color, electroweak, and any strongly broken interaction, the approximate global chiral symmetry group will be denoted G_f. When G_H becomes strong at Λ_H the chiral symmetries G_f are spontaneously broken down to some subgroup S_f. There must be Goldstone bosons for each of the broken G_f symmetries. In particular, if there are N flavors of hyperfermions transforming under the same complex representation of G_H there is a global symmetry

$$G_f = SU_L(N) \times SU_R(N) \times U(1) .$$

In analogy with QCD, when G_H becomes strong at the scale Λ_H the fermions acquire a dynamical mass which is associated with a chiral symmetry breakdown. The scale of masses generated is of order Λ_H, i.e. the full propagator of each hyperfermion, S(p), satisfies

$$\mathrm{Tr}[S^{-1}(p)] \sim \Lambda_H \neq 0 \quad \text{for } p^2 \leqslant \Lambda_H^2 . \tag{1}$$

The symmetry breaking is $G_f \to S_f = SU(N) \times U(1)$ with N^2-1 associated Goldstone bosons $\{\pi_a | a = 1, \ldots, N^2-1\}$. The respective decay constants are defined in terms of the associated spontaneously broken axial currents J^μ_{5a} by

$$\langle 0 | J^\mu_{5a} | \pi_b(q) \rangle = i\, q^\mu F_\pi \delta_{ab} \quad . \tag{2}$$

Because there is a remaining SU(N) vector symmetry all the decay constants are equal. Further it is expected that $F_\pi \sim \Lambda_H$ as there is only one scale. Although the general pattern of chiral symmetry breaking associated with non-Abelian gauge theories is still not completely understood, the assumptions made here are in agreement with recent progress in understanding this pattern.[11]

Armed with these two properties of the hypercolor interaction, we may calculate what happens when the weak interactions of hyperfermions are taken into account, and see how the dynamical models reproduce the successes of the W-S model. Some of the chiral symmetries in G_f are gauged by the electroweak interactions; thus the weak currents, $g_i J_i^\mu$, contain some combination of the broken chiral currents.

Let $J_i^\mu \equiv \sum_a A_{ia} J_{5a}^\mu$ + parts not associated with the broken generators.

Then the weak currents have non-zero matrix elements between the vacuum and the Goldstone bosons π_a given by

$$<0|J_i^\mu|\pi_a(q)> = i\, q^\mu F_\pi A_{ia} \qquad (A_{ia} \neq 0 \text{ for some a}). \tag{3}$$

And so a pole is generated in the polarization tensor of the weak gauge boson arising from the Goldstone boson intermediate states. The residue of the pole is the mass squared matrix of the gauge bosons M_{ij}^2 and is given by

$$M_{ij}^2 = g_i\, g_j \Big(\sum_{a=1}^{N^2-1} A_{ia}^* A_{ja} \Big) F_\pi^2 . \tag{4}$$

This is the mechanism by which the $W^\pm$ and Z^0 get their mass - the dynamical Higgs mechanism.

As mentioned in Section 1, it is known from experiments that

$$\rho \equiv \frac{M_W^2}{M_Z^2 \cos^2\theta_W} = 0.985 \pm 0.023 \text{ where } \sin^2\theta_W \equiv \frac{g_1^2}{(g_2^2+g_1^2)} = 0.218 \pm 0.020 \text{ and}$$

g_1 and g_2 are the $U(1)_W$ and $SU(2)_W$ weak coupling constants respectively. Susskind and Weinberg have shown,[9] independently, how to obtain $\rho = 1$ with this dynamical scheme.[12] The solution is simply to mimic the weak interactions of quarks and leptons, i.e. hyperfermions form left-handed doublets and right-handed isosinglets under $SU(2)_W$. To see how this works consider one "generation" of hyperfermions denoted by

$$\begin{pmatrix} U \\ D \end{pmatrix}_L , \quad U_R , \quad \text{and } D_R .$$

Then the chiral symmetry group $G_f = SU(2)_L \times SU(2)_R \times U(1)$ would be dynamically broken down to $S_f = SU(2) \times U(1)$, the vector subgroup. The three resulting Goldstone bosons, π_a, couple to the associated axial currents with equal decay constants, F_π,

$$<0|\bar{Q}(\gamma_\mu \gamma_5 J_a/2)Q|\pi_b(q)> = i\, q^\mu F_\pi \delta_{ab} , \qquad a,b = 1,2,3 \tag{5}$$

as a result of the residual SU(2) vector symmetry.

Using Eq. (4) and Eq. (5) we conclude

$$\frac{M_W^2}{M_Z^2} = \frac{\frac{1}{4} g_2^2 F_\pi^2}{\frac{1}{4}(g_1^2 + g_2^2) F_\pi^2} \equiv \cos^2\theta_W . \tag{6}$$

This recovers the successful result of the W-S model with Higgs doublets of elementary scalar fields. Furthermore, this provides the connection between the weak scale Λ_W and the hypercolor scale

$$M_W/g_2 = F_\pi/2 = G_F^{-1/2}\, 2^{-5/4} = \Lambda_W/2 . \tag{7}$$

Thus $F_\pi = 246$ GeV $= \Lambda_W$. If the ratio $\Lambda_H/F_\pi = \Lambda_{QCD}/f_\pi \sim 3$, then the scale at which G_H becomes strong (Λ_H) is approximately 1 TeV.[13] Thus the introduction of the hypercolor interactions which become strong near 1 TeV provides for a dynamical explanation of the origin of the $W^\pm$ and Z^0 masses and also naturally gives rise to the successful mass relation Eq. (6) above.

The new interaction as discussed so far is incomplete because hypercolor (G_H) commutes with the color interaction ($G_c \equiv SU(3)$) and the electroweak interaction ($G_W \equiv SU(2) \times U(1)$), and therefore the chiral symmetries associated with ordinary fermions are not broken when hyperfermions get mass. To generate current-algebra quark masses or lepton masses an interaction is required which couples fermions and hyperfermions. Within the assumed gauged nature of the fermion interactions this implies that some ordinary fermions and hyperfermions must belong to the same multiplet of this gauge interaction. Thus G_H is only a subgroup of the new gauge interaction required. The full gauge group is called sideways[8] or extended technicolor[14] and denoted G_S here.

The sideways group G_S must be partially broken (of course the G_H subgroup is unbroken); otherwise the existence of generators which couple hyperfermions to ordinary fermions would imply confinement of the flavor symmetries as well as color and hypercolor symmetries. The gauge bosons associated with the broken generators of G_S which couple ordinary fermions to hyperfermions (and thus gauge the flavor symmetries) must get mass spontaneously. I will not attempt to speculate on the origin of the dynamical symmetry breaking here. It is however tempting to assume that the gauge symmetry breaks internally to avoid adding any further interactions.[15]

Let F_S denote the scale associated with the dynamical G_S symmetry breakdown. The resultant masses for sideways gauge bosons are $\mu_s \approx g_S F_S$ where g_S is the G_S gauge coupling constant. Mass generation for ordinary fermions arises from the self-energy graph shown in Figure 1. For an external momentum much smaller than F_S, the graph simply gives the mass matrix for the ith left-handed fermion to couple to the jth right-handed fermion;

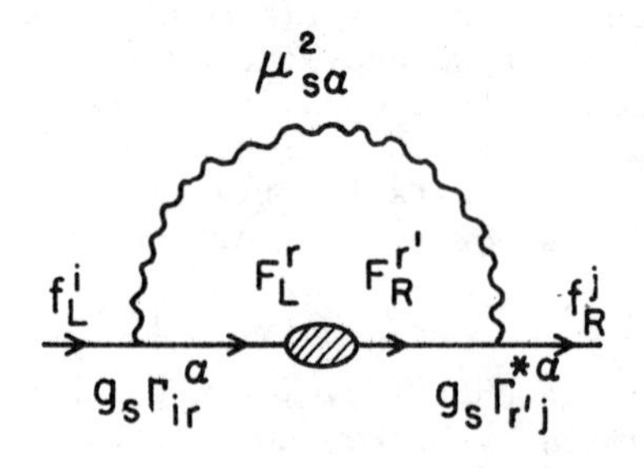

Figure 1. Graph showing the mechanism for generating ordinary fermion f mass from the dynamical hyperfermion F masses in the lowest order of sideways gauge boson exchange.

$$m_{ij} = \sum_{\alpha,r,r'} \int \frac{d^4p}{(2\pi)^4} \frac{g_S^2}{(p^2-\mu_{s\alpha}^2)} \mathrm{Tr}\left[\gamma^\mu \Gamma^\alpha_{ir} \left(\frac{1-\gamma_5}{2}\right) S^{rr'}(p) \gamma_\mu \Gamma^{*\alpha}_{r'j} \left(\frac{1+\gamma_5}{2}\right)\right], \tag{8}$$

where the summation runs over all sideways gauge bosons (labeled by α) and hyperfermion fields (labeled by r and r') which contribute to the i,j matrix element. Γ^α_{ir} and $\Gamma^{*\alpha}_{r'j}$ are Clebsch-Gordon coefficients of order one. Since the coupling of left-handed and right-handed hyperfermion fields is the result of dynamical symmetry breaking, it is expected to be soft - to vanish like $\Lambda_H^3/p^2 \times \log's^6$ for momentum much larger than Λ_H. Thus Eq. (8) gives

$$m_{ij} \approx \Lambda_H^3/F_S^2 \simeq (1 \text{ TeV})^3/F_S^2 . \tag{9}$$

Assuming typical elements m_{ij} to be of order 1 GeV

$$F_S \approx 30 \text{ TeV} .$$

Thus the energy scale at which G_S dynamically breaks down to a subgroup containing G_H is characterized by $F_S \approx 30$ TeV.

There are strong constraints on the fermion representation content of G_S which arise from the requirements imposed by low-energy phenomenology and particularly the absence of axions.[8] There are two immediate corollaries of these constraints:

(1) Both the color group G_C and the U(1) factor of the electroweak group G_W must be contained in G_S. In particular quarks and leptons must not be in entirely separate representations of G_S. There must be some sort of quark-lepton unification at the scale F_S.

(2) Grand unified gauge theories incorporating dynamical symmetry breaking cannot contain more than one irreducible representation of the full group involving light fermions.

To discuss the physics at 1 TeV and below, we may replace the full interaction associated with the broken generators of G_S by their effective four Fermi interactions. In lowest order of sideways gauge boson exchange this interaction is simply of a current-current form

$$\mathcal{L}_{\text{eff}} = \sum_\alpha \frac{g_S^2}{\mu_{S\alpha}^2} J_\alpha^\mu J_{\alpha\mu} \tag{10}$$

where the sum is over the broken generators Q_α whose current has a fermionic part J_α^μ and associated gauge boson has mass $\mu_{S\alpha}$. There are three types of terms in $\mathcal{L}_{\text{eff}}$ (shown in Fig. 2). They are (1) terms involving only ordinary fermion fields which provide new superweak

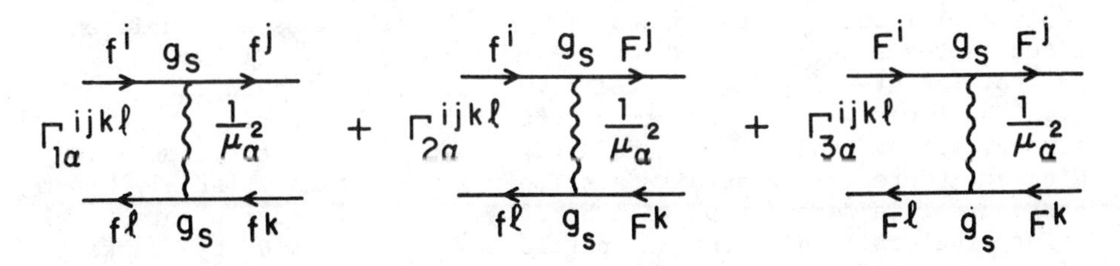

Figure 2. Contributions to the effective interaction $\mathcal{L}_{\text{eff}}$ associated with the exchange of the αth sideways vector boson of mass μ_α^2 and coupling g_S. Ordinary fermion fields are denoted by f and hyperfermion fields by F. Latin indices denote flavor quantum numbers. The Lorentz structure of the vertices has not been shown but is of the general form $A_L \gamma^\mu(1-\gamma_5)/2 + A_R \gamma^\mu(1+\gamma_5)/2$.

interactions among these fields, (2) terms involving two ordinary and two hyperfermion fields which are responsible for the current algebra masses of quarks and lepton masses, and (3) terms involving only hyperfermion fields which are responsible for breaking explicitly the hyperfermion chiral symmetries and provide masses for the PGB's of class 3 discussed in the next section. Of course $\mathcal{L}_{eff}$ is invariant under $G_H \times G_C \times G_W$.

To summarize we have found that to generate both masses for the $W^{\pm}$ and Z^0 and ordinary fermions a new gauge interaction called "sideways", G_S, must be introduced; the quantum numbers of the unbroken non-Abelian subgroup hypercolor, G_H, are confined while broken G_S quantum numbers correspond to the distinguishable fermion generations, and the gauge boson associated with these generators provide a new effective superweak interaction with strength $1/F_S^2 \approx 10^{-5} G_F$.

3. NEW PARTICLES: THEIR PROPERTIES AND THEIR DETECTION

When the hypercolor interaction becomes strong, the hyperfermions will bind into hyperhadrons which are hypercolor singlets. Some of these hyperfermions may also carry ordinary SU(3) color quantum numbers. Since the SU(3) color interaction commutes with the hypercolor interactions, there will in general be both ordinary color singlet and non-singlet hyperhadrons.

First consider hyperhadrons with net fermion number[16] which will be called hyperbaryons although they do not necessarily have half-integer spins. For example, if the hypercolor group G_H were SU(4) and the hyperfermions are in the 4 representation of SU(4), then the totally antisymmetric combination of four hyperfermion fields, denoted ψ_α^i with α the hypercolor index and i the remaining indices (color, flavor, and spin), form hypercolor singlet hyperbaryons

$$\varepsilon_{\alpha\beta\gamma\delta}\, \psi_\alpha^i \psi_\beta^j \psi_\gamma^k \psi_\delta^\ell$$

which have integral spins! These hyperbaryons form multiplets of definite spin and parity under the combined color and flavor symmetries of hyperhadrons. When color, electroweak, and broken sideways interactions are taken into account, some splitting within multiplets will occur. The masses of these hyperbaryons and their excited states should be of the order of 1-10 TeV. The following general features distinguish the lowest mass hyperbaryons:

(1) The lowest mass hyperbaryons should be stable against hyperstrong, strong, and electroweak decays. In addition to color SU(3) singlet, there will generally also be color non-singlet stable hyperbaryons. This second type will bind with ordinary quarks into total color singlets. These unusual particles have a "cloud" of quarks of size 1 $(\text{GeV})^{-1}$ around a hyperbaryon "nucleus" of size 1 $(\text{TeV})^{-1}$.

(2) Are hyperbaryons forever? This question cannot be fully answered without a specific model. However, the general possibilities are as follows:

(a) If hyperbaryons have non-integral electric charges then at least one of them must be absolutely stable.

(b) Assuming integral charges, they could still be absolutely stable; or they could decay through the broken sideways interactions given in Eq. (10). In the second case their lifetimes, τ_H, would be short. Let m_H denote the mass of the hyperbaryon; then $\tau_H \sim F_S^4/m_H^5 \sim 10^{-18}$ sec.

(c) Most likely, however, is that although none of the hyperbaryons are absolutely stable some decay only through interactions at a grand unified energy scale near the Planck mass. The lifetime in this case is very long. Since all possible interactions at this scale must preserve hypercolor, color, and electroweak symmetries, the decay cannot, in general, be mediated by a single superheavy gauge boson exchange (mass M_X). The process could be mediated by a single exchange only if a low energy (~10 TeV) effective four-fermion operator can be constructed which respects these symmetries.[17] Then the lifetime is

$$\tau_H \sim \frac{(m_X^2)^2}{m_H^5} \approx \left(\frac{m_P}{m_H}\right)^5 \tau_{proton} , \qquad (11a)$$

i.e. a factor of 10^{-15} to 10^{-20} times smaller than the proton lifetime. Otherwise

$$\tau_H \gtrsim \frac{(m_X^5)^2}{m_H^{11}} \gtrsim 10^{76} \text{ years} . \qquad (11b)$$

Essentially stable!

(3) How abundant would hyperbaryons be in the present universe assuming $\tau_H > 10^{10}$ years? If there were CP-violating interactions involving hyperfermions and ordinary fermions with roughly equal strength in the early universe, then the present excess of hyperbaryons would be comparable to the number of baryons. This is clearly ruled out as the total mass of the universe would be 10^3 times the observed mass. The other extreme is that there was no net excess of hyperbaryons created in the early universe. Then the production of hyperfermion-antihyperfermion pairs which failed to annihilate leads to a present average concentration, C, in ordinary matter given by[18] $C \approx 10^{-10} \times (m_H/m_P) \approx 10^{-7}$ to 10^{-6}. This is a lower limit on the possible average concentration of hyperbaryons. It is interesting to speculate on whether hyperbaryons could exist in terrestial matter at those concentrations and escape detection. A detailed analysis of this possibility has been presented by P. Frampton and

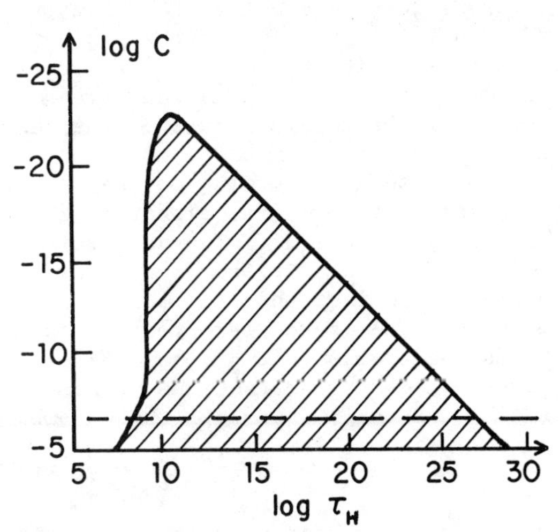

Figure 3. Concentration, C, versus lifetime τ_H for a hyperbaryon of mass ~3 TeV. The region under the solid curve is excluded by present experiments;[19] while the dashed curve is the lower limit on C.

S. Glashow.[19] They show that proton decay experiments put stringent limits on C/τ_H. The results are presented in Fig. 3. Note that the limits are restrictive only for theories in which the decay can be mediated by a single superheavy boson exchange.

Now consider the hyperhadrons which contain a hyperfermion and an antihyperfermion - the hypermesons.[16] There will be many similarities with ordinary mesons simply scaled up from 1 GeV to 1 TeV but also some important differences for the pseudoscalars: All states with $J^P \neq 0^-$ will have masses typical of the hypercolor interaction scale ~1 TeV. They will have hyperstrong decays when kinematically allowed; otherwise they will decay by electroweak and broken sideways interactions. The vector hypermesons (1^-) should be the lowest mass multiplet of these states. There is one unusual feature of the hyperstrong decays when the final state involves a 0^- state. As we will discuss below three of the 0^- states, denoted $\tilde{\pi}^+$, $\tilde{\pi}^0$, and $\tilde{\pi}^-$, are the unphysical Goldstone bosons which give masses to the W^+, Z^0, and W^- respectively. These states are simply the longitudinal components of the corresponding gauge particles. Thus the actual final state contains a $W^\pm$ or Z^0. For example $\tilde{\rho}^0 \to \tilde{\pi}^+\tilde{\pi}^-$ becomes $\tilde{\rho}^0 \to W^+W^-$. Turning this around, the W^+ and W^- will have hyperstrong interaction resonances with a typical mass scale about 1 TeV. Furthermore, the W^+ and W^- would be expected to lie on Regge trajectories associated with this new strong interaction.[20]

Finally we turn to the complex situation of the pseudoscalar states (0^-). These states will be very light compared to the 1 TeV scale. This is because these particles play a special role with regard to the dynamical breaking of the chiral global symmetry G_f of the hyperfermions. Ignore, for the moment, any explicit symmetry breaking due to color, electroweak, or broken sideways interactions. These mesons are exactly the Goldstone bosons $\{\pi_a\}$ associated with the dynamical breaking of the global hyperfermion chiral symmetries G_f discussed in Section 2. The ordinary pseudoscalars have exactly the same role in the dynamical breaking of quark chiral symmetries - they would be massless in the absence of electroweak interactions and explicit quark mass terms.

Now consider turning on the color, electroweak, and broken sideways interactions. The electroweak interactions gauge some of the broken G_f symmetries. The particular currents so gauged are not explicitly broken by either the color or broken sideways interactions of Eq. (10), since these interactions respect the electroweak symmetries. Therefore there remains one triplet of exactly massless Goldstone bosons $\{\tilde{\pi}^+, \tilde{\pi}^0$, and $\tilde{\pi}^-\}$ even in the presence of color and broken sideways interactions. This triplet gives masses to the $W^\pm$ and Z^0. Roughly speaking,[21] these combinations are

$$\tilde{\pi}^i \equiv \sum_a F_a^i \, \pi_a \, / \, [\sum_a {F_a^i}^2]^{\frac{1}{2}} \tag{12}$$

where F_a^i is the decay constant associated with the coupling of the ath Goldstone boson to the weak currents: $i = \pm$ are the charged $SU(2)_W$ currents while $i = 0$ is $1/\sqrt{2}$ times the difference of the neutral member of the $SU(2)_W$ currents and the $U(1)_W$ current. All the remaining Goldstone bosons may acquire mass through the explicit symmetry breaking interactions. The spin zero particles which acquire mass this way

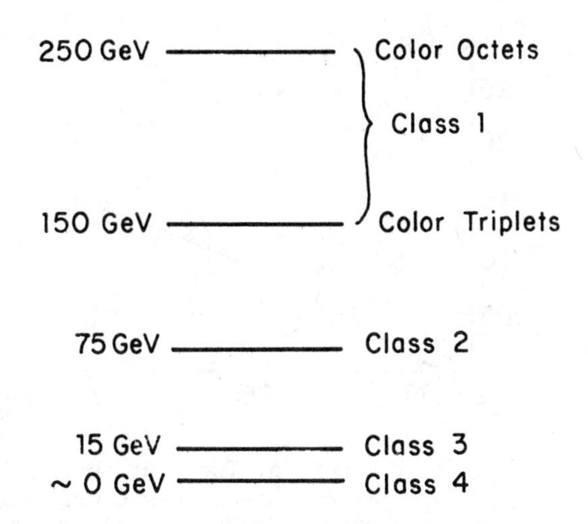

Figure 4. Approximate mass scales of PGB's of each class.

will be denoted here as pseudo Goldstone bosons (PGB).

The spectrum of PGB masses in these theories falls into distinct groups (or classes) depending on which interaction is responsible for the mass. The systematics of these masses have been worked out by a number of authors.[22-24] Ordering these contributions by decreasing strength gives four classes of PGB's (see Fig. 4):

Class 1. Those Goldstone bosons which are colored get mass by color gluon exchange in exactly the same manner as electromagnetic corrections give mass to the charged pions. An estimate[22-24] of the masses which result from these SU(3) color corrections gives ~150 GeV for a triplet state and ~250 GeV for a color octet.

Class 2. Of the remaining Goldstone bosons, some may get mass from the lowest order electroweak exchange (without including electroweak symmetry breaking effects). This will occur only if we allow real representations under G_H.[25] An estimate of the masses which result gives ~80 GeV.

Class 3. Those Goldstone bosons (which are not in classes 1 or 2) associated with the chiral symmetries broken by the sideways interactions of Eq. (10) will get a contribution to their mass estimated to be ~15 GeV. There is a significant variation in this mass estimate depending on exactly which terms in Eq. (10) contribute for each particular Goldstone boson. The range is approximately 5 to 40 GeV.[8]

Class 4. Any Goldstone bosons which still remain massless can only obtain mass from the full electroweak interaction.[8,22,23] In particular, by including the effects of the masses of the $W^{\pm}$ and Z^0, charge pseudoscalars in this class receive a mass contribution $= m_Z\sqrt{(3\alpha/4\pi)\ell n(\Lambda_H/m_Z)} \approx 5$ GeV while the neutral ones can get no mass from these interactions by a generalization of Dashen's theorem.[26] In fact neutral PGB in this class couple to a light fermion-antifermion pair like the axion[27] and are experimentally ruled out.

No realistic theory can have any PGB which fall into class 4. This result is the basis of the constraints on the fermion representation content of G_S mentioned in Section 2.

The phenomenology of PGB's of classes 1 and 2 has been discussed by S. Dimopoulos.[23] In general the signals are spectacular involving many jets or very hard (~50 GeV) photons. But, unfortunately, they require very high energies in e^+e^- accelerators and have at best marginally production rates in hadronic reactions at high energies.[28]

In any realistic theory the lightest mass PGB's have masses in

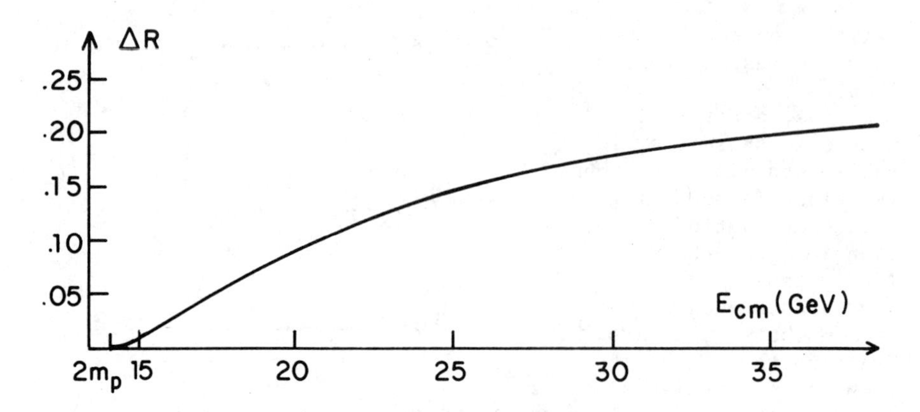

Figure 5. The contribution ΔR as a function of center-of-mass energy (E_{cm}) for a pair of PGB's of mass $m_\pm = m_p = 7$ GeV and unit charge.

the range 5 GeV to 40 GeV which puts them well in the range of accessible energies for e^+e^- colliding beam machines. The charged PGB's can be directly pair produced in e^+e^- annihilation and this seems the best reaction in which to detect them. Because these particles appear pointlike up to energies near a TeV they will contribute to R as an elementary scalar. The angular distribution of their production is $\sin^2\theta$ and

$$\Delta R = \tfrac{1}{4}Q^2(1 - 4m_\pm^2/E_{cm}^2)^{3/2} \tag{13}$$

where Q is the PGB's charge in units of e and $m_\pm$ its mass. E_{cm} is the total energy in the center-of-mass. The resulting ΔR for $Q = 1$ and $m_\pm = 7$ GeV is shown in Figure 5. It is clear from Figure 5 that the P wave threshold behavior and the small asymptotic ΔR make the direct observation of these thresholds in R very unlikely. It is necessary to look for some characteristic signal of the PGB's to have any hope of discovering these states. This is difficult without a specific model because the details of the decay of these PGB's are model dependent. However some general properties are:

(1) The decay of the lowest mass charged PGB's, $P^\pm$, proceeds through the broken sideways interactions only. The decay amplitude is of order

$$\frac{g_s^2}{\mu_s^2}\Lambda_H^2 = \left(\frac{g_s^2\Lambda_H^3}{\mu_s^2}\right)\Lambda_H^{-1} .$$

[See Eq. (10) and also Figure 2.] Now we notice from Eq. (9) that $g_s^2\Lambda_H^3/\mu_s^2 \sim m$, a typical current algebra quark mass or lepton mass matrix element ~ 1 GeV. Furthermore, $\Lambda_H^{-1} \sim G_F^{1/2}$ from Eq. (7) thus the decay amplitude $\sim m\,G_F^{1/2}$. The decays are semiweak.

(2) Somewhat more speculatively, the couplings to quark and lepton pairs within the same weak generation should be proportional to the mass scale of that generation.[10] Thus the decay amplitudes $P^+ \to u\bar{d}$, $P^+ \to c\bar{s}$ and $P^+ \to t\bar{b}$ as well as $P^+ \to \nu_e e^+$, $P^+ \to \nu_\mu \mu^+$, and $P^+ \to \nu_\tau \tau^+$ should be in the ratio 1:300:3000. The decays which

mix generations such as $P^+ \to c\bar{b}$ are harder to estimate. Naively they would be expected to be proportional to the sum of the masses of the final state quarks times a mixing angle. Since the mass of the P^+ must be at least 7 GeV,[29] and probably less than $m_t + m_b$ ($m_t + m_b \gtrsim 23$ GeV) the main decay modes should be $P^+ \to c\bar{s}$, $c\bar{b}$, $u\bar{b}$, and $\nu_\tau \tau^+$ with the hadronic modes probably dominating. If the mass is greater than $m_t + m_b$ then the mode $P^+ \to t\bar{b}$ should dominate.

The detection of the neutral PGB's is also most feasible in e^+e^- machines. The general decay properties of the lightest neutral PGB, denoted P^0, are the same as for the charged partner. The mass of the P^0, m_0, should be between 5 GeV and 40 GeV.[29] The best hope for observing P^0 is in the photon inclusive decays of a heavy quark-antiquark $^3S_1$ ground state [i.e. $\psi(c\bar{c})$, $\Upsilon(b\bar{b})$, or $\xi(t\bar{t})$] by $^3S_1 \to \gamma + P^0$. It is very unlikely that P^0 is light enough to be produced in ψ decay; there is a fair chance that it is light enough to be seen in Υ decays; and almost certainly it will be present in ξ decay. To estimate the branching ratio for this decay assume that $m_0 \ll M(^3S_1)$ [otherwise there is a phase space factor $(1 - m_0^2/M^2)$] and that the amplitude $P^0 \to Q + \bar{Q}$ is proportional to $2m_Q G_F^{1/2}$ (as in a standard Higgs model) then[30]

$$\frac{\Gamma(^3S_1 \to \gamma + P^0)}{\Gamma(^3S_1 \to \mu^+\mu^-)} \approx \frac{G_F M^2(^3S_1)}{4\sqrt{2}\,\pi\alpha} \tag{14}$$

This gives $B(\psi \to \gamma + P^0) \sim 10^{-2}\%$, $B(\Upsilon \to \gamma + P^0) \sim 3 \times 10^{-2}\%$ and $B(\xi(t\bar{t}) \to \gamma + P^0) \sim [M(\xi)/35 \text{ GeV}]^2 \times 1\%$, where the estimate $B(\xi(t\bar{t}) \to \mu^+\mu^-) \sim 10\%$ has been used. These rates allow some hope of seeing the P^0 in Υ decays and near certainty of seeing it in ξ decays.

4. NEW WEAK INTERACTIONS - RARE PROCESSES AND CP SYMMETRY VIOLATION

The effective four Fermi interaction, $\mathcal{L}_{eff}$ of Eq. (10), associated with the exchange of massive sideways gauge bosons gives rise to three new types of interaction terms as was illustrated in Figure 2. We have seen in Section 3 that currents linking different generations of hyperfermions were necessary to avoid unwanted pseudo-Goldstone bosons. Also currents linking ordinary fermions to hyperfermions are responsible for quark and lepton masses and provided the original reason for introducing G_S in Section 2. We have not yet discussed the last type of terms, those containing currents linking different generations of ordinary fermions.

This last type of current gives rise to flavor-changing neutral current interactions with an effective Fermi constant

$$\frac{1}{F_S^2} \simeq 10^{-5}\, G_F \,. \tag{15}$$

For example, one would expect a four Fermi effective interaction of the form

$$\frac{\Gamma}{4F_S^2}\left\{\bar{d}\,\gamma^\mu\left(\frac{1-\gamma_5}{2}\right)s + \bar{u}\,\gamma^\mu\left(\frac{1-\gamma_5}{2}\right)c\right\}\bar{\mu}\,\gamma^\mu\left(\frac{1-\gamma_5}{2}\right)e \tag{16}$$

where Γ is a dimensionless coupling associated with the particular model for G_S and fermion representations. Naively Γ would be expected to be of order 1. (We will return to this assumption.) Such an effective term allows the rare processes such as $K_L^0 \to \mu^\pm e^\mp$ and $K^+ \to \pi^+\mu^+e^-$. A whole new class of weak interactions are expected as a result of the broken sideways interactions between ordinary fermions with strengths determined by the scale F_S. Are these new interactions consistent with experimental bounds on such rare processes?

The experimental limits on rare processes have been investigated recently by a number of authors.[31-33] We employ the notation of R. Cahn and H. Harari[31] to classify the type of flavor-changing neutral current involved. Experimental limits exist only for lightest two generations of quarks G_1 = (u or d) and G_2 = (c or s) and the discussion will be limited to two generations of leptons $G_1^\ell = (\nu_e$ or e) and $G_2^\ell = (\nu_\mu$ or $\mu)$. The generation changing number associated with a current ΔG is defined by:

ΔG	Form for the Current
$\Delta G = +1$	$\bar{G}_2\,\gamma^\mu\,G_1$ or $\bar{G}_2^\ell\,\gamma^\mu\,G_1^\ell$
$\Delta G = 0$	$\bar{G}_2\,\gamma^\mu\,G_2$, $\bar{G}_1\,\gamma^\mu\,G_1$, $\bar{G}_2^\ell\,\gamma^\mu\,G_2^\ell$, or $\bar{G}_1^\ell\,\gamma^\mu\,G_1^\ell$
$\Delta G = -1$	$\bar{G}_1\,\gamma^\mu\,G_2$ or $\bar{G}_1^\ell\,\gamma^\mu\,G_2^\ell$

(17)

Thus the effective four Fermion interactions may have $|\Delta G|$ = 0, 1, or 2. Additionally, the $\Delta G = 0$ four Fermi interactions are distinguished to be diagonal (if $\Delta G = 0$ for each current as well) and non diagonal (if the two currents have equal and opposite non-zero ΔG). Thus, for example, the interaction display in Eq. (16) is a $\Delta G = 0$ non-diagonal type.

The experimental bounds on rare processes may be conveniently expressed as a limit on F defined by

$$\mathcal{L}_{int} = 1/F^2\, J_{1\mu}^{+}\, J_{2\mu} \tag{18}$$

with J_1 and J_2 the appropriate currents for the rare decay being considered. The results of Cahn and Harari[31] expressed this way are summarized below:

Process	Type	Limit on F
$K_L^0 \to \pi^+ e^- \mu^+$	$\Delta G = 0$ (non-diagonal)	$F \geqslant$ 55 TeV
$K^+ \to \mu^\mp e^\pm$	$\Delta G = 0$ (non-diagonal)	$F \geqslant$ 110 TeV
$\mu \to e^+\gamma$ [34]	$\|\Delta G\| = 1$	$F \geqslant$ 34 TeV
$\mu \to 3e$	$\|\Delta G\| = 1$	$F \geqslant$ 40 TeV
$\mu N \to eN$ [34]	$\|\Delta G\| = 1$	$F \geqslant$ 262 TeV
K_L^0-K_S^0 mass difference	$\|\Delta G\| = 2$	$F \geqslant$ 1200 TeV

(19)

Comparing these limits with our expectations two comments are appropriate:

(1) Although the value of F_S was determined to be 30 TeV in Section 2, remember that this estimate came from the assumption in Eq. (9) that the typical mass scale for ordinary fermions was 1 GeV. The small current algebra masses of u, d, and s and the small masses of e and μ suggest a more typical scale for these processes might be ~100 MeV.[10] The associated F_S would then be ~100 TeV.
(2) The factor Γ will in general involve Cabibbo-like mixing angles as well as group theory factors.

Thus the limits for $|\Delta G| = 0$ and $|\Delta G| = 1$ processes seem to be compatible with the experimental bounds. The K_L^0-K_S^0 mass difference however is in serious disagreement with the naive expectations of the dynamical models: a factor of at least 10 in F and thus a factor of 100 in the size of $\Delta m(K_L^0-K_S^0)$! This presents a serious challenge. How does one build a model which sufficiently suppresses these $|\Delta G| = 2$ interactions?

Finally consider the one known "super" weak interaction - the CP symmetry violating interaction. It has recently been shown that there is a natural mechanism for generating CP symmetry violation in dynamically broken electroweak gauge theories.[35] If we neglect the strongly broken sideways interactions $\mathcal{L}_{eff}$ (given in Eq. (10)) and the weak interactions (which play an inessential role in this whole discussion) the $G_H \times G_C$ invariant Lagrangian respects a global (chiral) flavor symmetry group G_f'. When hypercolor and color become strong at 1 TeV and 1 GeV respectively, G_f' is spontaneously broken to the subgroup S_f'. The ground state in this approximation is highly degenerate: the vacua parameterized by the coset space G_f'/S_f'. The perturbation $\mathcal{L}_{eff}$ lifts this degeneracy and picks out the true chiral perturbative vacuum. As Dashen[36] explained the correct vacuum is identified by minimizing

$$V(g) = \langle\Omega|U^{-1}(g)(-\mathcal{L}_{eff})U(g)|\Omega\rangle \ , \tag{20}$$

where $g \in G_f'$, $U(g)$ represents G_f' on the Hilbert space of states, and $|\Omega\rangle$ is the S_f' invariant vacuum. The true vacuum is $U(g_0)|\Omega\rangle$ where g_0 minimizes V.

It is assumed that (1) the fermion representation content under G_S is such that all G_S vacuum angles[37] can be simultaneously rotated to zero, and hence are unobservable, and (2) the breaking of G_S does not introduce CP nonconservation. Thus both $|\Omega\rangle$ and the perturbation $\mathcal{L}_{eff}$ are CP-invariant and the effective potential V is CP-symmetric. However the energy might be minimized by a CP-nonconserving vacuum $U(g_0)|\Omega\rangle$; then V(g) has a degenerate minimum and CP symmetry is spontaneously broken. The minimum of V(g) is determined once the pattern of G_S and G_f' symmetry breaking is known, and therefore so is the character of the spontaneous CP symmetry violation.

The fermions may be denoted $\psi_{Lr}^{(\rho)}$ and $\psi_{Rr}^{(\rho)}$ where the index identifies the irreducible representation $D^{(\rho)}$ (assumed comples) according to which $\psi^{(\rho)}$ transforms under $G_H \times G_C$. And the index r (r = 1,..., n_ρ) labels the various flavors of fermions which transform according to $D^{(\rho)}$.

The approximate chiral flavor group is then

$$G_f' = \prod_\rho [U_L(n_\rho) \times U_R(n_\rho)] \,/\, [U_A^H(1) \times U_A^C(1)] \tag{21}$$

which breaks down to $S_f' \equiv \prod_\rho [SU(n_\rho) \times U_V^\rho(1)]$. Under G_S' the fermions transform as

$$\psi_{Lr}^{(\rho)} \to W_{r'r}^{L(\rho)} \psi_{Lr'}^{(\rho)} \qquad \psi_{Rr}^{(\rho)} \to W_{r'r}^{R(\rho)} \psi_{Rr'}^{(\rho)} , \tag{22}$$

where $W^{L(\rho)}$ and $W^{R(\rho)}$ are unitary $n_\rho \times n_\rho$ matrices. If S_f is the diagonal subgroup with $W^L = W^R$ then the elements of G_f'/S_f' can be labeled by $W^{(\rho)} = W^{L(\rho)\dagger} W^{R(\rho)}$. The two axial U(1) currents which have the hypercolor and color anomaly, $U_A^H(1)$ and $U_A^C(1)$ respectively, can not be included in G_f; thus the unitary matrices are subject to two constraints

$$\prod_\rho [\det W^{(\rho)}]^{T_\rho^H} = \prod_\rho [\det W^{(\rho)}]^{T_\rho^C} = 1 , \tag{23}$$

where T_ρ^H (T_ρ^C) is the trace of the square of the hypercolor (color) generators in the representation $D^{(\rho)}$. Thus the effective potential in Eq. (20) can be expressed as a function of $W \equiv \{W^{(\rho)}\}$. The CP symmetry of V implies $V(W) = V(W^*)$.

V(W) may be minimized with respect to the matrices W subject to the constraints in Eq. (23) to determine the true vacuum. The extreme condition is

$$W_{rm}^{(\rho)} \frac{\partial V}{\partial W_{r'm}^{(\rho)}} - W_{rm}^{\dagger(\rho)} \frac{\partial V}{\partial W_{r'm}^{\dagger(\rho)}} = i(\nu_H T_\rho^H + \nu_C T_\rho^C)\mathbb{1} \tag{24}$$

where ν_H and ν_C are Lagrange multipliers associated with the constraints of Eq. (23).

The W which satisfies Eq. (24) and minimizes V(W) will satisfy one of these conditions:

(i) $W = W^*$ – CP is not spontaneously broken.

(ii) $W \neq W^*, \nu_C \neq 0$ – The effective Hamiltonian contains the CP nonconserving term $\frac{1}{2}(\nu_C/\Delta^q)\bar{q}\, i\gamma_5\, q$ where $\Delta^q \equiv \langle\Omega|\bar{q}q|\Omega\rangle$. A current algebra calculation[38] of the electric dipole moment D_N of the neutron gives $D_N \sim 4\times 10^{-16}(\nu_C/m_\mu \Delta^q)$ e cm. Since the natural scale of ν_C/Δ^q is of order the up quark mass m_u, $D_N \sim 10^{-15}$ e cm, which exceeds the experimental bound[39] by a factor of 10^9. In this case CP symmetry violation is strong and unacceptable.

(iii) $W \neq W^*$, $\nu_C = 0$ – In this case, CP symmetry is spontaneously broken but D_N is not unacceptably large.

To see more clearly the CP nonconservation in case (iii) note that the physics below 1 TeV can be described by an effective theory involving only quarks, leptons, gluons, electroweak bosons, and PGB. In integrating out all the other fields, a series of G_C invariant operators is generated.

$$\mathcal{L}_I = A_{r'r}\,\bar{q}_{Lr}\,q_{Rr'} + B_{r'r}\,\bar{q}_{Lr}\,\sigma_{\mu\nu}\,\lambda^a\,q_{Rr'}\,G^{\mu\nu a}$$
$$+ C_{rr'ss'}\,\bar{q}_r\,q_{r'}\,\bar{q}_s\,q_{s'} + \text{higher dimension operators}. \tag{25}$$

The coefficients A, B, and C depend on the $W^{(\rho)}$ for hyperfermions. The terms B and C have phases of order 1 and magnitude $[(\alpha_C(m_H))/2\pi]1/m_H^2$ for B, and $1/m_H^3$ for C. Thus the contribution to V(W) of the term B is suppressed by $\sim 10^{-9}$ relative to the quark mass term A and the term C is suppressed by $\sim 10^{-10}$. Therefore

$$W^{(q)}_{rm} \frac{\partial V}{\partial W^{(q)}_{r'm}} = \Delta_q \left(W^{(q)}A\right)_{rr'} \left(1 + O(10^{-9})\right) . \tag{26}$$

Using Eq. (26) the condition $\nu_C = 0$ implies that $W^{(q)}A$ is hermitian up to corrections of order 10^{-9}. The antihermitian part of the quark mass matrix as well as all other operators which contribute to D_N are of order 10^{-24}e cm.

The complex matrix $W^{(q)}$ of case (iii) also gives rise to other CP symmetry violating effects. For the electroweak interactions, these effects are entirely equivalent to the Kobayashi-Maskawa model.[40] Mixing matrices will also appear in the broken sideways currents. Typically phases in these matrices cannot be absorbed by redefining fields. If the operator $C\bar{s}d\bar{s}d$ occurs in the effective interaction, Im C must be suppressed. As we have already seen the real part must be suppressed. Naively such an operator would induce a CP non-conserving interaction in the K^0-$\bar{K}^0$ system which is 10^4 larger than observed.

5. SUMMARY

Dynamical schemes of electroweak symmetry breaking require:

(1) New spinless mesons (PGB) at accessible energies
   (a) The lowest mass states have masses ~7-40 GeV and can be produced directly or indirectly in e^+e^- annihilation.
   (b) Many new particles with unusual signals at masses ~70-300 GeV.

(2) The interactions at energies not directly accessible ($E \gtrsim 30$ TeV) have effects which can be seen through rare processes at low energies. In particular - flavor changing neutral currents.

(3) Weak CP symmetry violation can occur spontaneously in a natural way which implies an electric dipole moment of the neutron, $\sim 10^{-24}$e cm.

Observation of these signals would provide strong evidence for a dynamical alternative for electroweak symmetry breaking and in particular for a new strong interaction at 1 TeV.

Acknowledgements

I wish to thank my collaborators, K. Lane and J. Preskill, for many productive discussions. I would also like to thank V. Baluni, S. Dimopoulos, P. Frampton, S. Glashow, M. Peskin, and S. Weinberg for helpful discussions.

References and Footnotes

1. S. Weinberg, Phys. Rev. Lett. 19, 1264 (1967); A. Salam in *Elementary Particle Physics*, ed. N. Svartholm (Almquist and Wiksells, Stockholm, 1968), p. 367
2. See P. Langacker, *et al.*, University of Pennsylvania preprint COO-3071-243 (1979) and the references contained therein.
3. The definition of naturalness presented here is due to G. 't Hooft, "Naturalness, Chiral Symmetry, and Spontaneous Chiral Symmetry Breaking", Utrecht preprint (1980).

4. The necessity of fine tuning parameters in theories with elementary scalar fields has been pointed out by G. 't Hooft (Ref. 3) and L. Susskind (Ref. 9).
5. R. Jackiw and K. Johnson, Phys. Rev. D 8, 2386 (1973); J. M. Cornwall and R. E. Norton, *ibid.* D 8, 3338 (1973); E. Eichten and F. Feinberg, *ibid.* D 10, 3254 (1974).
6. K. D. Lane, Phys. Rev. D 10, 2605 (1974).
7. S. Weinberg, Phys. Rev. D 13, 974 (1976).
8. E. Eichten and K. D. Lane, Phys. Lett. 90B, 125 (1980).
9. L. Susskind, Phys. Rev. D 20, 2619 (1979); S. Weinberg, Phys. Rev. D 19, 1277 (1978).
10. The maximally attractive channel principle (MAC) is that the direction of the dynamic symmetry breakdown is signaled by the most attractive two-body fermion channel (under single gluon exchange). See S. Dimopoulos, S. Raby, and L. Susskind, "Tumbling Gauge Theories", Stanford preprint ITP-653 (1979) and the references contained therein.
11. Recent progress has been made in understanding the relation between chiral symmetry breaking and confinement in non-Abelian gauge theories. See S. Dimopoulos, L. Susskind, and S. Raby, "Light Composite Fermions", Stanford preprint ITP-662 (1980); S. Coleman and E. Witten, Phys. Rev. Lett. 45, 100 (1980); and Ref. 5.
12. M. Weinstein, Phys. Rev. D 8, 2511 (1973). To my knowledge, this was the first paper to point out that the value of ρ depends only on the representation content (under the weak gauge group) of the unphysical Goldstone bosons, not on whether these Goldstone bosons are elementary of dynamical.
13. In any realistic theory there will certainly be more than one generation of hyperfermions. Suppose there are N left-handed weak doublets of hyperfermions (generations) each of which may have a different decay constant F_r $(r = 1, \ldots, N)$. Then the relation between the weak scale Λ_W and the decay constants F_r is $F_\pi \equiv [\sum_{r=1}^{N} F_r^2]^{\frac{1}{2}} = \Lambda_W$.
14. S. Dimopoulos and L. Susskind, Nucl. Phys. B155, 237 (1979). The need for a strongly broken "sideways" interaction has also been realized by S. Weinberg, unpublished.
15. For a discussion of how this self-breaking might work, see Ref. 10.
16. This discussion assumes that a fermion number can be defined for hyperbaryons. This is trivial if all nontrivial representations of G_H are complex. It can also be done if there are real representations but here some of the real representations may have a zero fermion number. All hyperhadrons with zero "fermion" number should be considered hypermesons.
17. This analysis is an extension to hypercolor of the effective operator analysis for B-L conservation given by S. Weinberg, Phys. Rev. Lett. 43, 1566 (1979); F. Wilczek and A. Zee, Phys. Lett. 88B, 311 (1979).
18. The annihilation stops when the rate of annihilation equals the rate of expansion R/R of the universe. For the method of analysis see, e.g., G. B. Dover, F. K. Gaisser, and G. Steigman, Phys. Rev. Lett. 42, 1117 (1969).
19. P. Frampton and S. Glashow, Phys. Rev. Lett. 44, 1481 (1980).

20. The result that $W^{\pm}$ lie on Regge trajectories has been discussed within the context of elementary Higgs scalars by M. Grisaru and H. Schnitzer, Phys. Rev. D 20, 784 (1979).
21. Here the Goldstone bosons associated with the spontaneous breaking of the global chiral symmetries of quarks by the color interactions are ignored. Since $f_\pi/F_\pi \approx 10^{-3}$ this is a good approximation.
22. M. E. Peskin, "The Alignment of the Vacuum in Theories of Technicolor", Saclay preprint DPh-T/80/46 (1980).
23. J. P. Preskill, "Subgroup Alignment in Hypercolor Theories", Harvard preprint HUTP-80/A033 (1980).
24. S. Dimopoulos, Nucl. Phys. B168, 69 (1980).
25. For an example of a hypercolor model with real representations see E. Farhi and L. Susskind, Phys. Rev. D 20, 3404 (1979). Also see Ref. 24.
26. R. Dashen, Phys. Rev. 183, 1245 (1969).
27. S. Weinberg, Phys. Rev. Lett. 40, 223 (1978); F. Wilczek, Phys. Rev. Lett. 40, 279 (1978). A comprehensive review of experimental limits on axions is given by T. W. Donnelly, *et al.*, Phys. Rev. D 18, 1607 (1978).
28. G. L. Kane, "Could Higgs Bosons Be Found Before LEP", University of Michigan preprint UM79-37 (1979), and the references contained therein.
29. The total mass of $P^{\pm}$ includes a contribution from electroweak interactions $m_W \approx 5$ GeV as well as the sideways interaction contribution m_S. Thus $m_{\pm}^2 = \sqrt{m_S^2 + m_W^2}$ while $m_0 = m_S$.
30. F. Wilczek, Phys. Rev. Lett. 39, 1304 (1977).
31. R. N. Cahn and H. Harari, "Bounds on the Masses of Neutral Generation-Changing Gauge Bosons", LBL preprint LBL-10823 (1980).
32. G. L. Kane and R. Thun, "Searches for Effects of Flavor-Changing Neutral Currents", University of Michigan preprint UMHE80-8 (1980).
33. P. Herczeg, "Symmetry Violating Kaon Decays", Los Alamos preprint LA-UR-79-2616 (1979).
34. Four Fermi terms involving two leptons and two hyperfermions also contribute to the processes $\mu \to e^+\gamma$ and $\mu N \to eN$. Naively they might be expected to give much larger contributions to these processes than the interactions considered in the text. A detailed analysis however shows this not to be true.
35. E. Eichten, K. D. Lane, and J. Preskill, Phys. Rev. Lett. 45, 225 (1980).
36. R. Dashen, Phys. Rev. D 3, 1879 (1971).
37. G. 't Hooft, Phys. Rev. Lett. 37, 8 (1976); R. Jackiw and C. Rebbi, Phys. Rev. Lett. 37, 172 (1976); C. Callan, R. Dashen, and D. Gross, Phys. Lett. 63B, 334 (1976).
38. V. Baluni, Phys. Rev. D 19, 2227 (1979); R. Crewther, P. Di Vecchia, G. Veneziano, and E. Witten, Phys. Lett. 88B, 123 (1979).
39. W. B. Dress *et al.*, Phys. Rev. D 15, 9 (1977); I. S. Altarev *et al.*, Leningrad Nucl. Phys. Inst. preprint 430, 1 (1978); N. F. Ramsey, Phys. Rpt. 43C, 409 (1978).
40. M. Kobayaski and K. Maskawa, Prog. Theor. Phys. 49, 652 (1973).

A Review of D and B Meson Decays

Michael S. Chanowitz
Lawrence Berkeley Laboratory, Berkeley, California 94720

I. INTRODUCTION

There is a class of down and out members of the legal profession, known as "ambulance chasers", who materalize at automobile accidents hoping to find clients for an injury suit. Glashow has aptly used this term in referring to a widely diffused style of doing theoretical physics. In theoretical physics, if not in the law, chasing ambulances is not necessarily an ignoble practice. It is part of what distinguishes us from the mathematicians who do not have experimental colleagues to stimulate and guide them. Some noble discoveries have been made trying to fit a curve, a notable example being Planck's fit to the black body radiation spectrum.

The emerging experimental picture of charmed meson decays has brought the ambulance chasers out in force. There is a sense of disaster in the air, but the actual magnitude of the accident is not yet clear. There is still the possibility that we are dealing with a mere "fender-bender". While the simplest picture of charm decays[1] led us to expect that D^+ and D^o would have nearly equal lifetimes, $\tau_+ \simeq \tau_o$, data from emulsion experiments and from SPEAR make it likely that τ_+ is appreciably greater that τ_o. But we cannot tell yet whether the ratio is actually $\leqslant 3$ or $>> 5$. In the former case an explanation can probably be found in the context of the generally accepted theoretical framework. But if τ_+/τ_o is much larger than 5, I would say that the recent optimism about our understanding of all non-leptonic decays is called into question, including the basis of the $\Delta I = \frac{1}{2}$ rule for strange particle decays.

The plan of my talk is to review briefly the theoretical picture of K and D decays in order to explain why we did not expect large enhancements[1] in D decays. This expectation is contrasted with the available data on lifetimes and semileptonic branching ratios, which hint at large enhancements but are still ambiguous. I will then discuss some of the theoretical ideas proposed to explain the enhancement of D^o, and possibly F^+, nonleptonic decays. One of these, together with data presented earlier in this session, suggests a crazy way to try to detect the F^+ and a possible gluonium state in one fell swoop. Finally I will discuss what we learn about nonleptonic enhancements from decays into exclusive final states. In particular, the new data on $D \to \rho K/\pi K^*$ presented in this session have interesting implications for models which predict large enhancements.

In a briefer section, I will discuss two topics involving B decays which are related to the possibility of substantial

ISSN:0094-243X/80/620048-19$1.50

nonleptonic enhancements. The first is the worrisome prospect that though we need to know the Kobayashi-Maskawa angles to determine whether B decays are enhanced, it may be a practical requirement to first understand the pattern of nonleptonic enhancements in order to extract the K-M angles from the data. Second I will discuss the decays of B mesons into $J/\psi(3095)$ from the perspective of what these decays might teach us about the dynamics of nonleptonic enhancement.

II. D MESON DECAYS

A. Naive Expectations

The simplest imaginable picture of D meson decays is that the c quark decays into an s quark by bremsstrahlung of a virtual W^+ boson which materializes as a $u\bar{d}$ or $\nu\bar{\ell}$ pair. The light antiquark- $\bar{u}$ for D^o, $\bar{d}$ for D^+, $\bar{s}$ for F^+ - is a passive "spectator" to the decay. With this model we expect the D^o, D^+ and F mesons all to have lifetimes equal to that of the c quark itself,

$$\tau_+ = \tau_o = \tau_F = \tau_c \ . \qquad (2.1)$$

The branching ratio for μ semileptonic decays is just the fraction of W^+ bosons which materialize as $\nu_\mu\bar{\mu}$ pairs,

$$B(D \to \nu_\mu\bar{\mu}X) = \frac{1}{3+1+1} = \frac{1}{5} \qquad (2.2)$$

where the $u\bar{d}$ pair has a weight of three for color. The total width may be scaled from the rate for $\mu \to \nu_\mu e\bar{\nu}_e$:

$$\Gamma_{TOT} = 5\left(\frac{m_c}{m_\mu}\right)^5 \frac{1}{2}\Gamma(\mu \to \nu_\mu e\bar{\nu}_e)$$

$$= \left[(1.5^{+1.1}_{-0.7}) \cdot 10^{-12}\,\mathrm{sec}\right]^{-1} \qquad (2.3)$$

where I assume $m_c = 1.5 \pm 0.15$ GeV and $m_s/m_c = 0.3$. The uncertainly in (2.3) is just that which reflects the assumed spread in m_c; the factor $\frac{1}{2}$ is the reduction in the available phase space due to the strange quark mass. In the preceding I have discussed only Cabibbo allowed decays, which always give K's in the final state. The fraction of Cabibbo suppressed decays would be

$$B(D \to \text{no } K) \lesssim .08 \ , \qquad (2.4)$$

slightly larger than $\tan^2\theta_c$ because of the greater available phase space. In the four quark GIM model[2] we would expect the bound to be saturated while for six[3] or more quarks the ratio could be smaller.

There are two other lowest order Feynman diagrams to consider. In these the light antiquark is not a passive spectator. The F^+ may decay through a virtual W^+ in the s-channel into a $u\bar{d}$ pair: $F^+ \sim c\bar{s} \to "W^+" \to u\bar{d}$. The $D^o \sim c\bar{u}$ may decay by exchange of a W boson in the t-channel into an $s\bar{d}$ pair. No such mechanism is possible for the D^+. These diagrams were thought to be negligible for two reasons - a factor $(m_{u,d}/m_D)^2 \sim \frac{1}{25}$ due to helicity suppression (as in $\pi \to e\nu$) and a small factor $(F_D/m_D)^2 \sim \frac{1}{5} - \frac{1}{100}$ reflecting the probability for the initial quark pair to coincide in space as they must in these "annihilation" diagrams. Here F_D is the analogue of F_π for the pion, which in nonrelativistic models is proportional to the value of the wave function at the origin.

B. Nonleptonic Enhancement of K decays

Even if the annihilation diagrams are as small as presumed, it is not at all clear that the predictions (2.1) to (2.4) should be taken seriously. The point is of course that such a picture fails totally to account for the observed factor 400 enhancement in strange particle decays with $\Delta I = 1/2$. Our expectations for D decays are strongly coupled to our understanding of the K decays. In fact there was a wide-spread expectation among theorists that (2.1) - (2.4) would be a good zero'th order approximation. To explain this I have to make a slight detour to discuss the present understanding of the $\Delta I = 1/2$ rule.

If we consider the lowest order QCD corrections to the weak decay $s \to ud\bar{u}$, the diagrams with loops containing both a W boson and a gluon give rise to large factor $\alpha_s(\mu)\, \ell n \frac{M_W}{\mu} >> 1$, where μ is the renormalization point typically taken to be of order 1 GeV. The leading logs in this parameter may be summed to all orders using the renormalization group. The result[4] is that the effective four fermion interaction which in the absence of strong interactions is

$$\mathscr{H}_{\Delta S=1} \propto (\bar{s}u)_L(\bar{u}d)_L \qquad (2.5)$$

becomes the sum of two operators

$$\mathscr{H}_{\Delta S=1} \propto f_- O_- + f_+ O_+ \qquad (2.6)$$

where

$$O_\pm = \frac{1}{2}\left[(\bar{s}u)_L(\bar{u}d)_L \pm (\bar{s}d)_L(\bar{u}u)_L\right] . \qquad (2.7)$$

The abbreviated notation is that $(\bar{u}d)_L$ is the usual V-A weak current with an implied sum over color indices, $(\bar{u}d)_L \equiv \sum_{\alpha=1}^{3} (\bar{u}_\alpha d_\alpha)_L$. For $f_- = f_+ = 1$ (2.6) becomes the zero'th order interaction (2.5).

Now in O_- the u and d quark fields are arranged antisymmetrically, I = 0, so the net isospin of O_- is the I = (1/2) of the $\bar{u}$. In O_+ the ud pair has I = 1 so the net isospin of O_+ may be 1/2 or 3/2, $1 \otimes (1/2) = (1/2) \oplus (3/2)$. Therefore if f_- is much larger than f_+ we will have found an explanation for the $\Delta I = 1/2$ rule. The actual result of the calculation[4] is

$$f_- = \frac{1}{f_+^2} = \left[\frac{\alpha_s(\mu)}{\alpha_s(M_W)}\right]^{0.48} \cong 2.4 \ ,$$

$$f_+ \cong 0.65, \tag{2.8}$$

and

$$\left(\frac{f_-}{f_+}\right)^2 \cong 20. \tag{2.9}$$

Equation (2.9) is a generous estimate, but it falls far short of the needed 400.

A second possible source of $\Delta I = 1/2$ enhancement has been much discussed in recent years - the so-called penguin diagrams.[5] These occur by virtue of the strangeness - changing neutral currents which the GIM mechanism[2] was invented to suppress. In the GIM model they do still occur at a level given by the mass differences among the quarks, such as $m_c - m_u$. In the penguin diagrams, the loops contain the $Q = +(2/3)$ quarks and the W-boson, and the expansion parameter summed to leading-log order is $\alpha_s(\mu)\ell n \frac{m_c}{\mu}$. Because these logs are appreciably smaller than the log M_W effects discussed above, the leading log approximation must be taken with an even larger grain of salt in this case. The penguin diagrams are pure $\Delta I = 1/2$ because their net effect is to change an s quark into a d quark. The $\bar{u}$ or $\bar{d}$ quark in the K meson initial state interact in these diagrams only by gluon exchange which is flavor-preserving.

This unlikely mechanism has two large factors in its favor. First, there are large color factors of order 10. Second, because the $\bar{u}$ or $\bar{d}$ quark interact by the purely vectorial gluon interaction, the penguin diagrams give rise to four quark operators with left-right helicity structure rather than the usual Fermi left-left structure of Eqs. (2.5)-(2.7). The L-R structure is not susceptible to the suppression that occurs in Fermi decays of a pseudosular meson, such as $\pi \to e\nu$. For instance, a penguin induced operator is

$$\mathcal{H}^{(\mathrm{Penguin})}_{\Delta S=1} \propto (\bar{s}d)_L(\bar{u}u + \bar{d}d + \ldots)_R \tag{2.10}$$

A crude estimate of the helicity effect yields

$$\frac{\langle\pi\pi|(\bar{s}d)_L(\bar{u}u)_R|K\rangle}{\langle\pi\pi|(\bar{s}d)_L(\bar{u}u)_L|K\rangle} \sim \frac{m_\pi^2}{m_u m_s} \sim 30 \tag{2.11}$$

where $m_{u,s}$ are in this case the bare or current quark masses (because Eq. (2.11) is derived from the equations of motion and therefore uses the masses which appear in the Lagrangian rather that the effective constituent masses), $m_u \sim 5$ MeV and $m_s \sim 120$ MeV. A more careful estimate using the M.I.T. bag model gives a similar result.[6] The conclusion is that despite the considerable uncertainties, which mean these estimates are somewhere between qualitative and (semi)^n-quantitative with $n \geqslant 2$, it is plausible that the penguin mechanism may be the origin of the factor 400 enhancement in K decays. If the M.I.T. bag calculation is reliable, the factor 400 is not a synergistic combination of the penguin and the $\ln M_W$ effects due to the operator O_-: the two contributions interfere destructively so the enhancement must result from the overwhelming effect of the penguin mechanism alone.

C. D Decays with QCD Corrections

If we accept the penguin diagrams as the basis for the $\Delta I = (1/2)$ rule in K decays, it is easy to see why the free quark model was expected to be a reasonable guide to D decays. The effect of penguin diagrams should be vastly smaller in D decays than in K decays[7]. First, the penguin mechanism is Cabibbo-suppressed. This is no handicap in the K system but for D decays it costs a factor $\sim (1/20)$ in the rate. Second, the helicity enhancement of Eq. (2.11), which was a factor $m_\pi^2/m_u m_s \sim 30$ in the amplitude, becomes in the D case $m_\pi^2/m_u m_c \sim 2.5$. So relative to their importance in K decays, the penguins are demoted by roughly $\frac{1}{20} \cdot \left(\frac{2.5}{30}\right)^2 \sim \frac{1}{3000}$ in the D decays. It seems very unlikely that they could contribute a sizeable enhancement to the width of D mesons.

The analogue of the $\ln M_W$ effects summarized in Eqs. (2.6)-(2.9) are less important for D than K decays because $\alpha_s(m_c) \ \ln \frac{M_W}{m_c}$ is smaller than the analogous parameter in the K system. The calculation proceeds just as before.[8,9] The lowest order four quark interaction

$$\mathscr{H}_{\Delta C=1} \propto (\bar{c}s)_L(\bar{u}d)_L \tag{2.12}$$

becomes in leading log approximation

$$\mathscr{H}_{\Delta C=1} \propto f_- O_-(\Delta C=1) + f_+ O_+(\Delta C=1) \tag{2.13}$$

where

$$O_\pm = \frac{1}{2} \left[(\bar{c}s)_L(\bar{d}u)_L \pm (\bar{c}u)_L(\bar{d}s)_L\right] \tag{2.14}$$

$f_\pm$ are nearer the free quark values $f_+ = f_- = 1$ than in Eq. (2.8):

$$f_- = \left(\frac{\alpha_s(m_c)}{\alpha_s(M_W)}\right)^{0.48} \sim 2$$

$$f_+ = \frac{1}{\sqrt{f_-}} \sim 0.7 \tag{2.15}$$

The operator $O_-(\Delta C=1)$ is in the 6* of SU(3) while $O_+(\Delta C=1)$ is in the 15 of SU(3). So we expect a moderate "6-dominance." But no sizeable enhancement is expected in the rate of nonleptonic decays. The spectator ansatz for D decays together with Eqs. (2.13)-(2.15) implies that the total nonleptonic width is[10]

$$\Gamma_{NL}(D) = \frac{2f_+^2 + f_-^2}{3} \Gamma_{NL} \text{ (free quark)}$$

$$\sim \frac{5}{3} \Gamma_{NL}\text{(free quark)}. \tag{2.16}$$

That is, the color factor 3 is replaced by a factor 5.

QCD corrections to the width of the semi-leptonic decays are also expected to be moderate. No $\ln M_W$ effects occur because there are no loops containing both W-bosons and gluons. The $O(\alpha_s)$ correction to $c \rightarrow s\nu\bar{\ell}$, computed from the sum of virtual corrections and real gluon emission, is[11]

$$\Gamma_{SL} = (1 - \frac{1}{2}\alpha_s(m_c))\Gamma_{SL}\text{(free quark)}$$

$$\sim \frac{2}{3} \Gamma_{SL} \text{ (free quark)} \tag{2.17}$$

using $\alpha_s(m_c) \sim 0.7$ for $\Lambda \sim 0.5$ GeV. The perturbation expansion is not quantitatively believable here but may be a reliable qualitative guide.

In this framework of QCD corrections, the free quark model predictions Eqs. (2.1) - (2.4) are very little modified. We are still committed to the spectator ansatz

$$\tau_+ = \tau_o = \tau_F . \tag{2.18}$$

The estimate for the semi-leptonic branching ratio decreases by a factor 2, as a consequence of (2.16) and (2.17):

$$B(D \rightarrow \nu_\mu \bar{\mu} X) \sim \frac{2/3}{5 + \frac{2}{3} + \frac{2}{3}} \sim \frac{1}{10} \tag{2.19}$$

The estimate of the life-time decreases by only $\sim 30\%$,

$$\Gamma_{TOT} \sim \frac{5 + 2 \cdot 2/3}{5} \Gamma_{TOT}\text{(free quark)}$$

$$\sim \left[(1.2^{+0.9}_{-0.6}) \ 10^{-12} \text{sec} \right]^{-1} . \tag{2.20}$$

The estimate of the rate for Cabibbo-suppressed decays is unchanged from (2.4):

$$B(D \rightarrow \text{no } K) \lesssim .08 \tag{2.21}$$

D. Comparison with the Data

The experimental situation is different from these predictions but it is not yet clear how different. By SU(2) symmetry the semi-leptonic widths of D^+ and D^o must be equal, so that

$$\frac{B(D^+ \rightarrow \nu_\mu \bar{\mu} X)}{B(D^o \rightarrow \nu_\mu \bar{\mu} X)} = \frac{\tau_+}{\tau_o} . \tag{2.22}$$

One measurement[12] of the semileptonic branching ratios then gives $\tau_+/\tau_o \gtrsim 4$ (95% CL) and another[13] gives $\sim 3^{+7}_{-2}$. Early results[14] from the E-531 emulsion experiment at Fermilab with still sparse statistics, give $\tau_+/\tau_o \sim 10^{+20}_{-7}$. $B_{SL}(D^+)$ is measured at $\sim 0.16 \pm 0.05$ (Mark II[13]) or $\sim 0.24 \pm 0.04$ (DELCO)[12] while $B_{SL}(D^o)$ is $\sim 0.052 \pm 0.033$ (Mark II)[13] and $\lesssim 0.045$ (DELCO).[12] The emulsion data[14] give $\tau_+ \sim (1.00^{+0.89}_{-0.46}) \cdot 10^{-12}$ sec and $\tau_o \sim (0.93^{+0.52}_{-0.29}) \cdot 10^{-13}$ sec.

The branching ratios for Cabibbo supressed decays are reported[13] to be $B(D^o \to \text{no } K) \sim 0.25 \pm 0.11$ and $B(D^+ \to \text{no } K) \sim 0.41 \pm 0.16$.

The data suggest that Cabibbo allowed D^+ decays are not enhanced but that D^o nonleptonic decays are enhanced, perhaps substantially. For instance, Eq. (2.20) agrees remarkably with the emulsion measurement of τ_+ (even the reported experimental uncertainty is correctly predicted in (2.20)!). $B_{SL}(D^+)$ appears to agree with the free quark model prediction of 1/5; given the theoretical uncertainties in (2.16) and (2.17) the discrepancy with the expected 1/10, Eq. (2.19), is not unsettling. The branching ratios for $B(D \to \text{no } K)$ hint at the possibility that Cabbibo suppressed nonleptonic decays of the D^+ may share the enhancement of the Cabbibo allowed D^o decays. But a hard look at the quoted experimental uncertainties shows that the predictions (2.18) - (2.21) may yet survive at the level of a factor of two or better. Equations (2.19) and (2.20) should not be trusted at more than the factor two level in any case. Equations (2.18) and (2.21) are more reliable consequences of the assumed theoretical framework, but it is not hard to imagine effects which could also cause them to be modified by a factor two or three. So it is not yet clear whether theory and experiment are in a full scale collision or if the theorists will be able to walk away with only a dented fender and a few scratches to show.

E. Other Sources of Nonleptonic Enhancements

Regardless of the scale of the accident, it is clearly interesting to think now about additional sources of non-leptonic enhancements. One idea, which goes back to before the discovery of the J/ψ, is that decays into nonexotic channels are enhanced[15] or, equivalently, that the net quantum numbers of the final state of enhanced decays may be represented by the same number of quarks as the initial state.[16,17] D^o and F^+ decay into states with the net quantum numbers of $s\bar{d}$ and $u\bar{d}$ respectively so they are enhanced, but the D^+ decays into a final state with exotic quantum numbers

$s\bar{u}\bar{d}\bar{d}$. Cabibbo supressed decays of D^+ lead to a final state with the net quantum numbers of $u\bar{d}$, hence they may be a larger fraction of all D^+ decays. Had our understanding of K decays not "progressed" as described in Section B, this would probably have been the prevalent line of theoretical reasoning and $\tau_o << \tau_+$ would have seemed a likely prospect.

But this is only a rule or mnemonic. Even if it turns out to be correct we will still want to know its dynamical origin. One possibility is that hadronic final state interactions are the basis of the rule. Broad s-channel resonances could enhance the D^o and F^+ decay amplitudes. Hadronic final state interactions may have a tremendous effect on decays into particular final states (see Section F), but their effect on the total width should be less dramatic. Even if there were an s-channel $K\pi$ resonance around 1.86 GeV, its effect would be diminished by its probably large total width. We can get a feeling for how large the effect may be by looking at the structure in the e^+e^- total cross section between 1.5 and 2 GeV. My conclusion is that hadronic final state interactions are unlikely to contribute more than a factor two to τ_+/τ_o. If in the end we learn that $\tau_+/\tau_o \sim 2$, then hadronic f.s.i. could be part of the reason.

The "annihilation" diagrams are another possible dynamical explanation of the quark number conservation rule. It was argued in Section A that these are suppressed by a helicity factor $(m_u/m_c)^2$ and by the small probability for annihilation $(F_D/m_D)^2$. But in a first order calculation of the QCD corrections which assumes a nonrelativistic bound state model of the D, it is found that the annihilation diagrams might make a substantial contribution.[18] After bremsstrahlung of a gluon the initial state is no longer in an s-wave so that there is no helicity suppression of the subsequent weak decay. Furthermore the scale m_D in the factor $(F_D/m_D)^2$ is replaced by a light constituent quark mass m_u. The conclusion is that

$$\frac{\tau_+}{\tau_o} \sim 1 + 0.4\alpha_s(m_c)\left(\frac{m_D}{m_c}\right)^5 \frac{F_D^2}{m_u^2} \; . \tag{2.23}$$

A reasonable guess for F_D based on the e^+e^- decay widths of old and new vector mesons gives[37] $F_D \sim 150$ MeV. and then $\tau_+/\tau_o \sim 1.2$. Of course even if we know F_D precisely, (2.23) could be no more than a rough guide given the unreliability of the perturbation expansion for $\alpha_s(m_c) \cong 0.7$.

Another attitude[19] is more realistic but has less predictive power. As a relativistic bound state the D surely has a component of its wave function with one, two or many gluons. For this component the quark and antiquark need not be in an s-wave and can annihilate with no helicity suppression. Predictions are made by treating the magnitude of the gluonic component of the wave function as a free parameter. For reasonable values of this parameter the D^o and F annihilation amplitudes might be substantially larger than the spectator quark amplitude that yielded $\tau \sim 10^{-12}$ sec.

D^o annihilation proceeds by the t-channel exchange of a W-boson so the $c\bar{u}$ pair may be in a color octet and a single gluon is sufficient to make up the initial color singlet state. But F^+ annihilation proceeds through an s-channel W-boson so two gluons in a color singlet are needed to balance the color. Since the second gluon can be soft, this does not mean that F^+ annihilation is necessarily supressed relative to D^o annihilation. F^+ semileptonic decays may also proceed by this mechanism; hence these decays may be a source of gluon rich hadrons such as η' and possibly even of gluonium states.[20]

A good test of the quark number rule is given by the isospin structure of the final states.[21] The final states of D^o decays must be dominantly $I = \frac{1}{2}$ since they are formed from an $s\bar{d}$ pair plus $I = 0$ gluons. Therefore we expect

$$\frac{\Gamma(D^o \to \pi^+ K^-)}{\Gamma(D^o \to \pi^o \bar{K}^o)} = \frac{\Gamma(D^o \to \rho^+ K^-)}{\Gamma(D^o \to \rho^o \bar{K}^o)} = 2 \qquad (2.24)$$

and so on. For the $K\pi\pi$ mode we have[21]

$$\Gamma(D^o \to \bar{K}^o \pi^o \pi^o) = \frac{1}{2}\Gamma(D^o \to \bar{K}^o \pi^+ \pi^-) - \frac{1}{4}\Gamma(D^o \to \bar{K}\pi^+ \pi^o). \qquad (2.25)$$

There are also constraints on the decays into K plus n pions for any value of n.

Before leaving the subject of annihilation diagrams I want to make a crazy suggestion which is motivated by the experimental presentations of Coyne and Scharre in this session. We have heard that the anticipated F-associated rise in η production is not yet seen at the Crystal Ball, and we have also heard of the very large signal for $\psi \to \gamma + E$ (1420) seen at both the Mark II and the Crystal Ball. Since the E (1420) is not generally a prominent state in hadronic reactions and since $\psi \to \gamma X$ may be a copious source of gluon production, it is natural to speculate that the E may be a gluon-rich state, perhaps even a gluonium state. Unless the chain of reasoning is checked by a strong cup of coffee, it leads to the notion that the glueball _cum_ E may be a good tag for F production. $F^+ \to E\pi^+$ should be the dominant mode, since $F^+ \to E\rho^+$ is only permitted on the ρ tail. So by looking for $F^+ \to E\pi^+ \to (K\bar{K}\pi)\pi^+$ it might be possible to detect the F and a glueball in one fell swoop. If the annihilation amplitude is large this mode could be a very substantial fraction of F decays.

Another proposal[22,23] to explain $\tau_+ >> \tau_o$ requires assuming that f_-/f_+ is much larger than the leading log value, Eq. (2.15), and that gluon final state interactions do not change the color structure of the two $q\bar{q}$ color singlet clusters created according to the spectator quark ansatz. Recall that in leading log approximation

$$\begin{aligned} \mathcal{H}_{\Delta C=1} &= f_- O_- + f_+ O_+ \\ O_\pm &= \frac{1}{2}(O_1 \pm O_2) \\ O_1 &= (\bar{c}s)_L(\bar{d}u)_L \\ O_2 &= (\bar{c}u)_L(\bar{d}s)_L \end{aligned} \tag{2.26}$$

where for instance $(\bar{d}u)_L$ denotes the usual V-A color singlet current responsible for beta decay. Now it is easy to see that when O_1 acts on $D^+ \sim c\bar{d}$ it creates two color singlet quark-antiquark pairs, $(\bar{d}s)$ and $(\bar{d}u)$. O_2 acts on D^+ to create the same two pairs. But O_1 and O_2 create different color singlet pairs when they act on $D^o \sim c\bar{u}$: O_1 creates $(\bar{u}s)(\bar{d}u)$ while O_2 creates $(\bar{u}u)(\bar{d}s)$. Therefore in D^+ decays the contribution of $O_- = \frac{1}{2}(O_1 - O_2)$ may cancel coherently in the amplitude whereas in D^o decays no such cancellation occurs because O_1 and O_2 give rise to different final states which add incoherently. The result is that

$$\Gamma_{NL}(D^+) = \frac{4}{3} f_+^2 \Gamma_{NL}(\text{free quark}) \tag{2.27}$$

$$\Gamma_{NL}(D^o) = \frac{1}{3}(2f_+^2 + f_-^2)\Gamma_{NL}(\text{free quark}) \tag{2.28}$$

$$\frac{\tau_+}{\tau_o} = \frac{f_-^2 + 2f_+^2 + 4/3}{4f_+^2 + 4/3} \tag{2.29}$$

With the values obtained in leading log approximation, (2.15), $f_- = (f_+)^{-2} \sim 2$, the enhancement is $\tau_+/\tau_o \sim 2$. For the value $f_- = (f_+)^{-2} \sim 5$, chosen[22] to fit the observed ratio of $D^o \to K\pi$ decays as discussed in the next Section, the result is $\tau_+/\tau_o \sim 10$.

According to this approach, which I shall refer to as "enhanced 6 dominance", the effective $\Delta C = 1$ Hamiltonian is overwhelmingly dominated by the operator which is in the 6* of SU(3).

For the decays of D^o and F^+ into two pseudoscalars, this has the same consequences for the isospin of the final state as the quark number conservation rule. The symmetric product of two octets projects into the 27 and 8 of SU(3). The D^o is in the SU(3) 3* so the final state created by action of O_- is given by $3^* \otimes 6^* = 8 \oplus 10^*$. Therefore 6 dominance requires the D^o final state to be in an SU(3) octet and therefore in the $S = -1$ isodoublet. However 6 dominance does not require an $I = 1/2$ final state for decays of D^o into a pseudoscalar plus a vector, since in this case the antisymmetric 10^* is a permissible final state.

F. Exclusive Channels

In this section I want to give some examples of what can be learned from exclusive final states about the dynamical issues discussed in the preceding sections.[24] Consider first the measured rates[13] for $D^o \to K\pi$:

$$B(D^o \to K^-\pi^+) = 0.028 \pm 0.005 ,$$

$$B(D^o \to \bar{K}^o\pi^o) = 0.021 \pm 0.009 . \tag{2.30}$$

The $K\pi$ final state is a sum of $I = 1/2$ and $I = 3/2$ components

$$\mathcal{M}(D^o \to K^-\pi^+) = \sqrt{\frac{2}{3}}\, a_1 + \sqrt{\frac{1}{3}}\, a_3$$

$$\mathcal{M}(D^o \to \bar{K}^o\pi^o) = \sqrt{\frac{1}{3}}\, a_1 - \sqrt{\frac{2}{3}}\, a_3 \tag{2.31}$$

The quark number conservation rule requires the final state to be dominantly $I = 1/2$, so we expect

$$\frac{B(D^o \to K^-\pi^+)}{B(D^o \to \bar{K}^o\pi^o)} = 2 \tag{2.32}$$

which is consistent with the data (2.30).

The same model which led to Eqs. (2.27) and (2.23) was first applied[8,10] to the $D \to K\pi$ exclusive decays. The $q\bar{q}$ clusters are identified with the appropriate K and π mesons and again color rearrangement due to gluon final state interactions is neglected. The result is

$$\frac{B(D^o \to K^-\pi^+)}{B(D^o \to \bar{K}^o\pi^o)} = 2 \cdot \frac{(2f_+ + f_-)^2}{(2f_+ - f_-)^2} \tag{2.33}$$

Using the leading log estimate for $f_\pm$ the ratio is very large; for instance, substituting (2.15) I find 60 for the right hand side of (2.33). The choice[22] $f_- = f_+^{-2} \sim 5$, which gives $\tau_+/\tau_o = 10$ in (2.29), gives a ratio of 4 in (2.33), compatible with the data at the $1\frac{1}{2}\sigma$ level.

The prediction using the leading log estimate for $f_\pm$ appears to be dramatically excluded by the data. In fact the situation is not so simple, because of the effect of hadronic final state interactions[25,26] which shift the phases of the amplitudes in (2.31):

$$\mathcal{M}(D^o \to K^-\pi^+) = \sqrt{\frac{2}{3}}\, e^{i\delta_1} a_1 + \sqrt{\frac{1}{3}}\, e^{i\delta_3} a_3$$

$$\mathcal{M}(D^o \to \bar{K}^o\pi^o) = \sqrt{\frac{1}{3}}\, e^{i\delta_1} a_1 - \sqrt{\frac{2}{3}}\, e^{i\delta_3} a_3 \qquad (2.34)$$

The leading log prediction that the $K^-\pi^+$ mode is much more frequent than the $\bar{K}^o\pi^o$ mode means that in (2.31) we expect a cancellation, $a_1 \sim \sqrt{2}\, a_3$. But this delicate cancellation is easily undone if, as is not unlikely, $|\delta_1 - \delta_3|$ is large. If we assume for illustration that $a_1 = \sqrt{2}\, a_3$ exactly, then[25]

$$\frac{\Gamma(D^o \to K^-\pi^+)}{\Gamma(D^o \to \bar{K}^o\pi^o)} = \frac{1}{8}\left[9 \cot^2\left(\frac{\delta_1 - \delta_3}{2}\right) + 1\right] \qquad (2.35)$$

which can vary from infinity for $\delta_1 = \delta_3$ to 1/8 for $\delta_1 - \delta_3 = \frac{\pi}{2}$. The situation is further complicated by the inelasticity in the $K\pi$ channel, which is not included in (2.34). The conclusion is that we probably cannot learn much about the leading log approximation from the decays $D^o \to K\pi$. On the other hand, the predictions of the quark number rule and of enhanced 6 dominance are far less sensitive to these final state phases since they imply $|a_1| \gg |a_3|$.

Decays into a pseudoscalar plus a vector meson are of interest because, as discussed in the previous section, they offer more possibilities for distinguishing between the quark number rule and the enhanced 6 dominance hypothesis. The quark number rule always requires the D^o to decay to an $I = 1/2$ final state, so that

$$\frac{\Gamma(D^o \to \rho^+K^-)}{\Gamma(D^o \to \rho^o\bar{K}^o)} = \frac{\Gamma(D^o \to \pi^+K^{*-})}{\Gamma(D^o \to \pi^o\bar{K}^{*o})} = 2 \qquad (2.36)$$

Equation (2.36) need not hold for the enhanced 6 dominance hypothesis which does, together with SU(3) symmetry, imply that[27]

$$\Gamma(D^+ \to \rho^+\bar{K}^o) = \Gamma(D^+ \to \pi^+\bar{K}^{*o}) \tag{2.37}$$

Together with the SU(2) symmetry relations

$$\mathcal{M}(D^o \to \rho^+K^-) + \sqrt{2}\,\mathcal{M}(D^o \to \rho^o\bar{K}^o) = \mathcal{M}(D^+ \to \rho^+\bar{K}^o)$$

$$\mathcal{M}(D^o \to \pi^+K^{*-}) + \sqrt{2}\,\mathcal{M}(D^o \to \pi^o\bar{K}^{*o}) = \mathcal{M}(D^+ \to \pi^+\bar{K}^{*o})$$

(2.37) implies the inequality[28]

$$\frac{B(D^+ \to \pi^+\bar{K}^{*o})}{B(D^o \to \rho^+K^-)} \gtrsim \frac{\tau_+}{\tau_o}\left(1 - \sqrt{\frac{2B(D^o \to \rho^o\bar{K}^o)}{B(D^o \to \rho^+K^-)}}\right)^2 \tag{2.38}$$

Scharre in this session has reported measurements of these branching ratios. Two D^o decay modes are measured at

$$B(D^o \to \rho^+K^-) \gtrsim \frac{2}{3}B(D^o \to \pi^+\pi^oK^-), \tag{2.39}$$

$$B(D^o \to \pi^+K^{*-})\cdot B(K^{*-} \to \pi^-\bar{K}^o) \gtrsim \frac{2}{3}B(D^o \to \pi^+\pi^-\bar{K}^o), \tag{2.40}$$

from which we can deduce that

$$B(D^o \to \rho^o\bar{K}^o) \lesssim \frac{1}{3}B(D^o \to \pi^+\pi^-\bar{K}^o), \tag{2.41}$$

$$B(D^o \to \pi^+K^{*-})\cdot B(K^{*-} \to \pi^oK^-) \lesssim \frac{1}{3}B(D^o \to K^-\pi^+\pi^o). \tag{2.42}$$

Using the three body branching ratios

$$B(D^o \to K^-\pi^+\pi^o) = 0.085 \pm 0.032, \tag{2.43}$$

$$B(D^o \to \bar{K}^o\pi^+\pi^-) = 0.038 \pm 0.012, \tag{2.44}$$

we have the lower bound

$$\frac{B(D^o \to \rho^+K^-)}{B(D^o \to \rho^o\bar{K}^o)} \gtrsim 4.4 \pm 2.1, \tag{2.45}$$

where I have combined the uncertainties in (2.43) and (2.44) in quadrature. At the moment (2.45) is still just compatible at the 1σ level with the quark number conservation rule, Eq. (2.36), but a contradiction will develop if the bounds (2.39) - (2.42) are improved significantly. In this case the failure to observe $D^+ \to \pi^+\bar{K}^{*o}$ could also lead to a contradiction with the 6-dominance inequality, Eq. (2.38).

III. B Meson Decays

There is (fortunately) only enough time remaining to discuss two aspects of B meson decays, both of which are related to the issues discussed in the previous section. I will simply assume the b quark assignment of the standard Kobayashi-Maskawa model. I will discuss the problem of determining the K-M angles if there are significant enhancements in nonleptonic B decays and what we can learn about nonleptonic dynamics from the decays $B \to \psi X$.

A. Determining the K-M Angles

In the standard six quark model[3] the charged weak current is given by

$$J = (\bar{u}\ \bar{c}\ \bar{t})_L\ U \begin{pmatrix} d \\ s \\ b \end{pmatrix}_L , \tag{3.1}$$

where

$$U = \begin{pmatrix} c_1 & s_1c_3 & s_1s_3 \\ -s_1c_2 & c_1c_2c_3 - s_2s_3e^{i\delta} & c_1c_2s_3 + s_2c_3e^{i\delta} \\ s_1s_2 & -c_1s_2c_3 - c_2s_3e^{i\delta} & -c_1s_2s_3 + c_2c_3e^{i\delta} \end{pmatrix} . \tag{3.2}$$

The b quark decays by its coupling to the u quark

$$V_{ub} = s_1s_3 , \tag{3.3}$$

and to the c quark

$$V_{cb} = c_1c_2s_3 + s_2c_3e^{i\delta} . \tag{3.4}$$

Measurement of the K-M angles will be one of the most profound results of the study of B meson decays. The K-M angles are intimately connected with the origin of the quark masses. For instance, in the standard Higgs model the diagonalization of the Higgs-fermion coupling matrix yields both the fermion masses and the K-M angles. In general, models which offer an explanation of the quark masses will also predict or constrain the K-M angles.

Knowledge of the values of the angles is also an essential prerequisite to the study of weak interaction dynamics. For instance, it is necessary to know the K-M angles in order to extract dynamical enhancement factors from B lifetime measurements. Lepton yields which are not separated by B^o or B^- cannot necessarily tell us whether there are enhancements - as we are learning now from the D decays. Furthermore in B decays there are additional possibilities for confusion if $b \to u$ is a large mode. In principle, with measurements of the semileptonic to nonleptonic ratios $\Gamma_{SL}(b \to u)/\Gamma_{NL}(b \to u)$ and $\Gamma_{SL}(b \to c)/\Gamma_{NL}(b \to c)$, we can measure dynamical enhancements without knowing the K-M angles. But this is easier for theorists to imagine than for experimenters to do. I am doubtful that we will reliably know the strength of enhancements in B decays until we know the K-M angles. Unfortunately, as I will discuss below, it is not easy to measure the K-M angles if there are unknown nonleptonic enhancements.

The cleanest present constraints[29] are on θ_1 and θ_3. They are from beta decay

$$|c_1| = 0.9737 \pm 0.0025 \tag{3.5}$$

and from an analysis of $\Delta S = 1$ decays

$$|s_3| = 0.28^{+0.28}_{-0.21} \quad . \tag{3.6}$$

Equations (3.5) and (3.6) together imply that

$$|V_{ub}| = 0.06 \pm 0.06 \quad . \tag{3.7}$$

Thus V_{ub} is very small and is likely to be appreciably smaller than V_{cb}. Notice however in (3.4) that V_{cb} could also be extremely small if, for instance, all $|\theta_i|$ and $|\delta|$ are very small and if $s_2 \cong - s_3$. There are also constraints[30] on s_2, however these constraints require much more theoretical machinery and assumptions. No sacred principles would be violated if they turned out to be wrong.

The key to measuring V_{ub} and V_{cb} is lepton detection. One method is to use the single lepton spectrum[31] just above $B\bar{B}$ threshold, as at the Υ'''. Decays $b \to u$ have a harder endpoint and almost no soft secondary leptons, while $b \to c$ has an endpoint 15% softer than for $b \to u$ and a large number of soft secondary leptons from D decays. Another method[32] is to use the like-sign two lepton events generated by the cascade sequence

$$\begin{array}{llll} e^+e^- \to b & & + \ \bar{b} & \\ & \hookrightarrow c\ell^- X & \hookrightarrow \bar{c}X & \\ & & & \hookrightarrow \bar{s}\ell^- X \end{array}$$

to determine the ratio $\Gamma(b \to u)/\ (b \to c)$.

Both of these methods may be used to extract the K-M angles if there are no large nonleptonic enhancements. But the likelihood of significant enhancements seems greater today than a year or more ago when these analyses were first done - since the betting odds are coupled to whether there are large enhancements in D decays. And even if there are not large enhancements in D decays they might still occur in B decays: the enhancement mechanism for K decays might be specific to Cabibbo-suppressed decays (e.g. like penguins, which in particular should not be important for B if simple estimates are correct) in which case it could be important in B but not in D decays.

If there are significant but unknown enhancements in $\Gamma_{NL}(b \to u)$ and/or $\Gamma_{NL}(b \to c)$ then the method of like-sign dileptons can determine the ratio $\Gamma_{TOT}(b \to u)/\Gamma_{TOT}(b \to c)$. (Like-sign dileptons due to $B_o - \bar{B}_o$ mixing will both be primary or will both be secondary; their momenta will therefore help to separate them from the "usual" like-sign dileptons in which one is primary and the other secondary.) But without knowing the enhancements of $b \to u$ and $b \to c$ we cannot extract $|V_{ub}/V_{cb}|$ from this information.

The single lepton spectrum can in principle determine the ratio $|V_{ub}/V_{cb}|$ independent of enhancements. By measuring the yields of leptons near the endpoints for $b \to u\ell^- X$ and $b \to c\ell^- X$ one can unambiguously measure $\Gamma_{SL}(b \to u)/\Gamma_{SL}(b \to c)$ and extract $|V_{ub}/V_{cb}|$. But it may be impossible to accumulate enough statistics to carry this out in practice, because the signals near the endpoints are very small and because of a background from $e^+e^- \to \tau^+\tau^-$, $c\bar{c}$ which is a few times bigger than the signal at the $b \to (u/c)\ell^- X$ endpoints.

The problem requires more attention than it has been given. I do not know whether there is a practical method to extract $|V_{ub}/V_{cb}|$ from the data which would work given the most general possible pattern of enhancements and K-M angles.

B. What We Learn from $B \to \psi X$

I want to conclude this sketchy and disjointed discussion of B decays by mentioning what the rate for $B \to \psi X$ might teach us about the dynamics of nonleptonic decays. In principle we can learn from this inclusive rate about the importance of color rearrangement due to gluonic final state interations. These were assumed to be negligible in some of the models discussed in Section II, though some authors[33] have suggested that they play an important role in the formation of exclusive final states.

The lowest order effective Hamiltonian is

$$\begin{aligned} \mathcal{H}_{eff} &\propto (\bar{b}c)_L(\bar{c}s)_L \\ &\propto (\bar{b}_\alpha s_\beta)_L(\bar{c}_\beta c_\alpha)_L \qquad (3.8) \end{aligned}$$

where the second line is obtained by Fierz rearrangement and the color indices α,β are summed over. In calculating the rate for

$B \to \psi X$ the current $\bar{c}_\beta c_\alpha$ is in a color singlet for 1/9 of the final states and in the color octet for 8/9. Therefore there is a factor 9 difference in the predicted rate $B \to \psi X$ depending on whether we neglect color rearrangement due to gluon f.s.i. (i.e., keep the factor 1/9) or whether we assume that gluons rearrange the $c\bar{c}$ pair into a color singlet with probability one. Making the latter assumption a crude but plausible estimate gives[34]

$$B(B \to \psi X) \sim B(b \to \psi X)$$
$$\sim B(b \to c\bar{c}s)B(c\bar{c} \to \psi)$$
$$\sim 3 - 5\% \qquad (3.9)$$

where $|V_{cb}| >> |V_{ub}|$ is assumed. If color rearrangement is not assumed, the estimate would decrease by 1/9.

A more detailed estimate[35,36] uses the J/ψ wave functions at the origin as determined from $\Gamma(\psi \to e^+e^-)$. The result of this calculation is surprisingly large. Assuming no color rearrangement of the final state (i.e., which would have given 1/3 - 1/2% using the method of Eq. (3.9)), the result is

$$B(B \to \psi X) \sim (.018)(2f_+ - f_-)^2. \qquad (3.10)$$

For free quark values $f_+ = f_- = 1$, we get 1.8% from this estimate, which would become 16% if color rearrangement is permitted as in (3.9). At the other extreme if we use the renormalization group values for $f_\pm$ and do not allow color rearrangement, (3.10) yields[35] 0.2%. If enhanced values are taken for $f_\pm$ as in Ref. (22), then (3.10) yields 5% (as discussed in Section II, this approach assumes no color rearrangement).

What do we learn from all this? I invite the reader to tell me. It appears that the theoretical estimates are out of control at the level of the factor of 9 that we would like to be able to study.

IV Conclusion

I have tried to emphasize the importance of measuring with greater accuracy the lifetimes and semileptonic branching ratios for D^+ and D^o. It is clear that $\tau_+/\tau_o > 1$, but it is not yet clear whether the ratio is $\lesssim 3$ or $\gtrsim 10$ or somewhere in between. If $\tau_+/\tau_o \lesssim 3$ there need not be any drastic revision of our understanding of nonleptonic decays. Enhancements of this order could be explained by a combination of final state enhancements and QCD corrections. But if $\tau_+/\tau_o \gtrsim 10$ the conventional picture of nonleptonic decays is called into question. In my mind this would even raise doubts about the popular semiquantitative picture of the enhancements in $\Delta S = 1$ decays based on the penguin diagrams. If the enhancements in the D systems are not even understood at the level of an order of magnitude, then we are ignorant of important dynamical mechanisms which might also be important in $\Delta S = 1$ and $\Delta B = 1$ decays. I have also tried to emphasize that accurate measurements of exclusive final states can be a powerful probe of the dynamics, though any given prediction must be scrutinized carefully for sensitivity to complicating factors such as final state interactions and SU(3) symmetry breaking.

The possibility of large enhancements in D decays raises the spectre of the same possibility in B decays. One analysis indicates that annihilation diagrams eventually dominate (by a logarithm) as we scale to heavier flavors.[37] Even if it turns out that D decays are not drastically enhanced, caution requires that we take account of the possibility of significant enhancements in B decays. In view of their fundamental importance, we want to measure the K-M angles in a way that is independent of possible enhancements and which does not rely on the theoretical prejudice that the enhancements are small. Development of a practical program to accomplish this goal is a problem that merits careful consideration.

ACKNOWLEDGMENTS

I wish to thank Fred Gilman for several very helpful conversations, particularily concerning exclusive D decays. I have also profited greatly from discussions with Bob Cahn, Gerson Goldhaber Chris Quigg, Mahiko Suzuki, and Mark Wise. Research was supported by the the High Energy Physics Division of the U.S. Department of Energy under contract No. W-7405-ENG-48.

REFERENCES

1. For a review see M. K. Gaillard, Proceedings of the 1978 SLAC Summer Institute.
2. S. Glashow, J. Iliopolous, and L. Maiani, Phys. Rev. D2, 1285, 1970.
3. M. Kobayashi and T. Maskawa, Prog. Theor. Phys. 49, 652, 1973.
4. M. K. Gaillard and B. W. Lee, Phys. Rev. Lett. 33, 108, 1974; G. Altarelli and L. Maiani, Phys. Lett. 52B, 351, 1974.
5. M. Shifman, A. Vainshtein, and V. Zakharov, J.E.T.P. Lett. 22,553, 1975; Nuc. Phys. B120, 316, 1977.
6. J. Donoghue, E. Golowich, W. Ponce, and B. Holstein, Phys. Rev. D21, 186, 1980.
7. L. Abbott, P. Sikivie, and M. Wise, Phys. Rev. D21, 1768 (1980).
8. J. Ellis, M. K. Gaillard, and D. Nanopolous, Nucl. Phys. B100, 313, 1975.
9. G. Altarelli, N. Cabibbo, and L. Maiani, Nucl. Phys. B88, 285, 1975.

10. N. Cabibbo and L. Maiani, Phys. Lett. 73B, 418, 1978.
11. M. Suzuki, Nucl. Phys. B145, 420, 1978; N. Cabibbo and L. Maiani, Phys. Lett. 79B, 109, 1978.
12. J. Kirkby, Proceedings of the IX International Symposium on Lepton and Photon Interactions at High Energies, Batavia, IL.
13. V. Luth, ibid.
14. N.W. Reay Colloquium presented at the Lawrence Berkeley Laboratory, April, 1980.
15. M. Gaillard, B. Lee, and J. Rosner, Rev. Mod. Phys. 47, 277, 1975.
16. T. Hayashi, M. Nakagawa, M. Nitto, and S. Ogawa, Prog. Theor. Phys. 49, 351, 1973; 52, 636, 1974.
17. See also E. Ma, S. Pakvasa, and W. Simmons, University of Hawaii preprint UH-511-369-79, 1979; M. Matsuda, M. Nakagawa, and S. Ogawa, preprint, 1979; I. Bigi and L. Stodolsky, SLAC-PUB-2410, 1979.
18. M. Bander, D. Silverman, and A. Soni, Phys. Rev. Lett. 44, 7, 1980, and erratum, ibid, 44, 962 (1980).
19. H. Fritzsch and P. Minkowski, Phys. Lett. 90B, 455, 1980 and Bern University preprint, unpublished, 1980.
20. I. Bigi, CERN preprint TH 2798-CERN, 1979.
21. S. P. Rosen, Phys. Lett. 89B, 246, 1980.
22. B. Guberina, S. Nussinov, R. Peccei, and R. Rückl, Phys. Lett. 89B, 111, 1979.
23. M. Katuya and Y. Koide, Phys. Rev. D19, 2631, 1979; Y. Koide, Phys. Rev. D20, 1739, 1979.
24. SU(3) predictions for decays into two pseudoscalars in the 6 quark model are given by M. Suzuki, Phys. Lett. 85B, 91, 1979; L.-L. Wang and F. Wilczek, Phys. Rev. Lett. 43, 816 1979; C. Quigg, Z. Phys. C4, 55, 1980.
25. H. Lipkin, Phys. Rev. Lett. 44, 710, 1980.
26. J. Donoghue and B. Holstein, Phys. Rev. D21, 1334, 1980.
27. M. Einhorn and C. Quigg, Phys. Rev. D12, 2015, 1975.
28. I am grateful to F. Gilman for calling this relationship to my attention.
29. R. Shrock and L.-L. Wang, Phys. Rev. Lett. 41, 1692, 1978.
30. J. Ellis et. al., Nucl. Phys. B131, 285, 1977; V. Barger, W. Long, and S. Pakvasa, Phys. Rev. Lett. 42, 1585, 1979; R. Shrock S. Treiman, and L.-L. Wang, ibid. 42, 1589, 1979.
31. V. Barger, T. Gottschalk, and R. Phillips, Phys. Lett. 82B, 445, 1979; A. Ali, Z. Physik C1, 25, 1979; A. Ali and Z. Aydin, Nuc. Phys. B148, 165, 1979.
32. C. Quigg and J. Rosner, Phys. Rev. D19, 1532, 1979.
33. N. Deshpande, M. Gronau, and D. Sutherland, Phys. Lett. 90B, 431, 1980.
34. H. Fritzsch, Phys. Lett. 86B, 164, 1979; ibid, 343, 1979.
35. M. Wise, Phys. Lett. 89B, 229, 1980.
36. T. De Grand and D. Toussaint, Phys. Lett. 89B, 256, 1980.
37. M. Suzuki, LBL preprint Univ. of California, Berkeley pre-print, UCB-PTH-80/4, (submitted to Nucl. Phys. B), 1980.

PHOTON-PHOTON ANNIHILATION

Thomas A. DeGrand
University of California, Santa Barbara, CA 93106

ABSTRACT

I review recent progress (mostly theoretical) in our knowledge of photon-photon interactions at high energies, concentrating on resonance production and on reactions which expose the pointlike coupling of photons to quarks: deep inelastic scattering on a photon or electron target and $\gamma\gamma$-induced jet production at large transverse momentum.

I. INTRODUCTION

Photon-photon scattering is presently at a crossroads. The theoretical ideas which underlie it have been around for forty-six years.[1] And yet it is only in the last year or so that really meaningful experiments have begun to be done. The situation is steadily improving with the advent of the new generation of colliding beam machines, and many of the reactions which I want to describe today may actually soon be measured.

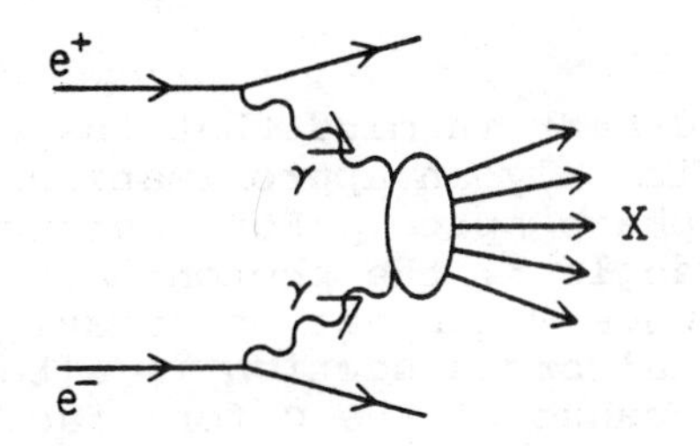

Fig. 1. The generic reaction $e^+e^- \to e^+e^-X$.

Photon-photon scattering is actually the reaction

$$ee \to e\gamma^* e\gamma^* \to e^+e^-X \qquad (1)$$

in which electrons of both incident beams emit virtual (space like) photons, γ^*'s, which in turn annihilate producing some final state X. Because the photon is a vector particle, the differential cross section for (1) peaks when the photons are of low mass, or equivalently, when the electrons scatter through a small angle. So most $\gamma\gamma$ experiments are built around detectors which are set at small angles to the beam axis, to "tag" the forward moving electrons and measure the photons' momenta.

ISSN:0094-243X/80/620067-21$1.50

Throughout this talk I will work in the context of the equivalent photon approximation,(2) which is useful for understanding the main qualitative features of the reaction $ee \to eeX$. In this approximation the leading behavior as $m_e^2/s \to 0$ of the reaction is given by

$$d\sigma(ee \to eeX) \simeq \int dx_1 dx_2 \, D_{\gamma/e}(x_1,s) D_{\gamma/e}(x_2,s) \times d\sigma_{\gamma\gamma \to X}(x_1,x_2) \quad (2)$$

where $d\sigma_{\gamma\gamma \to X}$ is the differential cross section for the annihilation of two oppositely directed real unpolarized photons of energies x_1E and x_2E $(s = 4E^2)$ into the state X. $D_{\gamma/e}(x,s)$ is the distribution of photons carrying (light-cone) fraction x of the electron's momentum

$$D_{\gamma/e}(x,s) = \frac{\alpha}{2\pi} \log \eta \, \frac{1 + (1-x)^2}{x} \quad (3)$$

and $\eta = s/4m_e^2$ if the electron is not tagged or $\theta^2_{max}/\theta^2_{min}$ if the electron is detected at an angle $\theta_{max} > \theta > \theta_{min}$.

Readers should bear in mind that the equivalent photon approximation is only an approximation valid in a limited region of phase space. For instance, if tagging is done at finite angle θ, the photons are not strictly massless-a consequence which has observable effects: for example, the $\gamma\gamma$ total cross section is altered when one photon is tagged because of the ρ form factor at $Q^2 \neq 0$.(3) A comprehensive study of the validity of the equivalent photon approximation has recently been carried out by Carimalo, Parisi, and Kessler.(4) I will use the equivalent photon approximation in this talk just for simplicity: I do not think that doing everything more correctly will change my answers much.

In many ways, photon-photon scattering complements ordinary e^+e^- annihilation. While all the action in e^+e^- annihilation takes place at one energy, that of the beams, in $\gamma\gamma$ scattering one measures many energy scales simultaneously. In e^+e^- annihilation only states of one spin parity are produced, while in $\gamma\gamma$ collisions many spins and parities are accessible.

The $\gamma\gamma$ total cross section is huge and rises logarithmically with energy. Integrating eq. (2) we have

$$\frac{d\sigma}{dM^2}\,(e^+e^- \to e^+e^-X) \simeq \left(\frac{\alpha}{2\pi}\ln\eta\right)^2 \ln\left(\frac{s}{M^2}\right) \frac{\sigma_{\gamma\gamma\to X}(M^2)}{M^2} \tag{4}$$

with $s = (p_{e_1} + p_{e_2})^2$, $M^2 = (p_{\gamma_1} + p_{\gamma_2})^2$ and $s >> M^2$. A simple vector dominance prediction for the hadronic $\gamma\gamma$ cross section gives $\sigma \sim 15$ nb for $E_{cm} = 30$ GeV, $M^2 \geq 1$ GeV^2. This is to be compared with the single-photon $e^+e^- \to \mu^+\mu^-$ cross section, 0.1 nb at that E_{cm}.

Photon-photon scattering bears much resemblance to proton-proton scattering. Most hadrons are produced with limiting fragmentation and small momentum transverse to the beam direction. $\gamma\gamma$ scattering possesses an extra flexibility in that the masses and linear polarizations of the photons can be tuned via electron tagging.

But that is not the whole story. A wide class of phenomena, typically characterized by the production of hadrons at large transverse momenta, are sensitive to the photon's pointlike coupling to the quark current. In fact, because of the special role of the photon in quantum chromodynamics (QCD), cross sections for jets and for deep inelastic scattering are sizable and are much larger than the extrapolation from hadron-hadron collisions.

Thus the study of $\gamma\gamma$ collisions at large $p_\perp$ provides a detailed laboratory for the study of QCD dynamics at short distances. The dominant hard scattering processes arise from the elementary field nature of the photon and are particular to γ-induced reactions: their observation will provide important tests of the basic subprocesses which govern large $p_\perp$ dynamics. While it is not proven that the perturbative component and couplings of the real photon survive confinement, we believe it is likely they do. In vector dominance language these contributions are represented by an infinite spectrum of massive vector mesons. Here I will approximate the hadronic interactions of the photon by combining the contributions of the lowest vector mesons (ρ, ω, φ) with the pointlike perturbative contribution. No serious double counting should occur since the pointlike interaction is equivalent to only very heavy vector mesons.[5]

Now let's turn to some examples. I want to concentrate on two topics which have become the "bread and butter" of colliding beam physicists: resonance production and jet physics.

II. RESONANCES

First, let us consider resonance production in $\gamma\gamma$ collisions. The cross section for production of a resonance R of spin J is given by the simple formula

$$\sigma(e^+e^- \to e^+e^-R) = \int dx_1 dx_2 D_{\gamma/e}(x_1) D_{\gamma/e}(x_2)\,\delta(x_1 x_2 s - M_R^2) \times \left[\frac{8\pi^2}{M_R}\,\Gamma(R\to\gamma\gamma)(2J+1)\right] \quad (5)$$

This equation is exactly the same that one would use in calculating massive meson production in hadronic reactions, except that the colliding partons are really photons. Performing the integral gives

$$\sigma(e^+e^- \to e^+e^-R) = \left(2\alpha \ln\frac{s}{M_e^2}\right)^2 f(\tau)\frac{(2J_R+1)}{M_R^3}\,\Gamma(R\to\gamma\gamma) \quad (6)$$

where $\tau = M_R^2/s$ and $f(\tau)$ arises from the convolution of the two photon distributions:

$$f(\tau) = \frac{1}{2}\,(2+\tau)^2 \ln\frac{1}{\tau} - (1-\tau)(3+\tau) \quad (7)$$

In $\gamma\gamma$ scattering the interesting physics is the measurement of $\Gamma(R\to\gamma\gamma)$. $f(\tau)$ is just a sort of "flux factor" which tells us how much phase space is available for the resonance.

The matrix element whose square gives $\Gamma_{\gamma\gamma}$ weights quark distributions in the meson according to the square of their charges

$$\langle\gamma\gamma|R\rangle \propto \Sigma\, e_i^2 \langle q_i\bar{q}_i|R\rangle \quad (8)$$

and as such is a very useful probe of the quark structure of hadrons. For example, the measured decay rates of π^0, η, and $\eta'\to\gamma\gamma$ are in nice agreement with simple SU(3) predictions based on fractional charged quarks and nonet symmetry.

Fred Gilman has prepared a nice table of resonance cross sections at SPEAR, PEP/PETRA, and LEP energies,(6) and I will just reproduce the PEP/PETRA part of it for reference here.

Table I Cross sections for production of various resonances in two photon collisions for an electron beam energy E = 15 GeV

Resonance	$\Gamma(R\to\gamma\gamma)$ (keV)	$\left\lvert 2\alpha \ln \frac{s}{4M_e^2}\right\rvert^2 f(\tau)$	$\sigma(ee\to eeR)$ (nb)
π^o	7.95×10^{-3} [a]	1.68	2.1
η	0.324 [a]	1.17	0.9
η'	5.9±1.6 [b]	0.97	2.6
A_2	1.8 [c]	0.86	1.3
f	5.0 [c]; 2.3±0.5 [d]	0.87	4.2
f'	0.4 [c]	0.81	0.18
η_c	6.4 [c]	0.57	5×10^{-2}
χ_0	1 [c]	0.53	5×10^{-3}
χ_2	4/15 [c]	0.51	1×10^{-3}
η_b	0.4 [c]	0.21	4×10^{-5}

[a] Ref. 7
[b] Ref. 8
[c] Ref. 9
[d] Ref. 3

The interesting "old physics" in my opinion are studies of $q\bar{q}$ mesons with one unit of orbital angular momentum: $J^{pc} = 0^{++}$, 1^{++}, 1^{+-}, and 2^{++}. Photon-photon reactions may help sort out which states in the wallet card correspond to which quark model multiplets. In particular, the $J^{pc} = 2^{++}$ states are well enough understood that one can make absolute (within a factor of 5!) predictions for decay rates such as $f \to \gamma\gamma$. Decay rates and angular distributions for these decays will test these predictions.

This mass range is also interesting because it is thought to be the home of "glueballs" - bound states of pure glue. Of course, because gluons are neutral, $\gamma\gamma$ collisions only measure the admixture of $q\bar{q}$ states into glueballs, $|\langle q\bar{q}|G\rangle|^2$. By definition, this is a number less than 1/2 (a state more than 50% $q\bar{q}$ isn't a glueball). No one knows how to make a reliable estimate of $\langle q\bar{q}|G\rangle$ and thus of $\Gamma(G \to \gamma\gamma)$. A simple argument which ignores essentially all important dynamics but is still amusing is provided by large N_c QCD (N_c colors, $N_c \to \infty$, but $\alpha_c N_c$ fixed). There one finds[10]

$$\frac{|\langle q\bar{q}|G\rangle|^2}{|\langle q\bar{q}|M\rangle|^2} \sim \frac{1}{N_c}$$

where M is an ordinary $q\bar{q}$ state. Perusing the Table and comparing $f \to \gamma\gamma$ we see that this (simple minded) prediction suggests $\Gamma(G \to \gamma\gamma) \sim \frac{1}{2}$ KeV or so - perhaps! I believe it is the consensus of theorists today that the best place to see glueballs is in ψ decay: $\psi \to \gamma X$, for example. $\gamma\gamma$ physics can do the Crystal Ball[11,12] a service by trying to see photon production of any suspicious new bumps they may find, thus helping to establish their $q\bar{q}$ structure. By the same token, if an "ordinary" meson has less $q\bar{q}$ in its wave function than expected, it may be because it has mixed with a glueball.

For heavy $q\bar{q}$ bound states the situation is clearer. To lowest order in the strong interactions

$$\Gamma(\eta_c \to \gamma\gamma) \propto e_c^4 \frac{|\psi(0)|^2}{m_c^2}$$

$$\Gamma(\psi \to e^+e^-) \propto e^2 e_c^2 \frac{|\psi(0)|^2}{m_c^2}$$

where $\psi(0)$ is the wave function at zero $q\bar{q}$ separation. This number cancels in the ratio so that, for example

$$\Gamma(\eta_c \to \gamma\gamma) = \frac{4}{3}\,\Gamma(\psi \to e^+e^-) = 6.4 \text{ KeV}$$

is probably a pretty reliable prediction. Similar results are obtained for the χ states: in particular, simple angular momentum considerations predict $\Gamma(^3P_0 \to \gamma\gamma)/\Gamma(^3P_2 \to \gamma\gamma) = 15/4$. This number is worth measuring because the corresponding prediction for total rates, calculated via decay into gluons, $\Gamma(^3P_0 \to gg)/\Gamma(^3P_2 \to gg)$ is in experimental difficulty and theoretically suspect[13] due to higher order corrections. As one can see from the table, the high mass of these charmed states makes their cross sections small in comparison to those of the ordinary $q\bar{q}$ states. One will probably have to get lucky in looking at specific decay channels. But they are certainly worth looking for. In my opinion confirming or denying theoretical predictions is much more clear-cut and hence useful here than anywhere else in the spectrum.

III. THE DIAGRAM $e^+e^- \to e^+e^-\mu^+\mu^-$

Now we come to the heart of recent $\gamma\gamma$ studies. It is the simple reaction $e^+e^- \to e^+e^-\mu^+\mu^-$ whose graph is shown in Fig. (2). "Dressing up" this reaction by exchanging muons for quarks and including gluon radiative corrections provides us with nearly all the recent QCD predictions of $\gamma\gamma$ physics.

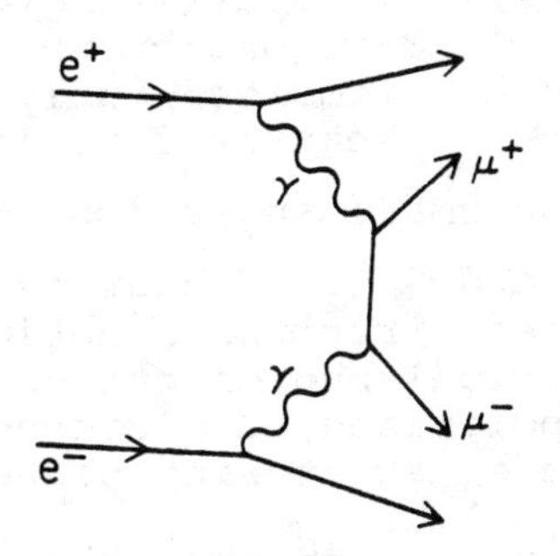

Fig. 2. The reaction $e^+e^- \to e^+e^-\mu^+\mu^-$.

First QED. Extensive calculations of $e^+e^- \to e^+e^-\mu^+\mu^-$ have been carried out by Bhattacharya, Grammer, Smith, and Vermaseren,[14] and I understand Vermaseren in particular has a nice set of computer programs "ready for use" by experimental groups.[15] One amusing observation they make is that $\gamma\gamma$ collisions can be a big background to $\gamma \to$ heavy leptons - or presumably to anything whose decays always involve neutrinos. This happens because the untagged electrons' missing energy in the $\gamma\gamma$ collision mocks the effect of the neutrino's undetectable energy in the lepton decay. They have shown that electron tagging is necessary to do a good job on the heavy leptons.

Now for the strong interactions. We pass from QED to QCD by relabelling all muons as quarks. This instantly gives us a parton model prediction: any differential cross section for jets is equal to a constant $R_{\gamma\gamma}$ times the same differential cross section for muons, where $R_{\gamma\gamma}$

measures the sum of fourth power of the quark charges, with a 3 for color,

$$R_{\gamma\gamma} = 3 \sum_i e_i^4 \quad . \tag{9}$$

Of course, this global prediction is only true as long as the reaction is "hard" - lots of momenta flowing through the diagram - and as long as one can successfully ignore the vector meson component of the photon. We will see that that happens often enough to expose useful dynamics.

To proceed further it is convenient to specialize by selecting particular regions of phase space and particular triggers.

First, deep inelastic scattering on a photon (or an electron) target.(16) Here one of the photons is far off shell and the other remains nearly massless. One detects the electron e_1 scattered at large angle, the other electron e_2 at small angle (to fix the momentum of the struck photon) and jets of hadrons from the collision. One can measure three structure functions for the real photon since the virtual photon can be either transversely or longitudinally polarized, and since the target photon is polarized in the e_2 scattering plane. The first structure function ($F_2^{\gamma}(x,Q^2)$) is largest, and proportional to log Q^2, and I will concentrate on it.

The photon structure function is defined through the inclusive reaction $e_1e_2 \to e_1'e_2'X$, with e_1 scattered through a large angle

$$\frac{Ed\sigma}{d^3pdx_\gamma}(e_1e_2 \to e_1'e_2'X) = \frac{4\alpha^2}{Q^4}(1 + \frac{u^2}{s^2}) \frac{s}{s+u} D_{\gamma/e}(x_\gamma, s) \times F_2^{\gamma}(x_{Bj}, Q^2) \tag{10}$$

where $Q^2 = -(p_{e_1} - p_{e_1'})^2$, $1-x_\gamma = |p_{e_2'}|/|p_{e_2}|$,

$x_R = 2|p_{e_1'}|/\sqrt{s}$, $u/s = -\frac{1}{2}x_R(1-\cos\theta)$, and $x_{Bj} = Q^2/x_\gamma(s+u)$.

(I've neglected F_L^{γ} but explicitly specified kinematics since conventions don't seem to be standardized.)

In QCD,(17) $F_2^{\gamma}(x,Q^2)$ consists of two pieces. The first involves the direct coupling of $q\bar{q}$ pairs to the photon, followed by their Q^2 evolution into the quarks and antiquarks seen by the large Q^2 photon. This term is in principle completely calculable order by order in perturbation theory since the coupling of the photon to a $q\bar{q}$ pair is known: it is just $e_q\gamma_\mu$. The second piece involves the Q^2 evolution of the quarks of the vector meson component of the photon. Just as in deep inelastic scattering on a proton, its Q^2 evolution is known but its shape is not calculable from first principles. The two terms have quite different behavior at large Q^2:

$$F_2^{\gamma}(x,Q^2)_{\text{pointlike}} \sim \log Q^2 f(x) \tag{11a}$$

$$F_2^{\gamma}(x,Q^2)_{\text{VMD}} \sim (1-x)^{c_1+c_2 \log(\log Q^2)} \tag{11b}$$

(the second equation resulting from an approximate analytic inversion).[18,19]

So at large Q^2 the calculable "pointlike" photon structure function dominates the less calculable hadronic part. The photon structure function is thus a candidate for a very good test of QCD.

How good a test is it? First of all, the factorized form (11a) is true even on the parton model, where $f(x)$ is just $\frac{\alpha}{2\pi} \Sigma e_i^4 x(x^2+(1-x)^2)$. Four years ago Witten[17] showed that (11a) is obtained as a lowest order in α_c QCD prediction where the $\log Q^2$ is really $1/\alpha_c(Q^2)$. The shape $f(x)$ is similar to that of the parton model except at very large and very small x. There the radiation of soft gluons by the fast quarks depletes the density of fast quarks and generates a "sea" of soft $q\bar{q}$ pairs.

Higher order corrections to F_2^{γ} have been computed by Bardeen and Buras[20] and recently by Duke and Owens.[21] They find

$$F_2^{\gamma}(x,Q^2) \sim f(x)\log Q^2/\Lambda^2 + g(x)\log\log Q^2/\Lambda^2 + h(x) \tag{12}$$

where the moments f, g, and h are all known. Scaling at fixed x versus Q^2 is not changed much, but the shape of F_2^{γ} at fixed, not enormous, Q^2 is even steeper. This suggests that the large x - or equivalently, the large moment - behavior of F_2^{γ} is poorly convergent and as such is not a good test of perturbative QCD.

Duke and Owens have recently recomputed F_2^{γ}. Including odd moments, they have extended the inversion of F_2^{γ} down to small values of x, where they make an appalling discovery: F_2^{γ} is negative at small x for small Q^2. This of course does not mean a negative cross section, just that perturbation theory is poorly convergent at small x and Q^2. By $Q^2 \sim 20$ GeV^2 or so everything has recovered, but corrections at small x are still sizable.

So far we have ignored the hadronic component of $F_2{}^{\gamma}$. Its magnitude may be estimated using vector dominance and the recent extractions of the pion structure function from Drell-Yan experiments.[21] I have plotted it and the pointlike $\frac{1}{\alpha} F_2{}^{\gamma}$ (using $\Lambda = 0.5$ GeV) at $Q^2 \approx 3$ GeV^2 together in Fig. (3). Where it is large

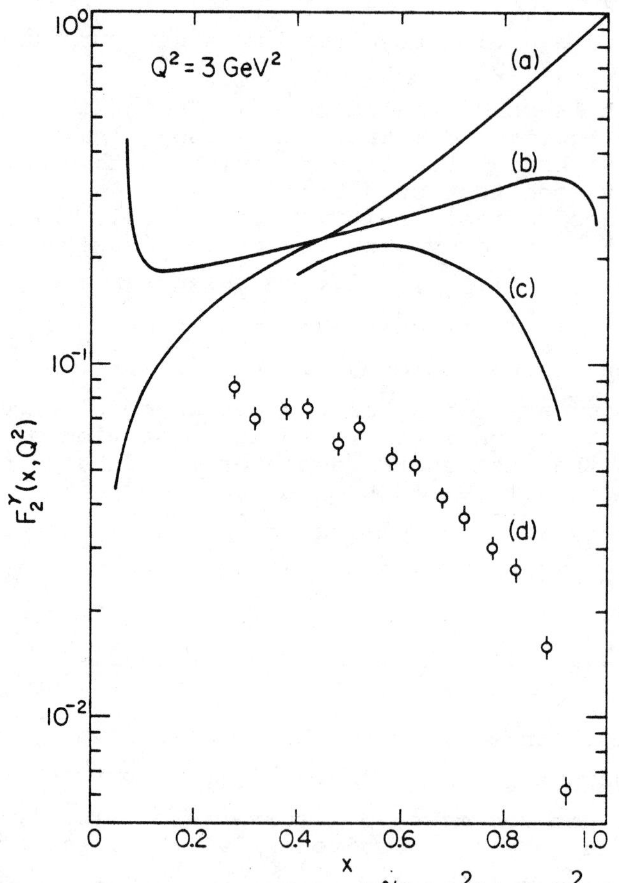

Fig. 3 The structure function $F_2{}^{\gamma}(x,Q^2)$ at $Q^2 = 3$ GeV^2 scaled by a factor of α: (a) parton model prediction, (b) lowest-order QCD, (c) higher order QCD, (d) vector dominance estimate using the πp Drell-Yan data of Ref. (21).

F_2^γ is not a good test of QCD: that is certainly true at small x. Equivalently, small moments of F_2^γ are badly contaminated by vector dominance.

So the best test of QCD in the photon structure function will be the moderate-j moments: $j = 4$ to 8, say. Their overall magnitude should show strong scale violation (proportional to log Q^2) but their ratios should show almost no Q^2 dependence. This prediction is probably valid for Q^2 greater than a few GeV^2.

JET PRODUCTION

Now we come to what I consider to be the most exciting QCD predictions for $\gamma\gamma$ annihilations: jet production.[22] In a hadron-hadron collision with a large-$p_\perp$ trigger, one always sees at least four jets: the trigger jet, the away-side jet which balances the transverse momentum, and two jets, forward and backward in the center of mass, arising from the hadronization of the constituents of the beam and target which did not participate in the hard scattering. But in $\gamma\gamma$ scattering the situation can be radically different. The photon is a pointlike object and can itself initiate the hard scattering, leaving no spectator partons in the beam or target to form a small-$p_\perp$ jet. Indeed, QCD perturbation theory predicts that the dominant source of very large-$p_\perp$ jets in $\gamma\gamma$ annihilations arises from the simple reaction $\gamma\gamma \to q\bar{q} \to 2$ jets. In addition, one also expects to see events with 3 jets or 4 jets, where one or both of the photons hadronizes or radiates spectator partons before participating in the hard scattering.

A good way for a theorist to discuss jets is to classify them according to their topology, since different numbers of jets reflect different physics of the $\gamma\gamma$ interaction.

A. Two jets. The reaction is just pair annihilation into fermions

$$\frac{d\sigma}{dt}(\gamma\gamma \to q\bar{q}) = \frac{2\pi\alpha^2}{s^2}\left(\frac{t}{u} + \frac{u}{t}\right)R_{\gamma\gamma} \tag{13}$$

with $R_{\gamma\gamma} = 3\Sigma e_i^4$ the ratio of jet production to μ-pair production. With double tagging the kinematics are completely constrained, but with single or no tagging it is just as convenient to think of the reaction as $e^+e^- \to$ jet+ X, where the inclusive cross section is obtained via the usual convolution formula

$$E\frac{d\sigma}{d^3p}(e^+e^- \to \text{Jet X}) = \int dx_1 dx_2 D_{\gamma/e}(x_1,s) D_{\gamma/e}(x_2,s) \times \frac{\hat{s}}{\pi}\, \frac{d\sigma}{dt}(\gamma\gamma \to q\bar{q})\Big|_{\substack{\hat{s}=x_1x_2s \\ \hat{t}=x_1t \\ \hat{u}=x_2u}} \tag{14}$$

Here s,t,u are the usual Mandelstam variables and $D_{\gamma/e}$ is the distribution function for photons in the electron beam. An approximate integration of eq. (14) yields

$$E\frac{d\sigma}{d^3p}(e^+e^- \to \text{Jet + X}) \simeq 2R_{\gamma\gamma}\frac{(1-x_R)}{p_\perp^4}\left(\frac{\alpha}{2\pi}\ln\eta\right)^2 \tag{15}$$

with $x_R = 2|p_{Jet}|/\sqrt{s}$

$$\sim 0.03\,\frac{(1-x_R)}{p_\perp^4}\ \text{nb - GeV}^{-2}$$

at $\sqrt{s}$ = 30 GeV. A graph of the predicted differential cross section is shown in Fig. (4).

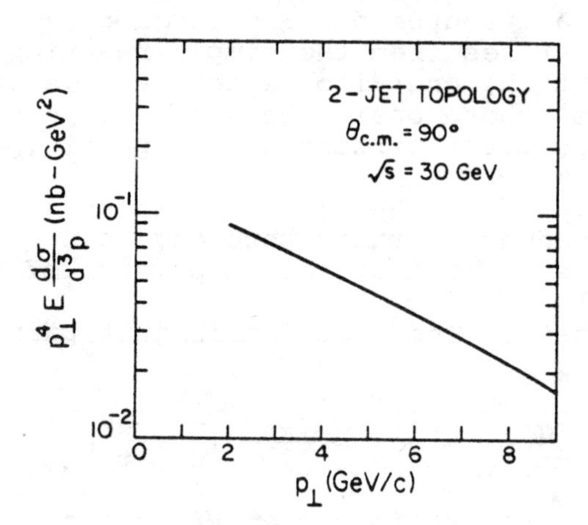

Fig. 4. The jet differential cross section $p_\perp^4(Ed\sigma/d^3p)$ vs $p_\perp$ at $\sqrt{s}$ = 30 GeV, θ_{cm} = 90° for reactions with a two-jet topology. From Brodsky, DeGrand, Gunion, and Weis, Ref. 22.

These jets may have already been seen at PETRA.[3] Future studies should expose their unmistakable signatures in colliding beam experiments. Since the distribution function $D_{\gamma/e}$ is essentially flat in rapidity, the distribution of jets per rapidity interval will be essentially flat away from the edge of phase space. Momentum conservation demands that a large-$p_\perp$ jet on one side of a reaction must be balanced by one or more large-$p_\perp$ jets on the away side. For the $\gamma\gamma \to q\bar{q}$ reaction the jets should be identical to those seen at SPEAR, though flavor correlations will be different due to the different charge weighting. Because the rapidity of the $\gamma\gamma$ system is generally not zero, the jet events range from nearly back-to-back to a "V" configuration along the beam axis.

This reaction is also amusing in that it may be the first place one can actually see the quark propagator $1/\not{p}$ at large p^2. (One also measures the quark propagator in reactions such as deep-inelatic scattering or $e^+e^- \to q\bar{q}g$ but there one must also disentangle the effects of the gluon spin: $\gamma\gamma \to q\bar{q}$ is much cleaner.)

The total cross section for quark and antiquark jets exceeding a minimum $p_\perp$ from the $\gamma\gamma \to q\bar{q}$ subprocess is

$$\sigma(p_\perp^{jet} > p_\perp^{min}) = \int_{p_\perp > p_\perp^{min}} d^2p_\perp \int dy \left[E\frac{d\sigma}{d^3p}(ee\to qX) + E\frac{d\sigma}{d^3p}(ee\to\bar{q}X)\right]$$

$$\simeq \frac{32\pi}{3}\alpha^2\left(\frac{\alpha}{2\pi}\ln\eta\right)^2 \frac{\left(\ln(s/p_{\perp min}^2) - 19/6\right)}{(p_\perp^{min})^2} R_{\gamma\gamma} \qquad (16)$$

which is about 0.5 nb $\mathrm{GeV}^2/(p_\perp^{min})^2$ at $\sqrt{s} = 30$ GeV (no tagging). This is not a small number: jets with $p_\perp > 4$ GeV/c contribute about 0.3 more units of R to the total cross section from this reaction alone. The cross section is shown in Fig. (5).

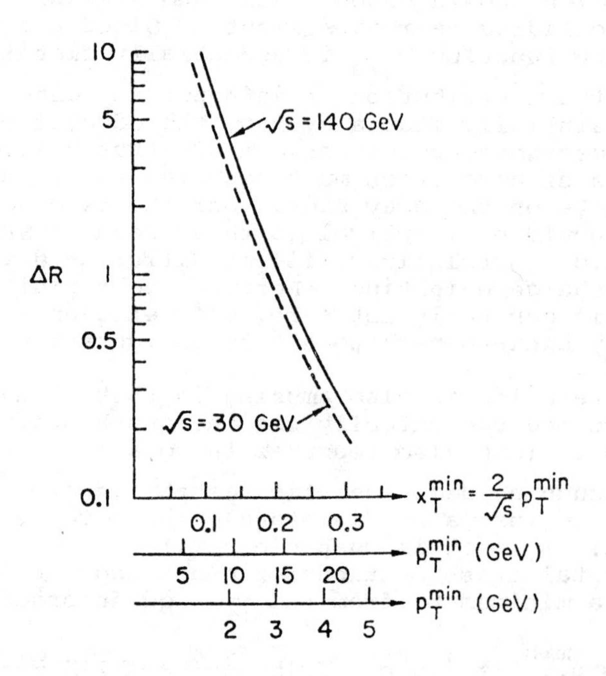

Fig. 5. The contribution to R from $\gamma\gamma \rightarrow q\bar{q}$ two jet processes at $\sqrt{s}$ = 30 and 140 GeV, K. Kajantie, Ref. 22.

Finally, Cahn and Gunion[23] have recently computed the reaction $\gamma\gamma \rightarrow gg$ - QCD light-by-light scattering. They find that it contributes to jet production at around the five to ten percent level of $\gamma\gamma \rightarrow q\bar{q}$.

B. Three jets. Several types of reactions may contribute to processes wherein the two large-$p_\perp$ jets are accompanied by a small $p_\perp$ jet: the third jet may arise because one of the photons vector dominates, while the hard scattering occurring between the other photon and one of the constituents of the vector meson, or (and these reactions dominate at large $p_\perp$) quarks or gluons arising from the Q^2 evolution of the photon's pointlike structure function may scatter off the other photon. Here one of the large-$p_\perp$ jets is a quark jet, the other a gluon jet. These reactions have the amusing property that, because the hard scattering differential cross section behaves as α_c, and the structure function of the photon behaves as $1/\alpha_c$, their overall magnitude is nearly independent of the value of the color coupling constant![24] This cancellation also has the result that violations to the naive $p_\perp^{-4}$ scaling form of $Ed\sigma/d^3p$ are much smaller than in hadron-hadron scattering.

Also present at this level one may find various "higher twist" QCD contributions where one of the "jets" is actually a meson or a resonance (compare Fig. (6)).

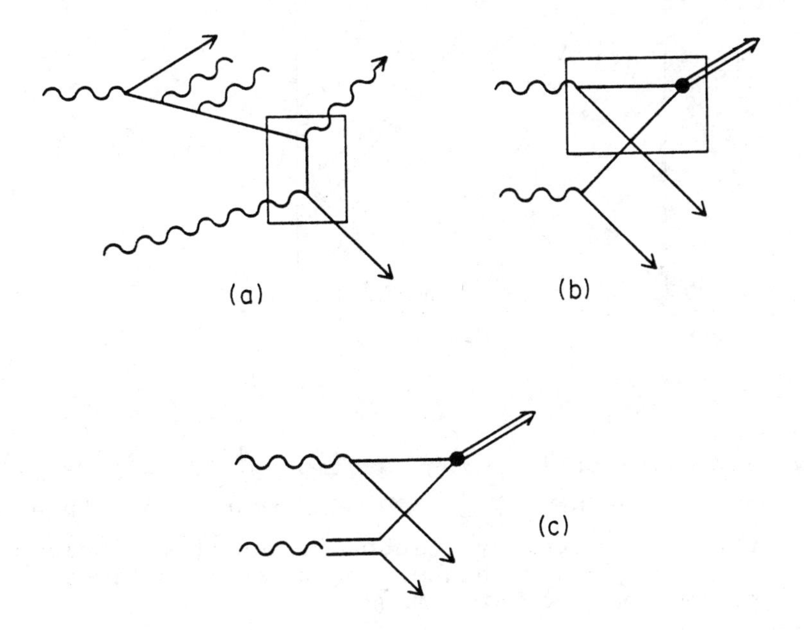

Fig. 6. Reactions giving rise to a 3-jet topology a) $\gamma\gamma \to q\bar{q}g$ showing the radiation of soft gluons which builds the structure function, and "higher twist" contributions, b) $\gamma\gamma \to Mq\bar{q}$ and c) $\gamma\rho \to Mq\bar{q}$. The hard scattering is shown in the box.

These contributions typically scale as $Ed\sigma/d^3p \sim p_\perp^{-6}$ due to wave function effects but if one parameterizes their normalization using hadronic data one finds that they may be as large as the $q\bar{q}$ g subprocess. Our predictions for inclusive jet cross-sections with a three-jet topology are shown in Fig. (7): the biggest high-twist reaction is $\gamma\gamma \to q\bar{q}$ + resonance.

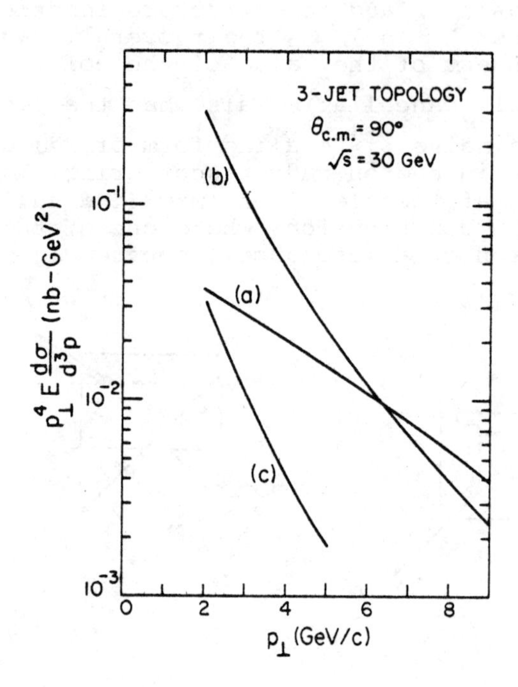

Fig. 7. Jet differential cross section $p_\perp^4\ Ed\sigma/d^3p$ vs $p_\perp$ at $\sqrt{s}$ = 30 GeV, θ_{cm} = 90° for reactions with a three-jet topology. Subprocesses shown include a) $\gamma\gamma \to q\bar{q}g$ and "higher twist" contributions b) $\gamma\gamma \to Mq\bar{q}$, c) $\gamma\rho \to Mq\bar{q}$.

There is also at this level the possibility of producing three large-$p_\perp$ jets and no small $p_\perp$ jets via the reaction $\gamma\gamma \to q\bar{q}g$. I am unaware if anyone has computed this process. I would expect that its magnitude is roughly in the same ratio to $\gamma\gamma \to q\bar{q}$ as $e^+e^- \to q\bar{q}g$ is to $e^+e^- \to \gamma \to q\bar{q}$.

C. Four jets. These reactions are most like hadronic large-$p_\perp$ scattering. The dominant QCD scattering mechanism $\gamma\gamma \to q\bar{q}q\bar{q}$, typically an order of magnitude smaller than $\gamma\gamma \to q\bar{q}$. As in the reaction $\gamma\gamma \to q\bar{q}g$, structure function and coupling constant cancellations leave a differential cross-section with nearly canonical $p_\perp^{-4}$ scaling. Here backgrounds can be estimated most easily by taking a combination of vector dominance and measured hadronic large-$p_\perp$ cross sections. These fall roughly like $p_\perp^{-8}$ and expose the pointlike QCD reactions for $p_\perp > 4$ GeV/c or so. In contrast to pp → Jet+X which is dominated by gluon-gluon scattering, the important subprocesses here are qq → qq since the gluon structure function of the photon is very small.

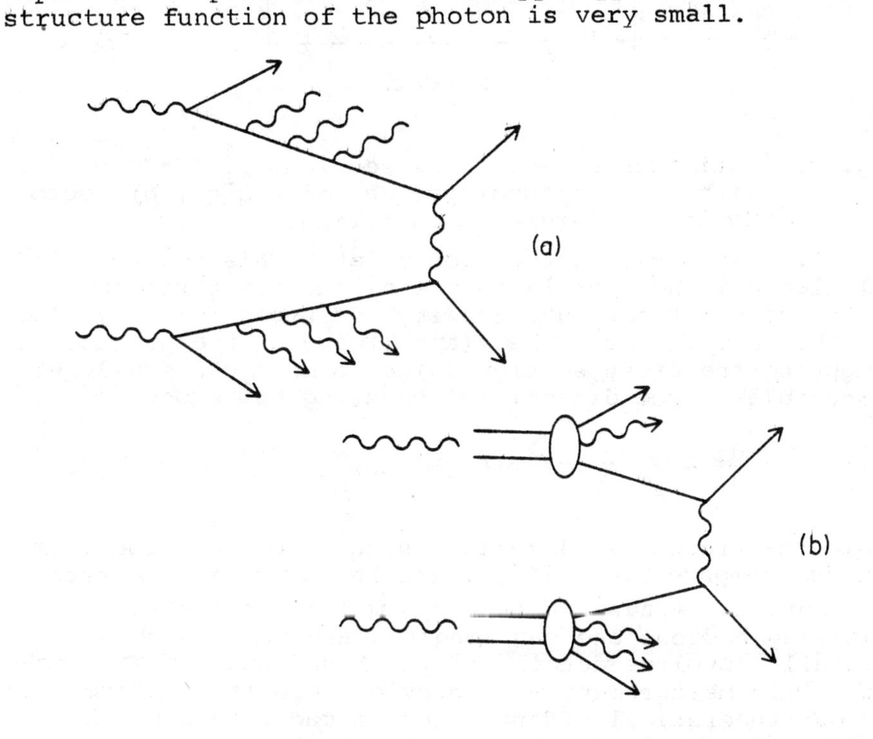

Fig. 8. Examples of four-jet topology reactions. a) $\gamma\gamma \to q\bar{q}q\bar{q}$, b) vector-dominance large-$p_\perp$ scattering.

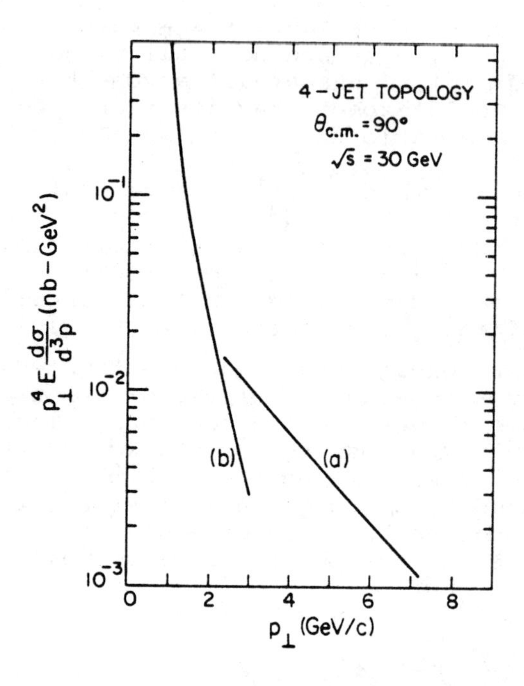

Fig. 9. Estimates of jet cross sections $p_\perp^4\, Ed\sigma/d^3p$ vs $p_\perp$ with a 4 jet topology: a) $\gamma\gamma \to q\bar{q}q\bar{q}$, b) vector dominance large-$p_\perp$ scattering.

D. $\gamma\gamma \to \pi\pi$. Not strictly jets, this reaction is calculable in QCD for large energy and scattering angle by the same methods that Brodsky and Lepage[25] have used for the form factor. They (and I) are in the process of computing the cross section which to me looks small but measurable. From dimensional counting one expects

$$\frac{d\sigma}{dt} \sim \frac{1}{s^4}\, \alpha_c(t)^2 f(\theta_{cm})$$

since the process is like the "square" of the pion form factor (compare Fig. (10)). The reaction is interesting to theorists because it is the simplest exclusive reaction involving hadrons one can imagine, and yet its shape crucially involves details of the bound state wave function. Its measurement will provide important information for our theoretical effort to begin understanding the bound state problem.

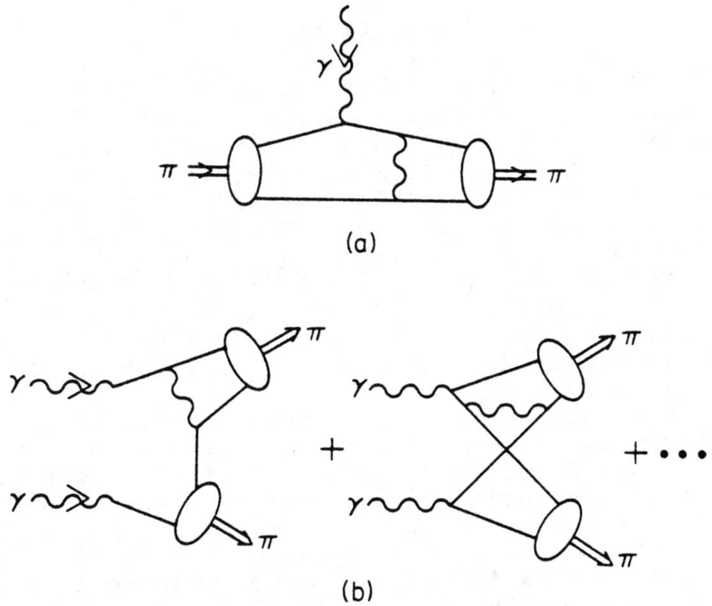

(a)

(b)

Fig. 10. The pion form factor (a) and contributions to large angle scattering $\gamma\gamma \to \pi\pi$ (b). The blobs are meson wave functions, the wiggly lines gluons which transfer the large momenta among the quark lines.

ENVOI

This concludes my summary of $\gamma\gamma$ physics. I hope that by the next Vanderbilt conference we'll have some data to confront the theoretical ideas I've been showing.

The customary finale to a $\gamma\gamma$ talk these days is a genuflection towards LEP. Photon-photon cross sections rise logarithmically with center of mass energy in contrast to single photon physics which falls like 1/s. At $\sqrt{s}$ = 200 GeV the integrated cross section for production of truly spectacular jets, $p_\perp > 10$ GeV/c, is about 9 units of R. So I think you had better start planning taggers for your LEP experiment right now, just so you don't die of boredom when you aren't running at the Z^0 pole!

ACKNOWLEDGEMENTS

I would like to thank Stan Brodsky, Al Eisner, Rollie Morrison, and Tom Walsh for conversations about $\gamma\gamma$ physics, and Bob Panvini and his colleagues at Vanderbilt for organising such a delightful conference. This research was supported by the National Science Foundation.

REFERENCES

1. E. J. Williams, Kgl. Danske Videnskab. Selskab, Mat.-Fys. Medd. 13 (No. 4) (1934); L. Landau and E. Lifshitz, Physik Z. Sowjetunion 6, 244 (134). See also S. J. Brodsky, T. Kinoshita, and H. Terazawa, Phys. Rev. D4, 1532 (1971) and the review articles of V.M. Budnev et. al. Phys. Reports 15C (1975) and H. Terazawa, Rev. Mod. Phys. 45, 615 (1973).
2. E. Fermi, Z. Physik 29, 315 (1924); C. Weizäker and E. J. Williams, ibid, 88, 612 (1934); L. Landau and E. Lifshitz, op.cit.; R. B. Curtis, Phys. Rev. 104, 211 (1956); R. H. Dalitz and D. R. Yennie, ibid, 105 1598 (1957).
3. cf. the contributions of G. Knies, these proceedings.
4. C. Carimalo, P. Kessler, and J. Parisi, Phys. Rev. D20, 1057, 2170 (1979), plus references therein.
5. For a comprehensive review of the vector-meson dominance model see T. H. Bauer, R. D. Spital, and D. R. Yennie, Rev. Mod. Phys. 50, 261 (1978). For an early discussion of the phenomenological necessity for a perturbative, point-like component for photon-induced reactions see S. J. Brodsky, F. E. Close, and J. F. Gunion, Phys. Rev. D6, 177 (1972). The relationship between point-like parton-model contributions and the infinite vector meson spectrum is discussed by S. J. Brodsky and J. Pumplin, Phys. Rev. 182, 1794 (1969); V. N. Gribov, "High Energy Interactions of Gamma Quanta and Electrons with Nuclei," preprint from the Fourth Winter Seminar of the Theory of the Nucleus and the Physics, Acad. Sci. (USSR), (October 1969); E. Etim, M. Greco, and Y. Srivastava, Nuovo Cimento Letters 16, 65 (1976); J. D. Bjorken, Proceedings of the International Symposium on Electron and Photon Interactions, Cornell (1971); and S. Brodsky, G. Grammer, G. P. Lepage, and J. D. Sullivan (in preparation).
6. F. Gilman, SLAC-PUB-2461 (to appear in the Proceedings of the 1979 International Conference on Two Photon Interactions, Lake Tahoe, California).
7. Particle Data Group, Phys. Lett. 75B, 1 (1978).
8. G. Abrams, et. al., Phys. Rev. Lett. 43, 477 (1979).
9. Theoretical estimate: either Ref. (6) or the text, below.
10. For a review, see E. Witten, Nucl. Phys. B160, 57 (1979).
11. D. Coyne's talk, these Proceedings.
12. D. Scharre's talk, these Proceedings.

13. E. D. Bloom, invited talk at the 1979 International Symposium on Lepton and Photon Interactions at High Energies, Batavia, Illinois, 23-29 August 1979, and R. Barbieri, et. al. Nucl. Phys. B154, 535 (1979).
14. R. Bhattacharya, J. Smith, and G. Grammer, Phys. Rev. D15, 3267 (1977), J. Smith, J. A. M. Vermaseren, and G. Grammer, Phys. Rev. D15, 3280 (1977), D19, 137 (1979).
15. J. Vermaseren, private communication.
16. S. Brodsky, T. Kinoshita, and H. Terazawa, Phys. Rev. Lett.
17. E. Witten, Nucl. Phys. B120, 189 (1977). Witten's result has also been rederived by summing ladder graphs by W. Frazer and J. Gunion, Phys. Rev. D20, 147 (1978) and by C. H. Llewellyn-Smith, Oxford preprint 67/78. See also Ref. (18).
18. Yu. L. Dokshitser, D. I. Dyakanov, and S. I. Troyan, Stanford Linear Accelerator Center translation SLAC-TRANS-183, from the Proceedings of the 13th Leningrad Winter School on Elementary Particle Physics, 1978.
19. T. DeGrand, Nucl. Phys. B151, 485 (1979).
20. W. Bardeen and A. Buras, Phys. Rev. D20, 166 (1979); D. Duke and J. Owens, Florida State preprint, April 1980 (unpublished).
21. C. B. Newman, et. al. Phys. Rev. Lett. 42, 951 (1979).
22. This subject was first discussed in the pioneering paper by J. D. Bjorken, S. Berman, and J. Kogut, Phys. Rev. D4, 3388 (1971). A discussion of this process for virtual γ reactions has been given by T. F. Walsh and P. Zerwas, Phys. Lett. 44B, 195 (1973). More recently, extensive studies of jet and single particle production have been carried out by S. Brodsky, T. DeGrand, J. Gunion, and J. Weis, Phys. Rev. Lett. 41, 672 (1978), Phys. Rev. D19, 1418 (1979), by K. Kajantie, Phys. Scripta 29, 230 (1979), K. Kajantie and R. Raitio, University of Helsinki preprint HU-TFT-79-13 (1979), and by C. H. Llewellyn-Smith, Phys. Lett. 79B, 83 (1978).
23. R. Cahn and J. Gunion, Phys. Rev. D20, 2253 (1979).
24. This observation was first made by Llewellyn-Smith, ref. (22).
25. Cf. S. Brodsky and G. P. Lepage, Phys. Lett. 87B, 359 (1979).

SOME DEVELOPMENTS IN $Q\bar{Q}$ SPECTROSCOPY*

Kurt Gottfried
Cornell University, Ithaca, NY 14853

ABSTRACT

The theory of the phenomenological $Q\bar{Q}$ interaction, and of the hadron transitions between $Q\bar{Q}$ states, is reviewed.

INTRODUCTION

In this report I shall not try to give a comprehensive review of recent developments in heavy quark-antiquark ($Q\bar{Q}$) spectroscopy. Instead, I confine myself to two topics: the static $Q\bar{Q}$ interaction, and transitions between $Q\bar{Q}$ states that are accompanied by the emission of light hadrons. A comprehensive review would, at the very least, also devote attention to the following items: i) The status of η_c, and in particular, the dispersive approach due to the ITEP/Moscow group, which correctly predicted its position several years ago[1]; ii) rigorous inequalities due to Martin[2], and Bertlmann and Martin[3], which test the flavor independence of the static $Q\bar{Q}$ interaction, and set bounds on the OZI threshold of any $Q\bar{Q}$ system; iii) the possible existence of vibrational states due to the gluon field which fall outside the Hilbert space of the conventional charmonium model[4]; and iv) recent work on the influence of decay channels on the $Q\bar{Q}$ states above the OZI threshold[5-7], where I would draw attention to Eichten's new calculations[7] which show that the 4^3S state in the Υ family has a shape that is very sensitive to its position with respect to the various B-meson thresholds.

Before I turn to the two topics that I shall discuss in detail, I wish to stress that the most elaborate model of the above-threshold region now in existence--that of the Cornell group[8], still suffers from the serious shortcoming of not having a complete set of decay channels. As we have just learned from Hitlin[9], both DORIS and SPEAR shall eventually produce data on the $c\bar{c}$ system of unprecedented quantity and quality. The existing model[8] was conceived and formulated before the charmonium P-state or the D-mesons were discovered, and it was therefore not intended to provide an accurate description of phenomena well above threshold. Indeed, it has been rather more successful than one would have anticipated, but that does not mean that it will serve the needs of the new generation of experiments. A considerable elaboration of the model (or its replacement by something altogether new) is now called for. This elaboration shall have to incorporate a complete set of final $Q\bar{q}$ + $\bar{Q}q$ states (q being a light quark), while retaining a proper rendition of thresholds, and if possible, it should also treat q and $\bar{q}$ relativistically.

THE STATIC $Q\bar{Q}$ INTERACTION

Since the very beginning, there has been a continuing effort to relate the $Q\bar{Q}$ interaction to more fundamental theory. Consequently

*Supported in part by the National Science Foundation.

ISSN:0094-243X/80/620088-13$1.50 Copyright 1980 American Institute of Physic

the first phenomenological potentials were taken to be Coulombic at short distances in rough accordance with asymptotic freedom, and as linear in r, as expected from the string picture of hadrons or, equivalently, Wilson's area law. In recent years, many authors[10] have modified the short distance potential by incorporating logarithmic corrections arising from asymptotic freedom. Unfortunately, in the absence of a complete understanding of confinement in QCD, no one knows how to connect this short distance regime to the confinement force at long distances. Nevertheless, during the last year several studies of lattice gauge theory, in particular the numerical work by Creutz, have given rather clear indication that there is a very rapid transition from the perturbative regime at short distances, to the simple string-like situation at long distances. Consequently, one can entertain the hope that there exists a simple semi-empirical formula for the $Q\overline{Q}$ interaction which amalgamates the knowledge that we have about the two asymptotes. One can even expect this interaction to have only one adjustable constant, because QCD is parametrized by one arbitrary length.

This hope appears to have been realized in a potential put forward by Richardson.[11] Quite aside from any theoretical considerations, it is clear that Richardson's potential provides a remarkably accurate and economical description of both the $c\overline{c}$ and $b\overline{b}$ spectra. His potential has only one parameter, Λ_R, which can be naively viewed as the parameter that describes scaling violation in perturbative QCD. Very recently, the relationships between Λ_R and the QCD scaling violation parameter, as well as the Regge slope, have been elucidated by Buchmüller, Grunberg and Tye.[12] I will now describe these developments.

As I have said, the goal is to relate, in as economical a fashion as possible, the Regge slope α' of "old" meson spectroscopy, the scaling violations in deep inelastic lepton-hadron scattering, and $Q\overline{Q}$ spectroscopy, via a potential that describes that spectroscopy. To this end we write the potential in momentum space as

$$V(Q^2) = -\frac{64\pi^2}{3}\frac{\rho(Q^2)}{Q^2}, \qquad (1)$$

where $|\vec{Q}|$ is the 3 momentum transfer, and

$$\rho(Q^2) \equiv \frac{1}{4\pi}\alpha_s(Q^2),$$

$\alpha_s(Q^2)$ being the running coupling constant of QCD.

At short distances the dominant correction to the $1/Q^2$ interaction comes from vacuum polarization due to particles that are <u>light</u> compared to m_Q; in other words, due to u, d, and s quarks, and to gluons. In this respect our problem is actually reminiscent of μ-mesic atoms, where e^+e^- vacuum polarization is the leading correction to the non-relativistic spectrum.[13] This well known, but nevertheless remarkable circumstance arises because the Bohr radius of a μ-mesic atom is of comparable size to the electron's Compton wavelength, and it is the latter which characterizes the dimensions of the vacuum polarization cloud.

These leading corrections due to vacuum polarization are summarized by the famous formula

$$\rho(Q^2) = \frac{\rho(\Lambda^2)}{1 + b_o(N_f)\rho(\Lambda^2)\ell n(Q^2/\Lambda^2)} , \tag{2}$$

where

$$b_o(N_f) = 11 - \frac{2}{3} N_f , \tag{2'}$$

with N_f being the number of quark flavors that are light compared to m_Q. In the case of the $c\bar{c}$ and $b\bar{b}$ systems we shall take $N_f = 3$, or $b_o = 9$. Note that at very short distances

$$\rho(Q^2) \xrightarrow[Q^2\to\infty]{} \frac{1}{b_o(N_f)\ell n(Q^2/\Lambda^2)} . \tag{3}$$

At long distances the interaction in coordinate space must have the form

$$V(r) \to \frac{1}{2\pi\alpha'} r . \tag{4}$$

The appearance of α' in this relationship cannot be inferred from the low lying states of heavy $Q\bar{Q}$ systems. This is quite obvious, because α' is a characteristic of the light mesons, which are relativistic systems. Consequently, the familiar flat Regge trajectory will only make its appearance in heavy $Q\bar{Q}$ spectra when the angular momentum is large.[14] Nevertheless, Eq. (4) is correct even in the problem at hand because the connection between the string tension and the Regge slope has been established on quite general grounds.[15]

Comparing Eqs. (4) and (1) we see that $\rho(Q^2)$ must have the following behavior as Q^2 tends to zero:

$$\rho(Q^2) \to \frac{3}{16\pi^2 Q^2 \alpha'} . \tag{5}$$

Richardson's Ansatz "follows" from a comparison of the two asymptotic conditions, Eqs. (3) and (5):

$$\rho(Q^2) = \frac{1}{b_o(N_f)\ell n(1 + \frac{Q^2}{\Lambda_R^2})} . \tag{6}$$

By construction

$$\rho(Q^2) \xrightarrow[Q^2\to\infty]{} \frac{\Lambda_R^2}{b_o Q^2} , \tag{7}$$

and therefore the parameter Λ_R is related to the Regge slope by

$$\alpha' = \frac{3b_o(N_f)}{16\pi^2} \frac{1}{\Lambda_R^2} . \tag{8}$$

or $\alpha' = (27/16\pi^2)\Lambda_R^{-2}$ for $N_f = 3$.

Richardson has used his Ansatz to compute the ψ and Υ spectra. He determines $\Lambda_R = 400$ MeV from the ψ-ψ' splitting and this yields $\alpha' = 1.07$ GeV^{-2}. This is in remarkable agreement with the familiar value of the Regge slope. Furthermore, Richardson's potential gives an excellent rendition of the ψ spectrum, and accurately predicted the position of the 3S state in the Υ spectrum (see Table I). It is

Table I. The ψ and Υ Spectra

		Expt.	Richardson	$\Lambda_{\overline{ms}}$		
				500	400	250
ψ	$1P_{cog}$	427	419	425		
	2S	589	--	595	590	507
	1D	677	704	715		
	3S	990	1.001	1.015		
Υ*	2S	560	555	--	528	
	3S	891	886	890		
	4S	1113	1.16	1.16		

*Experimental data from D. Andrews *et al.*, Phys. Rev. Lett. **44**, 1108 (1980); T. Böhringer *et al.*, *ibid.*, **44**, 1111 (1980).

to be noted that in contrast to the Cornell Coulomb + linear potential[16], Richardson did not have to engage in any flavor-dependent fine tuning of parameters; the only input that changes in going from the $c\bar{c}$ to $b\bar{b}$ system is the quark mass!

At first sight, one is also tempted to say that $\Lambda_R = 400$ MeV is in excellent agreement with what is known about scaling violation in deep inelastic scattering. But as Buchmüller *et al.* point out, the QCD parameter Λ is not defined if $\rho(Q^2)$ is only used at the one loop level. That is to say,

$$\rho(Q^2/\Lambda_1^2) - \rho(Q^2/\Lambda_2^2) = \frac{1}{b_o} \ln \left(\frac{\Lambda_2}{\Lambda_1}\right)^2 \left|\ln \frac{Q^2}{\Lambda_1^2} \cdot \ln \frac{Q^2}{\Lambda_2^2}\right|^{-1},$$

which is of order the two loop contribution. Hence one may ask whether there is any significance in the fact that Λ_R is compatible with the Λ inferred from scaling violations.

To answer this question, one must incorporate vacuum polarization at the two loop level. For this purpose we define

$$\beta(\rho) \equiv Q^2 \frac{\partial}{\partial Q^2} \rho(Q^2) .$$

From perturbative QCD we know that

$$\beta(\rho) \xrightarrow[\rho\to 0]{} -b_o \rho^2 - b_1 \rho^3 + \dots . \tag{9}$$

Here the terms are respectively the one- and two-loop contributions; b_o was already given in Eq. (2'), and

$$b_1 = 102 - \frac{38}{3} N_f ,$$

which equals 64 for $N_f = 3$. In the confinement regime, where we know that ρ must behave like $1/Q^2$ (cf. Eq. (5)) we must have

$$\beta(\rho) \xrightarrow[\rho\to\infty]{} -\rho[1 + O(\frac{1}{\rho})] . \tag{10}$$

Since QCD is characterized by only one length, there must be a relationship between the Regge slope α' and the scaling violation parameter. For the latter we take $\Lambda_{\overline{ms}}$, which is the currently popular parameter defined by a modified form of the dimensional regularization scheme. As Buchmüller et al. have shown, the two asymptotic forms of β, as given by Eqs. (9) and (10), can be related, and this then leads to the following relation between α' and $\Lambda_{\overline{ms}}$:

$$-\ln(\alpha' \Lambda^2_{\overline{ms}}) = \ln(12\pi^2) + \frac{1}{b_o}(\frac{34}{3} - \frac{10}{9} N_f) + \frac{b_1}{b_o} \ln b_o$$
$$+ \int_0^1 dx \left[\frac{1}{b_o x^2} - \frac{b_1}{b_o^2 x} + \frac{1}{\beta(x)}\right] + \int_1^\infty dx \left[\frac{1}{x} + \frac{1}{\beta(x)}\right] . \tag{11}$$

In these integrals the pieces of β that would lead to singularities have been explicitly removed.

We are now in a position to make a somewhat more sophisticated interpolation Ansatz than Richardson's. This is done in terms of the function β. The form guessed by Buchmüller et al. is

$$\frac{b_o}{\beta(\rho)} = -\frac{1}{\rho^2}\left(1 - e^{-\frac{1}{b_o\rho}}\right)^{-1} + \frac{b_1}{b_o}\frac{1}{\rho} e^{-\lambda\rho} . \tag{12}$$

Here the first term is precisely the same as Richardson's, whereas the second is new, and supposedly incorporates the two loop contributions at short distances. (A plot of β is shown in Fig. 1.) When substituted into Eq. (11), (12) leads to

$$\ln(\alpha' \Lambda^2_{\overline{ms}}) = -\frac{7}{9} + \ln(\frac{3b_o}{16\pi^2}) + \frac{b_1}{b_o^2}(0.577 + \ln\frac{\lambda}{b_o}) . \tag{13}$$

Eq. (12) shows that $\lambda \to \infty$ retrieves Richardson's Ansatz. As we see from Eq. (13), this limit does not allow any comparison between the Regge slope and the scaling violation parameter.

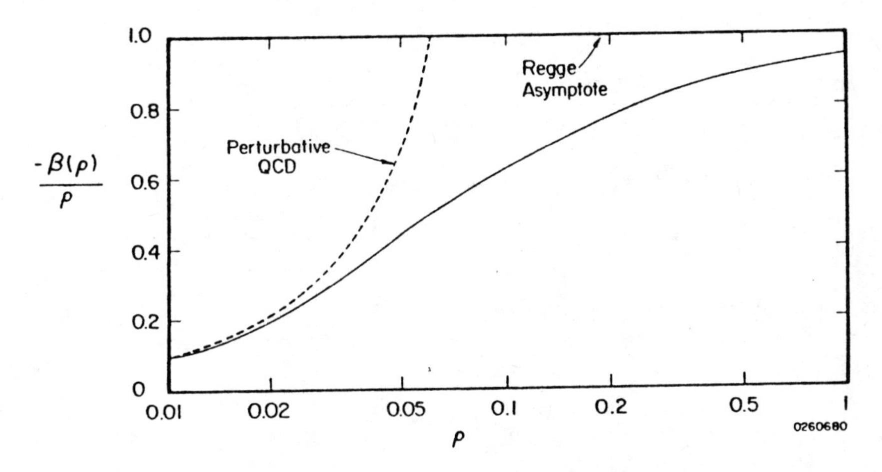

Fig. 1. The function $\beta(\rho)$.

We are now in a position to compare this second Ansatz with the data. From scaling violations we know that $\Lambda_{\overline{ms}} \leq 0.5$ GeV. Buchmüller et al. find that for $\lambda = 24$, which corresponds to $\Lambda_{\overline{ms}} = 0.5$ GeV, the fit to both the $c\bar{c}$ and $b\bar{b}$ spectra is as excellent as Richardson's. This value of $\Lambda_{\overline{ms}}$ corresponds to a Regge slope of 1.04 GeV^{-2}, which is in excellent agreement with the "old spectroscopy". This is hardly surprising since such a large value of λ makes the second term of Eq. (12) virtually negligible. When λ decreases, so does $\Lambda_{\overline{ms}}$, and the fit to the spectra deteriorates quite rapidly. Thus for $\Lambda_{\overline{ms}} = 0.4$ ($\lambda = 13$), the fit is still acceptable but for $\Lambda_{\overline{ms}} = 0.25$ ($\lambda = 4$) it is already terrible. A more detailed description of this behavior is provided by Table I. When examining this table one should remember that the observed state $\psi(3772)$ is $^3D_1$ $(c\bar{c})$, whereas this spin independent potential only provides the energy of the center of gravity of the ^3D multiplet.

The values of the lepton widths in the Υ family, expressed as ratios to the width of the 1S state, are given in Table II. Fig. 2 shows the Richardson potential, as well as the asymptotic forms at short and long distances. Also shown are the mean radii of various $Q\bar{Q}$ states.

To summarize, we now have a very simple formula for the $Q\bar{Q}$ interaction--in essence due to Richardson--which combines our knowledge of the short distance interactions, as described by perturbative QCD, with the confining force that correctly describes the Regge slope of the light hadrons. It achieves this in terms of one parameter Λ_R that can be associated with the scaling violations in deep inelastic scattering.

Table II. Υ leptonic width ratios

	$\Gamma_{ee}(\Upsilon^n)/\Gamma_{ee}(\Upsilon)$		
	Exp.*	Richardson	Buchmüller et al. ($\Lambda_{\overline{ms}} = 0.5$)
Υ(2S)	0.44±0.06	0.42	0.45
Υ(3S)	0.35±0.04		0.32
Υ(4S)	0.21±0.06 0.25±0.07		0.27

*For references, see Table I.

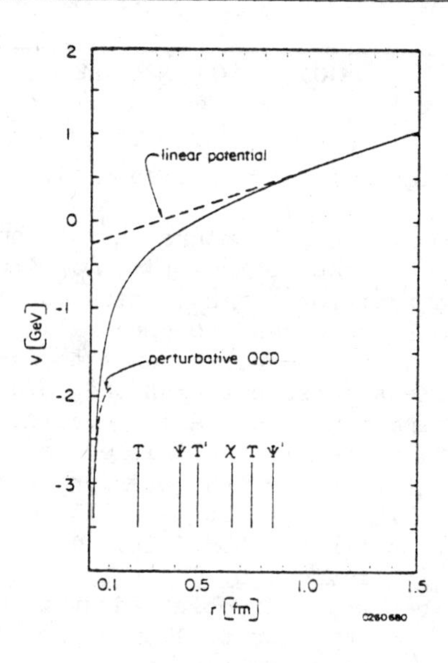

Fig. 2. The potential constructed from Eq. (12), via Eq. (1). The asymptotes corresponding to the Regge and perturbative QCD regimes are also shown, as are the mean radii of various Υ and ψ states.

RADIATION OF LIGHT HADRONS IN $Q\bar{Q}$ SYSTEMS

At the moment the only known examples of this phenomenon are $\psi' \to \psi\pi\pi$ and $\psi' \to \psi\eta$. There is indirect evidence[17] that the rate for $\Upsilon' \to \Upsilon\pi\pi$ is at least an order of magnitude smaller than the corresponding one in the ψ family.

Since the light hadrons carry off a small amount of momentum, it is clear that straightforward perturbative QCD cannot describe these processes. Nevertheless, they are characterized by a small parameter, and the theory that has been developed seeks to exploit this fact to the utmost. The small parameter is the velocity v of the $Q\bar{Q}$ motion (in units of c). In the case of the ψ's, $v^2 \sim 1/4$, whereas it is of order 1/10 in the Υ's. Once a system is nonrelativistic, the wave length of radiations emitted by it are large compared to the dimensions $\bar{r}$ of the source. In atomic and nuclear physics one takes advantage of this by carrying out a multipole expansion of the coupling to the electromagnetic field. By the same token, it is possible to multipole-expand the coupling between the heavy quarks and the color gauge field. Thus in the Υ system $k\bar{r}$ is $\sim 1/5$, where k is the characteristic momentum of a virtual gluon which, subsequently, converts into light hadrons. In contrast to QED, in QCD one must retain all terms in the coupling constant power series that contribute to a given multipole. Furthermore, because of the non-linear nature of the gauge field, a complex set of interactions occur in QCD that have no counterpart in the theory of electromagnetic radiations.

In the original formulation of the QCD multipole expansion[18], the non-Abelian couplings were handled in a clumsy, and in one respect even incorrect, fashion. Subsequently[19-21] Peskin, Voloshin, Bhanot, Fischler and Rudaz re-examined the problem in the limit of very large m_Q, where perturbation theory is valid because the $Q\bar{Q}$ level splittings become very large, and the $Q\bar{Q}$ spectrum tends to the hydrogenic limit. Peskin, in particular, showed that the multipole expansion can be cast into the language of the operator product expansion. Unfortunately, it was not clear how this large m_Q limit was to be applied to the systems of current interest, but a naive application[21] of the theory led to results that were very encouraging. Recently, Yan[22] has discovered a technique which exploits the gauge invariance of QCD to achieve a formulation which incorporates the essential features of the work of Peskin *et al*., and at the same time is sufficiently transparent to permit a very plausible application to systems where $k\bar{r}$ is small, but where the spectrum is still far from hydrogenic. I shall now describe the essential features of Yan's work.

The objective of any multipole expansion is to approximate the intricate coupling between the source and the object radiated into a sequence of terms, each of which is a product of two operators, one of which refers solely to the source, the other to the emitted objects. If the expansion is successful, the sequence can be ordered so that selection rules imply that only a few terms contribute to a particular transition. In our problem the $Q\bar{Q}$-h couplings fall into two categories: the first has a direct analogue in QED, the second is intrinsically non-Abelian (here h is the emitted light hadron system). This latter category is illustrated by Fig. 3, where two transverse gluons are emitted by instantaneous Coulomb interaction. The central point of Yan's treatment is that in the leading multipoles all terms of this type can be eliminated by a canonical transformation of the heavy quark field.

This transformation has the following form. Let $A_\mu \equiv 1/2\ \lambda_a A_\mu^a$ be the gluon field in Coulomb gauge, and $\vec{x}$ be any point measured with respect to the $Q\bar{Q}$ center of mass. One then defines the unitary

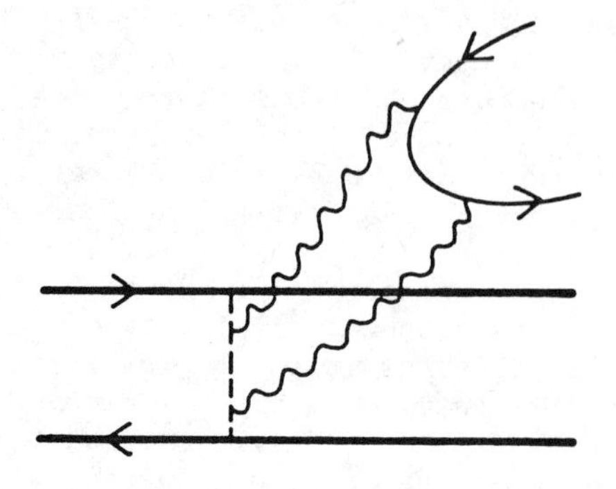

Fig. 3. A non-Abelian contribution to the emission of a light hadron. The heavy (light) solid lines represent heavy (light) quarks, whereas dashed (wavy) lines are Coulomb (transverse) gluons.

operator

$$U(\vec{x}) = P \exp\{ig \int_0^{\vec{x}} d\vec{s}\cdot\vec{A}(\vec{s})\}, \tag{14}$$

where P is the path ordering operator. If $\psi(\vec{x})$ is the original Q field, one then defines a "constituent" field by

$$\Psi(\vec{x}) \equiv U^{\dagger}(\vec{x})\psi(\vec{x}) . \tag{15}$$

The essential point is then the following: If $V(\vec{x})$ is the unitary operator that affects an <u>arbitrary</u> local gauge transformation, the constituent field transforms as

$$\Psi(\vec{x}) \to V(\vec{0})\Psi(\vec{x}) , \tag{16}$$

i.e., as if the transformation were <u>global</u>.

The instantaneous $Q\bar{Q}$ Coulomb interaction, which, in terms of the original variable ψ, has an infinity of non-linear couplings to the A-field, takes on the familiar electromagnetic form

$$H_C = -\frac{1}{2}\frac{g^2}{4\pi}\sum_a \int d^3x\, d^3y\, \frac{\rho_a(\vec{x})\rho_a(\vec{y})}{|\vec{x}-\vec{y}|} \tag{17}$$

provided that the color densities ρ_a are written in terms of the constituent fields:

$$\rho_a(\vec{x}) = \Psi^{\dagger}(\vec{x})\,\frac{1}{2}\,\lambda_a\Psi(\vec{x}) . \tag{18}$$

In (17) all explicit dependence on the A-field has disappeared!

In terms of the constituent field one can write a gauge invariant expression for any color singlet $Q\bar{Q}$ state:

$$|\chi\rangle = \int d^3x\, \chi(\vec{x})\, \Psi^{\dagger}(\vec{x})\Gamma\, \Psi(0)|vac\rangle ;$$

here Γ is a Dirac matrix which depends on the spin of the state, and χ is its spatial wave function.

Since the entire Hamiltonian has now assumed a form similar to the one in electrodynamics, the multipole expansion is straightforward. The interaction between the $Q\bar{Q}$ source and the gauge field has the following structure:

$$H_I = Q_a A_o^a(0) - \vec{d}_a \cdot \vec{E}_a(0) - \vec{m}_a \cdot \vec{B}_a(0) + \ldots \quad (19)$$

Here the gauge field makes its appearance through the potential and the field strengths evaluated at the $Q\bar{Q}$ c.o.m. $\vec{x} = 0$. The $Q\bar{Q}$ system is characterized by the multipoles Q_a, $\vec{d}_a$, $\vec{m}_a$. The first term, a monopole, only contributes to octet-octet $Q\bar{Q}$ transitions--i.e., it will only appear in virtual states. The electric and magnetic dipole operators are

$$\begin{aligned} \vec{d}_a &= g\int d^3r\, \Psi^\dagger(\vec{r})\, \frac{1}{2}\lambda_a \Psi(\vec{r})\vec{r} \\ &\to \frac{1}{4} g\vec{r}(\lambda_a - \bar{\lambda}_a) , \end{aligned} \quad (20)$$

and

$$\begin{aligned} \vec{m}_a &= \frac{1}{2} g\int d^3\vec{r}\; \vec{r}\times[\Psi^\dagger(\vec{r})\vec{\alpha}\, \frac{1}{2}\lambda_a\Psi(\vec{r})] \\ &\to \frac{g}{4m_Q}\frac{1}{2}(\lambda_a - \bar{\lambda}_a)(\vec{\sigma}_Q - \vec{\sigma}_{\bar{Q}}) . \end{aligned} \quad (21)$$

In these equations the first expression still contains the relativistic field Ψ, and the second shows the nonrelativistic limit. As we see, the E1 term is of order kr, whereas the M1 is of order k/m_Q.

The full Hamiltonian now takes the form

$$H = H_o + \sum_i M_i , \quad (22)$$

where the "unperturbed" part is:

$$H_o = H_{Q\bar{Q}} + H_{qA} + Q_a A_o^a . \quad (23)$$

Here $H_{Q\bar{Q}}$ is the Hamiltonian of the naive charmonium model, H_{qA} is the Hamiltonian of light quarks and transverse gluons, and the last term is the monopole of Eq. (19). Note that H_o only contains singlet-singlet and octet-octet $Q\bar{Q}$ operators. The "perturbation" terms in Eq. (22) are the multipole operators; M_1 is the electric, M_2 is the magnetic dipole, etc.

An arbitrary transition amplitude now takes the form

$$A(T\to hT') = \sum_n \sum_{i_1\ldots i_n} \langle T'h|M_{i_1}\frac{1}{E_o - H_o^8}M_{i_2}\cdots\frac{1}{E_o - H_o^8}M_{i_n}|T\rangle , \quad (24)$$

where T is an arbitrary $Q\bar{Q}$ state, and H_o^8 is the projection of H_o

into the octet $Q\bar{Q}$ subspace. The leading amplitudes are therefore

$$A_{E1;E1} = \langle T'h|\vec{d}_a \cdot \vec{E}_a \frac{1}{E_o - H_o^8} \vec{d}_b \cdot \vec{E}_b|T\rangle , \tag{25}$$

which has the selection rules $\Delta S = 0$, $\Delta L = 0$ or 2, whereas the next-to-the-leading term is

$$A_{E1;M1} = \langle T'h|\vec{d}_a \cdot \vec{E}_a \frac{1}{E_o - H_o^8} \vec{m}_b \cdot \vec{B}_b|T\rangle , \tag{26}$$

with the selection rules $\Delta S = \Delta L = 1$. In order of magnitude, these amplitudes are

$$A_{E1;E1} \sim (k\bar{r})^2 , \quad A_{E1;M1} \sim (k\bar{r})(\frac{k}{m_Q}) . \tag{27}$$

As in atomic and nuclear physics, the Wigner-Eckart theorem can be used to great advantage. At the moment the most interesting example is the set of $\pi\pi$ transitions between the 2^3P and 1^3P $b\bar{b}$ states. There are nine of these E1-E1 transitions, and because of the Wigner-Eckart theorem, only three reduced matrix elements. Consequently there are relationships between the various rates. For example, the $\pi\pi$ spectra are related by[22]

$$\frac{d\Gamma(2\to2)}{dM_{\pi\pi}} = \frac{d\Gamma(0\to0)}{dM_{\pi\pi}} + \frac{3}{4}\frac{d\Gamma(0\to1)}{dM_{\pi\pi}} + \frac{7}{20}\frac{d\Gamma(0\to2)}{dM_{\pi\pi}} ,$$

where phase space corrections have been ignored.

As we see from Eq. (24), the transition amplitude involves the operator $(E - H_o^8)^{-1}$. Since the ψ and T spectra have very similar spacings, these energy denominators should be roughly constant. Therefore the dominant variation in rates will stem from the change in $\bar{r}$ and m_Q. Each multipole operator M_i has a definite dependence on $\bar{r}$, and on m_Q, and one therefore expects scaling rules which relate a particular transition in one $Q\bar{Q}$ family to its counterpart in another family. If one is prepared to assume that the multipole expansion is applicable to the $c\bar{c}$ family, one can use these scaling rules to estimate rates in the T system. It is probably not true that the multipole expansion, when expressed as a power series in r, converges well in the ψ family. On the other hand, it is known from nuclear physics that a multipole expansion in terms of Bessel functions works very well even when kr is of order 1, and gives results that do not differ markedly from the naive expansion.

The $2S \to 1S + \pi\pi$ transitions have the selection rules $\Delta L = \Delta S = 0$, and therefore [cf. Eq. (27)] have amplitudes that vary like $\bar{r}^2$. Consequently

$$\frac{\Gamma(\psi \to \psi\pi\pi}{\Gamma(T' \to T\pi\pi)} \sim \left(\frac{r_\psi}{r_T}\right)^4 \sim 10 . \tag{28}$$

Here it should be noted that if the gluon had spin 0, this ratio would be 1, in gross disagreement with the observed $pA \to e^+e^-X$

excitation curve in the T region.[17]

At first sight, the transitions 2S → 1S + η have the same scaling law as the ππ case. That is not so.[23] Since η is a pseudoscalar, the amplitude must be proportional to $\hat{\varepsilon}'\times\hat{\varepsilon}\cdot\vec{p}_\eta$, where $\hat{\varepsilon}'$ and $\hat{\varepsilon}$ are the ψ' and ψ polarization vectors, and this structure cannot arise from the operator $(\vec{d}_a\cdot\vec{E}_a)\ \ldots\ (\vec{d}_a\cdot\vec{E}_a)$. Consequently the amplitude for η emission has, to leading order, contributions from M1-M1 and E1-M2. Both of these have roughly the same scaling law, which can be expressed by[22]

$$\frac{\Gamma(\psi' \to \psi\eta)}{\Gamma(T' \to T\eta)} \sim \left(\frac{m_b}{m_c}\right)^4 \left|\frac{p_\eta(\psi)}{p_\eta(T)}\right|^3$$

where the last factor takes into account the difference in p-wave phase space. This ratio is ~400, and therefore the η transition is expected to be negligible in the T family.

If the radiated pions have small momenta, one might expect that soft pion theorems could also be exploited in these phenomena. This is indeed the case, as has been shown by Yan.[22] When combined with the multipole expansion, the soft pion technique leads to a variety of predictions. From among these I single out the ππ spectra in the transitions $^3D \to T\pi\pi$ and $2^3S \to T\pi\pi$, as shown in Fig. 4. The characteristic difference between these spectra could be of importance in deciding whether a e^+e^- resonance near the $B\bar{B}$ threshold is a ^3D state, or the vibrational state that has been predicted there[4], and which should decay like a ^3S state.

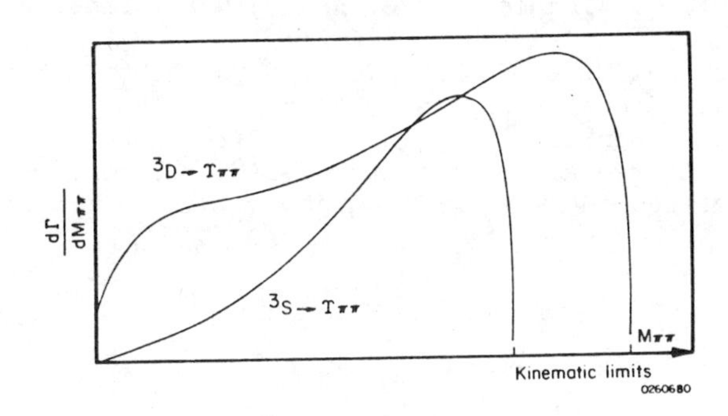

Fig. 4. 2-pion mass spectra.

REFERENCES

1. M. A. Shifman, Z. Phys. C4, 345 (1980). For the original predictions of this group, see references cited therein, and for a detailed exposition of their method, see V. A. Novikov et al., Physics Reports 41C, 1 (1978).
2. A. Martin, Phys. Lett. 88B, 133 (1979).
3. R. A. Bertlmann and A. Martin, TH.2772-CERN (1979).
4. W. Buchmüller and S.-H. H. Tye, Phys. Rev. Lett. 44, 850 (1980).

5. S. Ono, Aachen preprint PITHA 80/2 (1980).
6. A. Bradley and D. Robson, Manchester preprint (1980).
7. E. Eichten, Harvard preprint HUTP-80/A027 (1980).
8. E. Eichten, T. Kinoshita, K. Gottfried, K. D. Lane and T.-M. Yan, Phys. Rev. D17, 3090 (1978); 21, 203, 313(E) (1980).
9. D. Hitlin, these Proceedings.
10. See, for example, W. Celmaster, H. Georgi and M. Machacek, Phys. Rev. D17, 879 (1978); W. Celmaster and F. Henye, ibid., D18, 1688 (1978); A. Billoire and A. Morel, Nucl. Phys. B135, 131 (1978); J. Rafelski and R. D. Viollier, Ref.TH.2673-CERN (1979); R. D. Viollier and J. Rafelski, MIT preprints CTP 791, 819 (1979).
11. J. L. Richardson, Phys. Lett. 82B, 272 (1979).
12. W. Buchmüller, G. Grunberg and S.-H. H. Tye, Cornell preprint CLNS 80/453 (1980).
13. E. Eichten and K. Gottfried, Phys. Lett. 66B, 286 (1978).
14. See Fig. 1 of Ref. 4.
15. See, for example, Celmaster and Henye, Ref. 10.
16. See Note Added in Proof to E. Eichten et al., Phys. Rev. D26, 203 (1980).
17. K. Gottfried, Proc. 1977 International Symposium on Lepton and Photon Interactions at High Energies, Hamburg, F. Gutbrod, ed. (DESY, Hamburg, 1978). R. N. Cahn and S. D. Ellis, Phys. Rev. D17, 2338 (1978).
18. K. Gottfried, Phys. Rev. Lett. 40, 598 (1978).
19. M. E. Peskin, Nucl. Phys. B156, 3651 (1979). See also M. B. Voloshin, ibid. B154, 365 (1979).
20. G. Bhanot, W. Fischler and S. Rudaz, Nucl. Phys. B155, 208 (1979).
21. G. Bhanot and M. E. Peskin, Nucl. Phys. B156, 391 (1979).
22. T.-M. Yan, Cornell preprint CLNS 80/451 (1980). See also K. Shizuya, Berkeley preprint LBL-10714 (1980).
23. I failed to recognize this in Ref. 18, and consequently gave an incorrect scaling law for η emission.

C E L L O - DESIGN AND PERFORMANCE

CELLO Collaboration:

Deutsches Elektronen-Synchrotron, HAMBURG
Universität und Kernforschungszentrum, KARLSRUHE
Max-Planck-Institut für Physik und Astrophysik, MÜNCHEN
Laboratoire de l'Accelerateur Lineaire, ORSAY
Université de Paris VI, PARIS
Centre d'Etudes Nucleaires, SACLAY

(Presented by V. Schröder, DESY)

ABSTRACT

A description of the CELLO detector at PETRA is given as well as first results showing the performance of the major components under beam conditions.

INTRODUCTION

CELLO is a 4π magnetic detector sited at one of the interaction regions of the e^+e^- storage ring PETRA at DESY. CELLO is designed to identify and measure simultaneously leptons, photons and hadrons with high precision over almost the entire solid angle.

The primary physics goals are:

- measurement of the hadronic cross section and particle production characteristics
- search for new leptonic and hadronic states
- study of weak current effects, QED-processes and 2-photon physics

A schematic view of the detector is shown on Fig. 1. CELLO consists of the following main components:

- central tracking device consisting of interleaved cylindrical proportional and drift chambers and four pairs of planar endcap proportional chambers
- superconducting magnet providing a magnetic field up to 1.4 T parallel to the beam axis; thickness is only .5 rad.length (including the cryostat)
- two additional superconducting coils compensating the influence of the main coil on the PETRA beams
- lead liquid argon calorimeter consisting of a cylindrical part containing 16 modules and two symmetric endcaps containing 2 modules each
- μ-detectors consisting of a hadron filter (80 cm iron) and 32 planar drift chambers mounted outside
- a pair of small angle forward spectrometers consisting of drift chambers, scintillator hodoscopes and lead glass shower detectors
- fast programmable hardware trigger using signals from the various detectors; it makes a fast

ISSN:0094-243X/80/620101-12$1.50

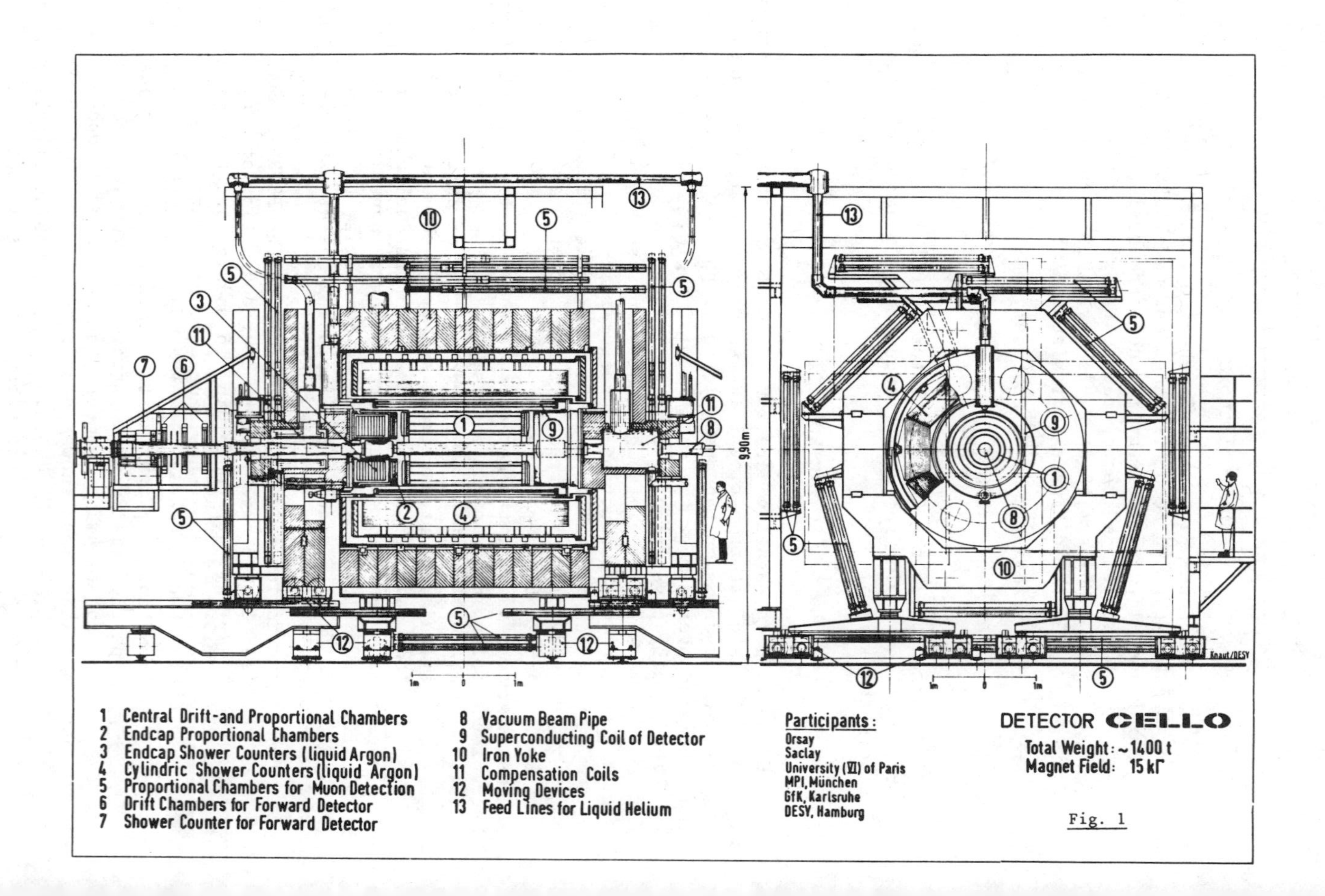
1 Central Drift-and Proportional Chambers
2 Endcap Proportional Chambers
3 Endcap Shower Counters (liquid Argon)
4 Cylindric Shower Counters (liquid Argon)
5 Proportional Chambers for Muon Detection
6 Drift Chambers for Forward Detector
7 Shower Counter for Forward Detector
8 Vacuum Beam Pipe
9 Superconducting Coil of Detector
10 Iron Yoke
11 Compensation Coils
12 Moving Devices
13 Feed Lines for Liquid Helium
Participants:
Orsay
Saclay
University (VI) of Paris
MPI, München
GfK, Karlsruhe
DESY, Hamburg
DETECTOR CELLO
Total Weight: ~1400 t
Magnet Field: 15 kΓ
9,90 m
1m
0
1m
Knaut/DESY

Fig. 1

decision on tracks coming from the interaction point or on particles which exceed a certain amount of energy deposit in the calorimeters
- a computer network with CAMAC readout for data acquisition and reduction and monitoring; it consists of minicomputers (one for each subsystem), a PDP11/45 which mainly does data acquisition, a PDP11/55 for monitoring and supervision and an online link to the DESY IBM-system

The characteristic figures for the various subdetectors are shown on Table I. Some main components are described in more detail below.

CYLINDRICAL PART OF THE CENTRAL TRACKING DEVICE

The charged particle tracking device is designed so that it provides direct accurate measurement of ϕ, the angular coordinate in the plane normal to the beam axis and containing the interaction point, and z, the coordinate along the beam axis. Besides this it provides signals to trigger on charged tracks coming from the interaction region, and is safe to operate in a high magnetic field.

The device consists of seven separated cylindrical groups concentric with respect to the beam axis. The groups contain: 2 proportional wire chambers (PWC's), 2 drift chambers (DC's), 1 PWC, 3 DC's, 1 PWC, 2 DC's and 1 PWC (from the innermost chamber to the outermost). The radii vary between 170 mm and 700 mm, and the layers are spaced nearly uniformly in radius. The cylindrical part covers 87 % of the entire solid angle. Its resolution is $.015 \cdot p$ (GeV/c) at a magnetic field of 1.4 T. The total thickness of the device is 1 % of a rad.length typically. All wires are orientated parallel to the beam axis.

CYLINDRICAL PROPORTIONAL WIRE CHAMBERS

The proportional wire chambers (PWC's) provide full spatial information on charged tracks without ambiguities.

The wire spacing of the 5 chambers varies between 2.10 mm and 2.87 mm and had been chosen so that trigger signals corresponding to equally sized angular bins in the $r\phi$-projection can be obtained. There is a total amount of 5120 wires with a diameter of .020 mm. The gap width is 4 mm. The anode resolution is $\sigma = .26 \cdot$wire spacing (mm), a value which is consistent with the theoretical value of .29. All anodes are readout digitally with a gate width of 50 nsec. This value has been chosen so that drift time and jitter is taken into account and synchrotron radiation background (which is delayed by about 45 nsec) is reduced.

The cathodes of the PWC's are made of free standing mylar cylinders with silver painted cathode strips with an orientation of 90° and 30° with repect to the beam axis. Analog measurement of the induced charge collected on the strips is done for each of the 4400

TABLE I: CHARACTERISTIC FIGURES FOR THE CELLO DETECTOR

Detector component	# of modules	Solid angle ($\Delta\Omega/4\pi$)	Characteristic dimensions (cm)	Characteristic quantities
Tracking device	5 cyl. PWC 7 cyl. DC	0.87	length = 220, radius = 17-70	$\sigma_p/p = .015 \cdot p$ (GeV/c) at 1.4 T, total thickness typ. 1 % X_o
	8 plan.PWC	0.10	radius = 21-66	semicircular shaped
				correlated space points by charge measurement on all PWC-cathodes
Central solenoid	1		length = 400, radius = 80	superconducting, up to 1.4 T, thickness = .49 X_o
Compensating solenoid	2			superconducting, ip to 3.5 T
Lead-liquid argon calorimeters	16 cyl. 4 endcaps	0.96		$\sigma(E)/E = .085/(E(GeV))^{1/2}$, ang. res. 4 mrad typ., detection eff. for low energy e.m. showers = 50 % at 60 MeV and 100 % at 110 MeV, e/h sep. = .2 % at 1 GeV, π,K/p sep. up to 1.1 GeV
μ-chambers	32	0.92	tot.area 200m^2	correlated space points by cathode readout, hadron punch through at 15 GeV < .8 %
Forward detector	24 DC 96 scint. 112 Pb-gl.	25mr<θ<50mr		$\sigma(E)/E = .05/(E(GeV))^{1/2}$, e/h sep. 1 % down to 150 MeV
Charged trigger		0.97		$p_T$cut = 350 MeV/c, cell size = $4^o - 6^o$, software programmable
Neutral trigger		0.96		Ecut = 1 GeV, cell size = $.05 \cdot 4\pi$

channels. The width of the strips is about 4 mm; the width of the 90°-cathodes varies so that trigger signals corresponding to equally sized angular bins in the rz-projection are provided. The precisión of position determination in the longitudinal projection is $\sigma_z = .4$ mm:

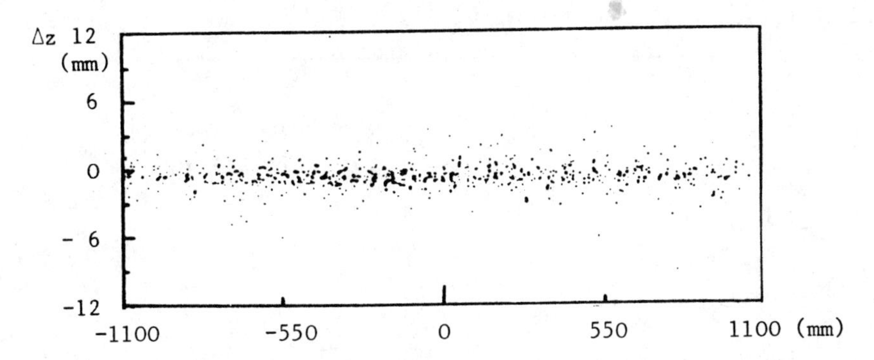

Fig. 2. Spatial resolution of the cathodes of the cyl. PWC's

Together with the informations of the anodes and the 30°-cathodes a space point reconstruction is possible with a precision of $\sigma = (.5 \text{ mm})^2$.

CYLINDRICAL DRIFT CHAMBERS

The 7 drift chambers (DC's) provide an accurate measurement of the position of charged tracks in the rϕ-projection. The chambers consist of nearly equally sized drift cells. Two sense wires have a distance of 15 mm and are separated by 2 cathodes and 1 potential wire. The gap width is 5 mm. This small cell size has the advantage that no field shaping electrodes are needed and that simple single hit electronics can be used. Time digitalisation is done with a resolution of about 3 nsec.

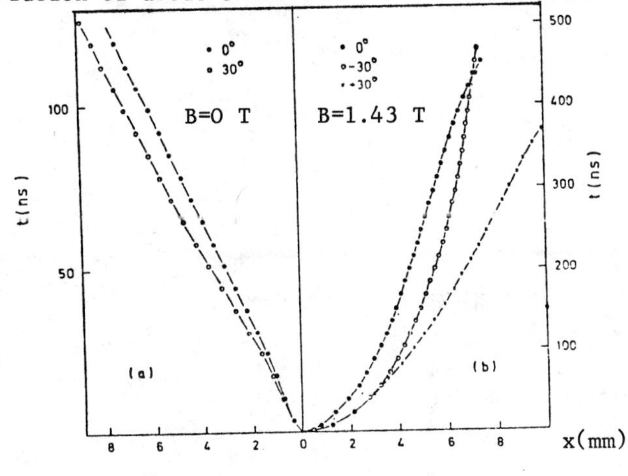

Fig. 3.
Space-time relationship for the cylindrical drift chambers without (a) and with (b) magnetic field of 1.43 T applied

The inclination of the tracks, which is up to 30° in the rϕ-projection due to the high magnetic field as well as the magnetic field itself cause a substantial amount of nonlinearity in the space-time relationship (Fig. 3). Using a phenomenological model of the drift process the nonlinear space-time relations are parametrized by cubic spline functions thus giving a sufficient amount of precision:

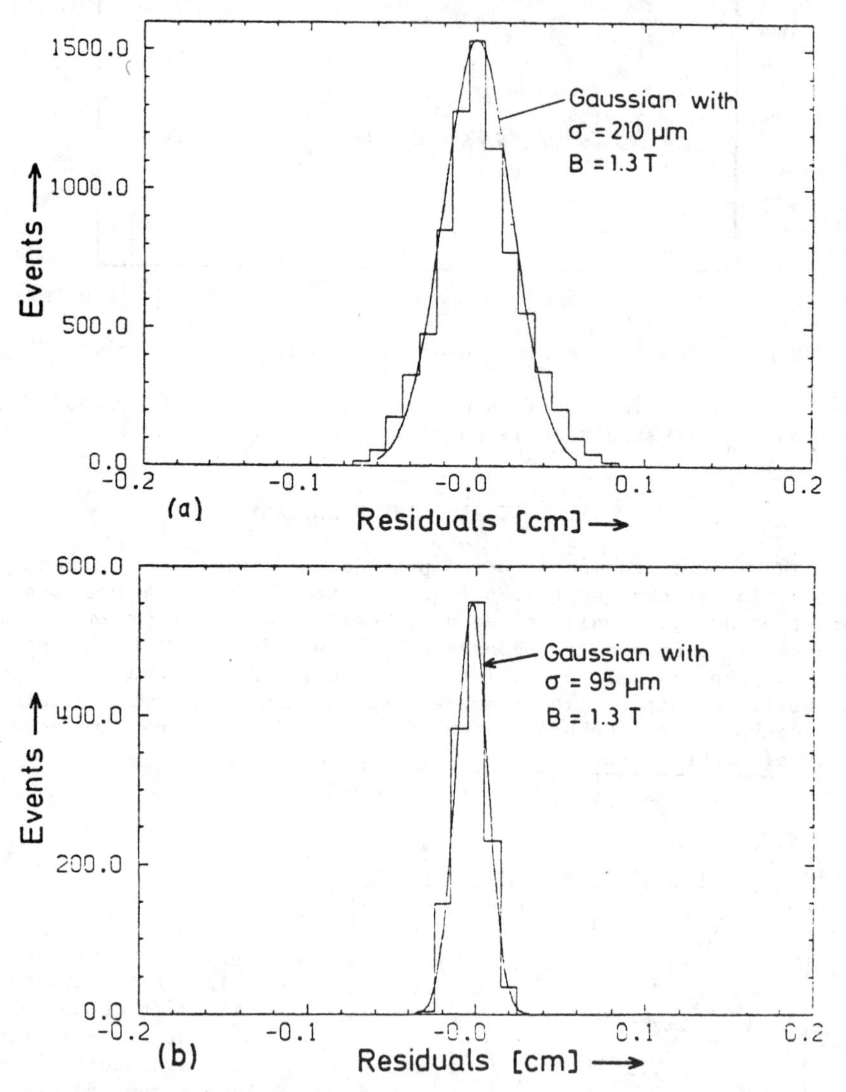

Fig. 4: Spatial resolution of the cyl. drift chambers averaged over the entire drift cell and for angles of incidence between ± 30° (Fig. 4a) and nonaveraged (Fig. 4b)

CYLINDRICAL LEAD-LIQUID ARGON CALORIMETER

The lead-liquid argon calorimeter provides an accurate position and energy measurement of electromagnetic showers and a good electron/hadron separation.

The cylindrical part of the calorimeter is built mirror symmetric with respect to the z=0-plane and consists of 16 equally sized modules in a single cryostat of 25 m^3 volume. The modules are arranged so that an octagonal symmetry is obtained. Each module is trapezoidal shaped and is a stack of 41 layers which are divided in 2 cm wide strips orientated at 0°, 45° or 90° relative to the beam axis. In front of each stack there are two additional layers which serve as dE/dx gaps. These layers are built of copper foils glued on epoxy plates instead of lead plates. The thickness of the lead plates is 1.20 mm, the ionisation gap width is 3.6 mm and the total thickness of one stack is more than 20 rad.lengths, which is large enough to contain electromagnetic showers at even the largest PETRA energies.

The wiring scheme of the strips is optimized to achieve a good resolution in space and energy and even good detection efficiency for low energy showers while keeping the number of electronic channels small (384 per stack). The strips in the same layer are grouped so that an angular resolution of 4 mrad typically (<20 mrad) is obtained. Electronic sampling in depth is 17-fold and is optimized to get a good electron/hadron separation. Information on the energy deposit in the stacks is available for triggering (see below).

Before being installed at CELLO one full size module has been tested in hadron and electron beams at CERN and in electron and photon beams at DESY. Fig. 5 shows the linearity measured at DESY with electrons with energies between 100 MeV and 7.2 GeV.

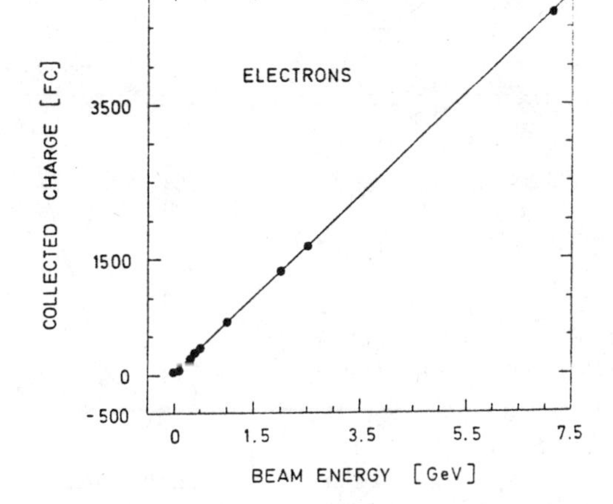

Fig. 5. Linearity of the lead-liquid argon calorimeter

The energy resolution measured in these tests (Fig. 6) is affected by some material in front of the calorimeter. The amount of material is .5 rad.length for the CERN test and 1.0 rad.length at DESY. The resolution obtained by extrapolating the measured values to zero passive material is $\sigma_E/E = .085/(E)^{1/2}$.

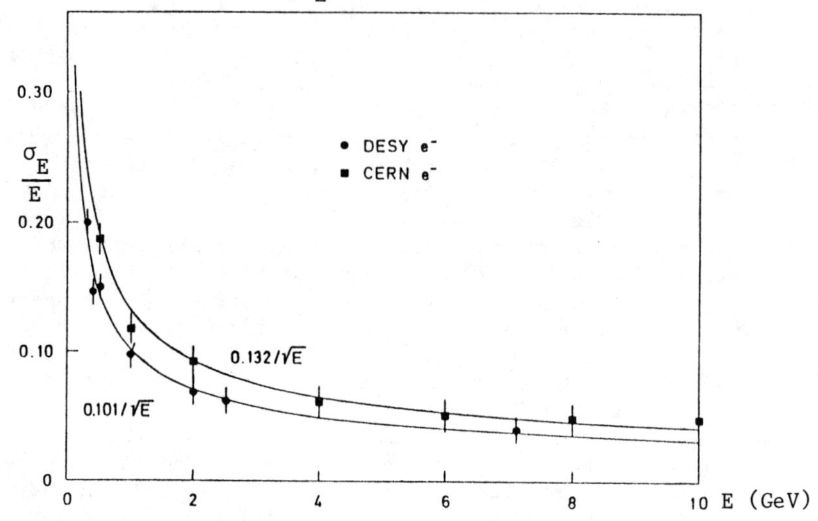

Fig. 6. Energy resolution of the lead-liquid argon calorimeter

Fig. 7 shows the efficiency curve for low energy photons with two different thresholds applied to eliminate noise. It shows virtually full efficiency down to an energy of approximately 110 MeV:

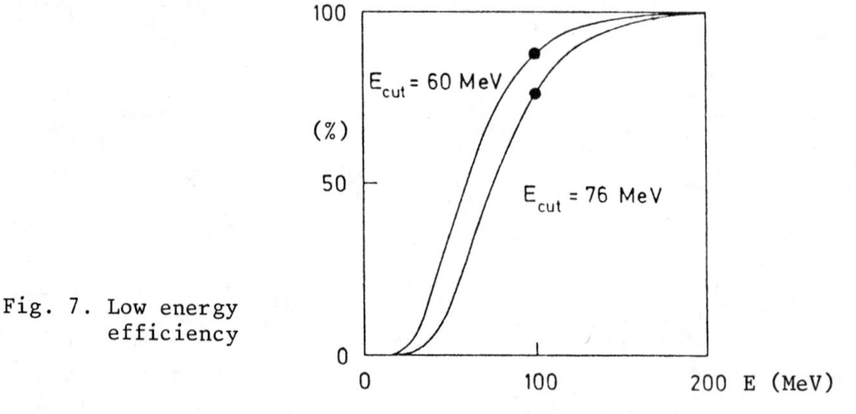

Fig. 7. Low energy efficiency

The spatial width of the charge distribution of showers measured in different sampled depths varies between 5 mm and 20 mm. It is nearly independent on energy for a depth less than 5 rad.lengths.

The center of a shower can be determined with an overall resolution of $\sigma = 4$ mm for electrons of 1 GeV/c.

The CERN test data show, that pion/proton separation by means of ionization measurement in the dE/dx-gaps is possible up to energies of 1 GeV.

MUON CHAMBERS

Muon identification is provided by a hadron filter of 5 - 8 absorption lengths followed by 32 large planar drift chambers with digital anode and cathode readout. The chambers cover an active area of about 200 m^2 total.

The anode wire spacing of the chambers is 12.7 mm, the anodes are separated by one field wire. The gap width is 8 mm. The cathodes are made of 10.6 mm wide copper foil strips on mylar glued on a honeycomb structure with a thickness of 60 mm. The orientation of the cathode strips is $\pm 34^{o}16'$ with respect to the anode wires.

The informations of the anodes and cathodes of only one chamber can be used for space point reconstruction. The resolution hereby obtained is $\sigma = 12$ mm. This value has to be compared to the uncertainty caused by multiple scattering in the hadron filter, this uncertainty is about ± 8 mm for muons of 15 GeV/c.

The signals of each muon chamber are ored to be used in coincidence with other trigger signals.

FORWARD DETECTOR

The angular region between 25 mrad and 50 mrad with respect to the beam axis is observed by two special spectrometers. They are designed to measure bhabhas, to study 2-photon events by single and double tagging and backgrounds in 1-photon events.

Each quarter of the detector is mounted on pivoting supports and contains three groups of planar drift chambers with 11 sense wires orientated at 45° or 135°. The first and the third group contain large scintillators which have the same size as the active area of the drift chambers. The middle group contains two planes of 12 small overlapping finger scintillator hodoscopes which mirror the wire spacing of the drift chambers. The scintillators are used to cope with synchrotron radiation background.

A lead glass shower counter arrangement is mounted behind the drift chambers. Each quarter of the forward detector contains 6 vertical, 6 horizontal and 16 longitudinal lead glass blocks. The total thickness of the shower counter is 20 rad.lengths. Parts of the lead glass counters are used as bhabha luminosity monitor in coincidence with two additional pairs of small scintillation counters mounted near the beam pipe.

The spatial resolution of the drift chamber assembly is $\sigma = .3$ mm obtained with reconstructed tracks. The energy resolution of the lead glass counters is $\sigma_E/E = .05/(E)^{1/2}$. By means of sampling the showers in depth a pion/electron discrimination of about 1 % down to energies of 150 MeV is obtained.

TRIGGER

The fast hardware trigger of CELLO makes the decision whether to readout an event by the online computer or not in less than 1.8 μsec (that is less than the time between two PETRA bunches). The trigger is an arrangement of the following components:

- A trigger on charged particles in the central tracking device pointing to the interaction region. It demands a coincidence of tracks in the rϕ-projection as well as in the rz-projection; the minimum numbers of tracks and the lower limit of the transverse momentum in the rϕ-projection can be preselected individually. Later on the endcap region will be included in the trigger.
- Trigger on showers in the cylindrical lead-liquid argon calorimeter. It demands an energy deposit of showers which exceeds a preselected value in individual stacks or in the entire calorimeter.
- A trigger on showers in the endcap lead-liquid argon calorimeters. It is a track trigger pointing to the interaction region and is built up of overlapping energy sums in individual parts of the calorimeter.
- A forward trigger which uses signals of the scintillators and lead glass counters of the forward detector. It is a threefold trigger and enables to trigger on bhabhas, on tagged events or on two photons only.

CHARGED PARTICLE TRIGGER

The charged particle trigger is a software programmable hardware trigger. As the technical solution is a very interesting one it is described in more detail. The trigger uses the signals of the anodes and the 90°-cathodes of proportional wire chambers and the sense wires of two layers of the drift chambers of the cylindrical part of the tracking device.

In the rϕ-projection the signal wires are grouped so that equally sized angular bins (sectors) of 5.625° (=360°/64) are obtained. The hitpatterns in all sectors are compared in parallel with the shapes (masks) of all possible charged tracks coming from the interaction region which exceed a predetermined lower limit of the transverse momentum (>.35 GeV/c) in the rϕ-projection. The trigger acceptance has a radius of up to 10 mm around the beam axis.

A similar system is used in the rz-projection. It receives the signals of the 90°-cathodes which are grouped so, that 37 sectors are obtained. The acceptance of the rz-part of the trigger is up to ±150 mm along the beam axis.

The mask scheme to be used in the trigger is stored in a random access memory (RAM, size about 5 Mbit) by means of a minicomputer. The signals in the sectors of the tracking device are used to address different parts of the RAM in parallel, the contents of the addressed

RAM-words are used to make the trigger decision. It is possible to store up to 48 masks for each sector, that means that the equivalent RAM-word is set nonzero. Masks found in individual sectors can be counted individually or even OR-ed. Trigger decision is done on the total numbers of masks which have to exceed preselected values in the $r\phi$-projection and in the rz-projection separately.

The trigger combines an efficiency of nearly 100 % with a background reduction to the order of 1 Hz. In addition the RAM concept offers a great flexibility: chamber inefficiencies can be taken into account, the acceptance region can be altered e.g. in the rz-projection to determine beam-gas background, or the transverse momentum limit can be raised by simply reprogramming the RAM. Besides this the trigger delivers a lot of information about triggered events to the online computer. This information concerns trigger types, mask shapes and sector numbers and is used in online event flagging and reconstruction.

FIRST ROUND OF DATA TAKING

The trigger configuration used during the first round of data taking is the following one:

- charged particle trigger, which demands a coincidence of at least 2 tracks in the $r\phi$-projection and 1 or more tracks in the rz-projection of the cylindrical part of the tracking device
- trigger on showers in the cylindrical lead-liquid argon calorimeter with an energy deposit of at least 2 GeV in one or more stacks or an energy deposit of at least 5 GeV in the entire calorimeter
- trigger on showers in the endcap lead-liquid argon calorimeters pointing to the interaction region
- trigger on two showers in opposite parts of the forward lead glass shower counters; this trigger is scaled down by a factor of 100

With this trigger configuration a trigger rate of about 3 Hz is obtained at a luminosity of 10^{30} $cm^{-2}sec^{-1}$.

An online scanning of the triggered events is done. It is based on the informations the hardware triggers deliver to the online computer which flags the events according to different event types. By means of this online scanning about 85 % of all hadronic events are identified.

During the first round of data taking of CELLO an energy scan was done at PETRA energies between 17.50 GeV and 17.63 GeV in 10 MeV-steps, each with an integrated luminosity of 30 nb^{-1}. The average hadronic ratio obtained by online scanning is consistent with a value of about 4, but data analysis has just started.

APPENDIX: Physicists participating in the CELLO collaboration

DESY

H.-J. Behrend
J. Field
U. Gümpel
V. Schröder
H. Sindt

IKP Karlsruhe

D. Apel
J. Bodenkamp
D. Chrobaczek
J. Engler
D.C. Fries
G. Flügge
H. Müller
H. Randoll
G. Schmidt
H. Schneider

MPI München

W. de Boer
G. Buschhorn
G. Grindhammer
P. Grosse-Wiesmann
B. Gunderson
C. Kiesling
R. Kotthaus
H. Lierl
D. Lüers
T. Meyer
L. Moss
H. Oberlack
P. Schacht
M.J. Schachter-Radig
A. Snyder

LAL Orsay

G. Carnesecchi
A. Cordier
M. Davier
D. Fournier
J.F. Grivaz
J. Haissinski
V. Journé
F. Laplanche
F. Le Diberdet
J.-J. Veillet
A. Weitsch

Universite Paris VI

R. George
M. Goldberg
B. Grossetete
F. Kapusta
F. Kovacs
P. Limon
G. London
L. Poggioli
M. Rivoal
Ph. Villeneuve

CEN Saclay

R. Aleksan
J. Bouchez
G. Cozzika
Y. Ducros
A. Gaidot
J. Pamela
J.P. Pansart
F. Pierre

RESULTS FROM THE JADE COLLABORATION *

presented by

R.D. Heuer

Physikalisches Institut der Universität Heidelberg

*W. Bartel, D. Cords, P. Dittmann, R. Eichler, R. Felst, D. Haidt, S. Kawabata, H. Krehbiel, B. Naroska, L.H. O'Neill, J. Olsson, P. Steffen, W.L. Yen

(Deutsches Elektronen-Synchrotron DESY, Hamburg)

E. Elsen, M. Helm, A. Petersen, P. Warming, G. Weber

(II. Institut für Experimentalphysik der Universität Hamburg)

H. Drumm, J. Heintze, G. Heinzelmann, R.D. Heuer, J. von Krogh, P. Lennert, H. Matsumura, T. Nozaki, H. Rieseberg, A. Wagner

(Physikalisches Institut der Universität Heidelberg)

D.C. Darvill, F. Foster, G. Hughes, H. Wriedt

(University of Lancaster)

J. Allison, J. Armitage, A. Ball, I. Duerdoth, J. Hassard, F. Loebinger, H. McCann, B. King, A. Macbeth, H. Mills, P.G. Murphy, H. Prosper, K. Stephens

(University of Manchester)

D. Clarke, M.C. Goddard, R. Hedgecock, R. Marshall, G.F. Pearce

(Rutherford Laboratory, Chilton)

M. Imori, T. Kobayashi, S. Komamiya, M. Koshiba, M. Minowa, S. Orito, A. Sato, T. Suda, H. Takeda, Y. Totsuka, Y. Watanabe, S. Yamada, C. Yanagisawa

(Lab. of Int. Coll. on Elementary Particle Physics and Department of Physics, University of Tokyo)

ISSN:0094-243X/80/620113-27$1.50

ABSTRACT

Results from the JADE experiment at PETRA in the energy range between 27.7 and 35.8 GeV are reported. QED is tested using the reactions $e^+e^- \to e^+e^-$, $\mu^+\mu^-$, $\gamma\gamma$ and lower limits on hypothetical cut-off parameters of up to 100 GeV are obtained. Predictions of the standard electro-weak theory are compared with the data and limits are derived on the values of the vector and axial vector coupling constants. Extended gauge models containing more than one neutral weak boson are also tested and lower limits are placed on the masses of such bosons.

For the reaction $e^+e^- \to$ hadrons topological distributions of the hadrons are studied. At $\sqrt{s}$ of about 30 GeV it is shown in a model independent way that the observed planar events actually possess three-jet structure strongly suggesting gluon bremsstrahlung as the origin of these events. By comparison of the data with the $q\bar{q}g$ - model we obtain a value for the strong coupling constant of $\alpha_s (q^2) = 0.17 \pm 0.02 \pm 0.05$.

At the highest energy of about 35 GeV a search for new flavour production is performed. An average value of $R = 3.9 \pm 0.2$ is obtained, compatible with the production of quarks with only the known flavours (u, d, s, c, b). Also the study of the topological distribution of the final state hadrons at this energy shows no evidence for new flavour production.

In the energy range between 12 and 35 GeV a measurement of the neutral energy fraction in multihadronic events is reported.

Free quarks with charge $Q = 2/3$ are searched for and upper limits for the production have been derived from dE/dx measurements which are of the order of $0.01 \cdot \sigma_{\mu\mu}$. An upper limit of the lifetime of B-mesons of about $2 \cdot 10^{-9}$s is obtained.

DETECTOR

Fig. 1 shows a sectional view of the JADE detector in a vertical plane containing the beam axis. A 7 cm thick aluminium coil, 3.5 m long, 2 m in diameter, produces a uniform solenoidal magnetic field of 0.5 T parallel to the beam axis. An array of 42 scintillation counters is mounted immediately inside the coil serving for trigger purposes and for time of flight measurements. The trajectories of charged particles are measured by a cylindrical drift chamber, called "jet-chamber", which is operated with a gas mixture of argon-methane-isobutane at a pressure of 4 atm. Details about the principle of the jet-chamber, its construction and the electronics have already been published [1,2]. 48 points in space are measured along each track in the polar angular range $|\cos\theta| < .83$ (θ being the angle with respect to the direction of the incident positron) and at least 8 points on

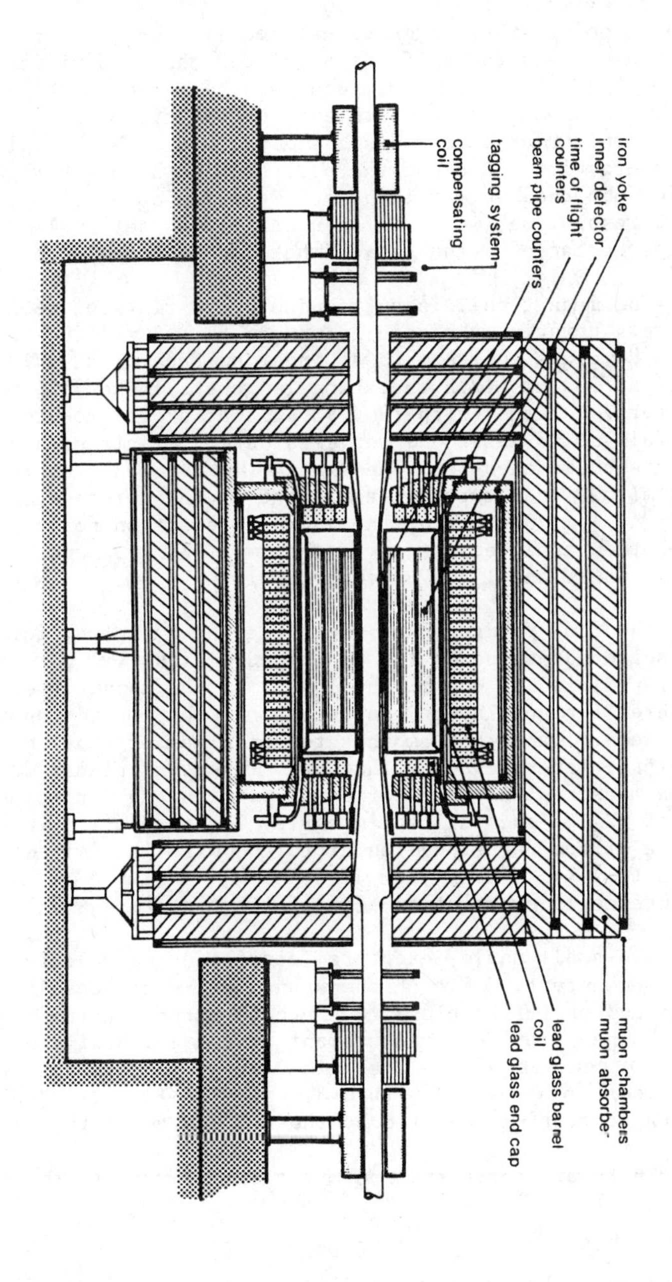

Fig. 1 Sectional view of the JADE - detector.

a track are obtained over a solid angle of 97% of the full sphere. At each point, three dimensional coordinates r, ϕ, z are given by the wire number and the drift time and the charge division measurements. The charge division method requires the measurement of the integrated charge from each hit at both ends of the signal wire. The ratio of these amplitudes determines z and the sum of both amplitudes measures the energy loss dE/dx of the particle in the chamber gas. The resolution obtained so far is σ_ϕ = 180 μm and σ_z = 1.6 cm. The double track resolution (which is adjustable) was set to 7 mm. Details about the performance of the jet-chamber can be found elsewhere [3].

The magnet coil is surrounded by 30 rings of lead glass shower counters covering the angular range $35^o < \theta < 145^o$. Each ring contains 84 glass wedges with an inner surface of 85 x 102 mm^2 and a depth of 300 mm (12.5 radiation lengths). These 2520 barrel shower counters, together with the 192 end cap shower counters cover 90 % of the full solid angle and serve to detect electrons and photons and measure their energies. The fine granularity of the lead glass counters allows an accurate measurement of the emission angles of typically $\Delta\theta = \pm 0.4^o$ using the known interaction point. The energy resolution achieved so far is $\sigma_E/E = \pm$ 7%, 4.3% and 3.8% at $\sqrt{s}$ = 1 GeV, 6 GeV and 15 GeV, respectively.

The flux return yoke, including the end caps, forms a rectangular box surrounding the so far described cylindrical part of the detector. It is utilized as one of the layers of the muon filter and is followed by three further layers consisting of iron loaded concrete. The total thickness of absorber amounts to a minimum of 785g/cm^2 (6 absorption lengths), interspersed with 4 or 5 layers of planar drift chambers which measure the trajectories of penetrating particles with a coverage of 92% of the full solid angle. The resolution in these chambers in the direction perpendicular to the wire is limited by the clock width to 2 mm. Parallel to the wire $\sigma \approx$ 15 cm is obtained using the difference in propagation time of the signals.

Two small angle detectors, consisting of an array of scintillation counters, drift chambers and lead glass modules record electrons and positrons close to the beam direction (35 mrad $\leq \theta \leq$ 75 mrad). They provide a measurement of the luminosity and tag the two photon processes ($e^+e^- \rightarrow e^+e^-$ + hadrons). The luminosity measured with this monitor agreed to within 3% with the luminosity determined from Bhabha scattering detected by the end cap counters.

There are three types of trigger relevant to the results presented here.

1) The "charged-particle trigger" required

   a) at least two time-of-flight counters to have fired

   b) total shower energy (sum of the barrel and end caps) to be more than 1 GeV

c) at least one track recognized by hard-wired logic based on the hit pattern of the jet-chamber.

2) The "two-prong trigger" required

a) two or three time-of-flight counters to have fired where the angle between two of the counters has to be $180^o \pm 30^o$,

b) two tracks recognized by the hard-wired logic.

3) The "shower energy trigger" only required a total energy of more than 4 GeV recorded in the shower counters.

ELECTROMAGNETIC AND WEAK INTERACTION

In the first part of this chapter a test of the QED-reactions $e^+e^- \rightarrow e^+e^-$, $\mu^+\mu^-$, $\gamma\gamma$ is reported. The deviation of the data from theory is expressed in terms of hypothetical cut-off parameters.

In the energy range of PETRA ($Q^2 \approx 1000$ GeV^2) it should be possible to study the effect of the weak interaction in the purely leptonic processes $e^+e^- \rightarrow e^+e^-$ and $e^+e^- \rightarrow \mu^+\mu^-$. Part 2 of this chapter describes the determination of the neutral current parameters in the standard model with one neutral weak boson. In part 3 the data are compared with extended gauge models containing two neutral weak bosons.

The selection criteria for events from the reactions $e^+e^- \rightarrow e^+e^-$ and $e^+e^- \rightarrow \gamma\gamma$ were as follows :

a) at least two clusters of energy in the lead glass each having more than one third of the beam energy,

b) the two lines joining the clusters to the centre of the interaction region have to be parallel within 10°.

The $\gamma\gamma$-events were then distinguished by demanding that at least one of the energy clusters was not related to any track in the jet-chamber. More details can be found elsewhere[4].

Events of the type $e^+e^- \rightarrow \mu^+\mu^-$ were selected from the event sample taken with the "two prong trigger" according to the following criteria :

a) two tracks seen in the jet-chamber with collinearity better than 11°

b) both tracks must enter the fiducial volume of the lead glass shower detector and each deposit less than one third of the beam energy

c) both tracks must penetrate the muon filter

d) to reject cosmic rays, the difference in flight time measurement from the TOF counters must be less than 4 nsec and the flight time of each track must be within ± 3 nsec of the central value.

e) The sum of the momenta of the two tracks determined by the inner detector must be greater than 40% of the summed beam energies. This loose cut removed the background from two-photon processes.

I. TEST OF QED

Fig. 2 shows the measured scattering angle distribution for the reactions $e^+e^- \rightarrow e^+e^-$ and $e^+e^- \rightarrow \gamma\gamma$. The curves in Fig. 2 were calculated from QED. Radiative corrections [5] and, in the case of Bhabha scattering, a correction for hadronic vacuum polarization [6] have been applied. There is excellent agreement between the data and the

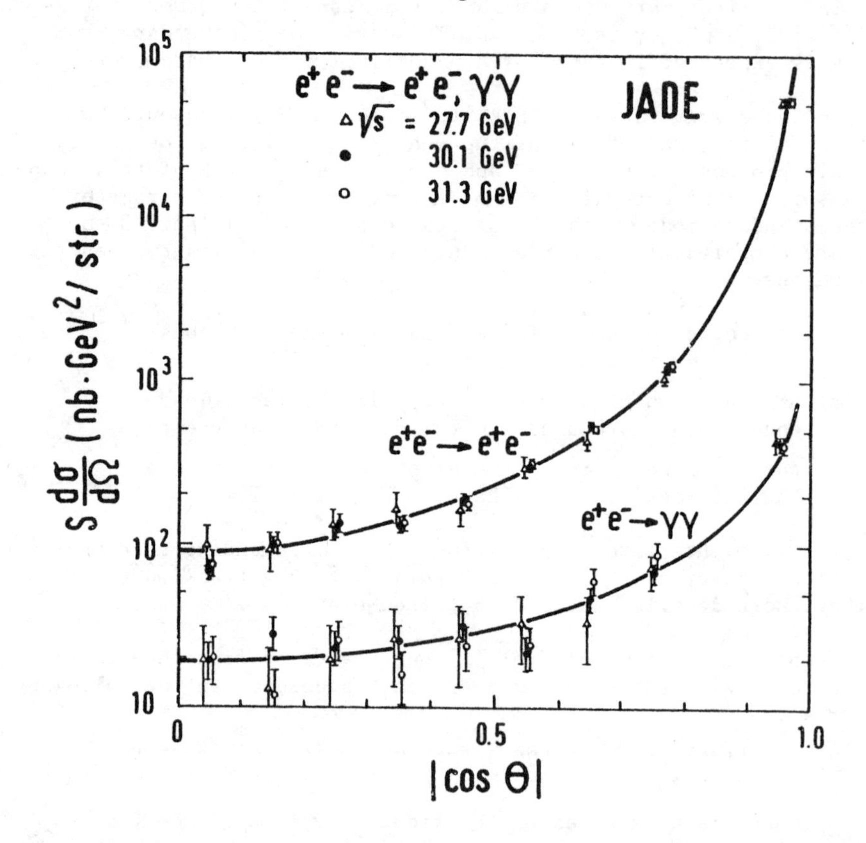

Fig. 2. Angular distributions for the reactions $e^+e^- \rightarrow e^+e^-$ and $e^+e^- \rightarrow \gamma\gamma$ at $\sqrt{s}$ = 27.7, 30.1 and 31.3 GeV. The curves are the predictions of QED.

QED predictions. In order to express the results in terms of a cut-off parameter Λ, for the reaction $e^+e^- \rightarrow e^+e^-$ a hypothetical modification [7] of the photon propagator $1/q^2$ is introduced by multiplying it with a form factor

$$F(q^2) = 1 \pm q^2/(q^2 - \Lambda_\pm^2)$$

For the reaction $e^+e^- \rightarrow \gamma\gamma$ we use [8]

$$\frac{d\sigma}{d\Omega} = \frac{\alpha^2}{s} \frac{1 + \cos^2\theta}{1 - \cos^2\theta} \left(1 \pm \frac{s^2}{2\Lambda_\pm^4} (1 - \cos^2\theta)\right)$$

and for the reaction $e^+e^- \rightarrow \mu^+\mu^-$ [7]

$$\sigma_\Lambda = \sigma_{QED} \left(1 \pm \frac{2s}{\Lambda_\pm^2} \right) .$$

χ^2-fits of the theoretical cross sections to the scattering angle distributions of Fig. 2 and to the measured total cross section of the process $e^+e^- \rightarrow \mu^+\mu^-$ were performed. The lower limits on the parameters Λ at the 95% c.l. obtained in this way are listed in Table 1.

TABLE 1

Lower limits on the cut-off parameters (in GeV) (95% confidence level)

	Λ_+	Λ_-
$e^+e^- \rightarrow e^+e^-$	104	87
$e^+e^- \rightarrow \gamma\gamma$	45	38
$e^+e^- \rightarrow \mu^+\mu^-$	136	91

II. DETERMINATION OF NEUTRAL CURRENT PARAMETERS

The processes $e^+e^- \rightarrow e^+e^-$ and $e^+e^- \rightarrow \mu^+\mu^-$ are assumed to proceed via the following graphs :

$e^+e^- \rightarrow e^+e^-$

$e^+e^- \rightarrow \mu^+\mu^-$

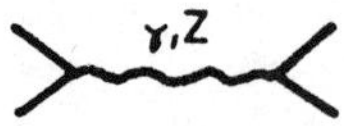

A general form for the differential cross-sections is as follows [9] :

$e^+e^- \rightarrow e^+e^-$:

$$\frac{8s}{\alpha^2}\frac{d\sigma}{d\Omega} = 4B_1 + (B_3+B_2)\ (1+\cos^2\theta) + 2(B_3-B_2)\cos\theta \qquad (1)$$

with

$$B_1 = (\tfrac{s}{t})^2 |1+(v^2-a^2)Q|^2$$

$$B_2 = |1+(v^2-a^2)R|^2$$

$$B_3 = \tfrac{1}{2}\ \{|\tfrac{s}{t}(1+(v+a)^2Q) + 1+(v+a)^2R|^2 + |\tfrac{s}{t}(1+(v-a)^2Q + 1+(v-a)^2R|^2\}$$

v and a are the vector and axial vector coupling constants respectively and the notation is such that in the Salam-Weinberg (S-W) model, $v = 1-4\sin^2\theta_W$ and $a = 1$. s and t are the kinematic factors, $s = 4E^2_{beam}$ and $t = -4E^2_{beam}\ (1-\cos\theta)$ where θ is the polar scattering angle.

In the most general form, Q and R are arbitrary, although in the case of a single gauge boson of mass M_z and width Γ :

$$Q = \frac{\sqrt{2}G_F M_z^2 t}{4e^2(t-M_z+iM_z^2\Gamma)} \qquad\qquad R = \frac{\sqrt{2}G_F M_z^2 s}{4e^2(a-M_z+iM_z^2\Gamma)}$$

where $G_F = 1/(293\ \text{GeV})^2$ and $e^2 = 4\pi\alpha$.

$e^+e^- \rightarrow \mu^+\mu^-$:

$$\frac{8s}{\alpha^2}\frac{d\sigma}{d\Omega} = (B_4+B_2)\ (1+\cos^2\theta) + 2(B_4-B_2)\cos\theta \qquad (2)$$

with B_2 the same as for e^+e^-

and $$B_4 = \frac{1}{2}\{|1+(v+a)^2R|^2 + |1+(v-a)^2R|^2\}$$

The angular dependence of the forward-backward asymmetry in muon pair production is then given by

$$A(\cos\theta) = 2\frac{(B_4-B_2)}{(B_4+B_2)} \frac{\cos\theta}{1+\cos^2\theta} \quad (3)$$

Our value for the asymmetry integrated over the angular range $|\cos\theta| < 0.75$ is

$$A = -0.06 \pm 0.10 \text{ at } \langle s\rangle = 1014 \text{ GeV}^2 .$$

Radiative effects [10], which cause a positive asymmetry of + 0.05 for the collinearity cut of 11° used in our analysis, were corrected for in the data.

The e^+e^- differential cross-section and the $\mu^+\mu^-$ asymmetry measurement were compared with equations (1) and (3) to obtain information about the values of v^2 and a^2. The fit results are

$$a^2 = 1.04 \pm 1.7 \qquad (a^2 = 1.0 \text{ in S-W})$$

$$v^2 = -0.1 \pm 0.6 \qquad (v^2 = 0.006 \text{ in S-W})$$

The 95% c.l. contour derived from the combined fit in the v^2-a^2 domain is shown in Fig. 3. The limits from the separate fits to electron pair and muon pair data are also shown in Fig.3. The determination of the vector and axial vector couplings is still much more inaccurate than that provided by measurements of νe-scattering. However, the electro-weak theory is being tested completely independently in a different reaction and at values of Q^2 (≈1000 GeV^2) which are larger than any used for neutral

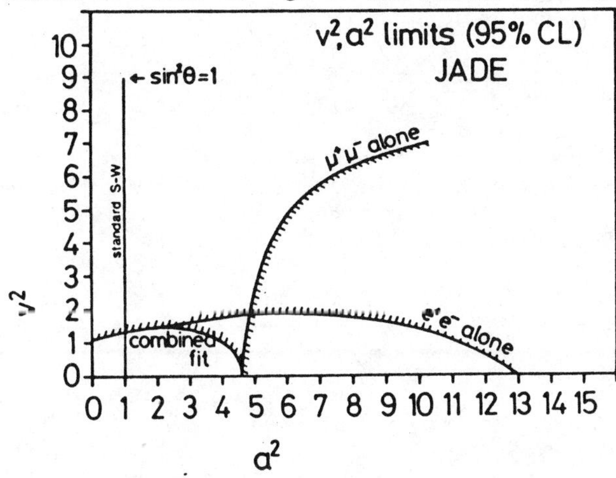

Fig. 3. The 95% c.l. limits for the coupling constants v^2 and a^2 derived from the reactions $e^+e^- \to e^+e^-$ and $e^+e^- \to \mu^+\mu^-$.

current studies to date. Furthermore, the values of v^2 and a^2 can be extracted in a model independent way, i.e. without assuming the validity of the S-W (or any other) model in the first place.

The Salam-Weinberg model ($a^2 = 1.0$) is shown in Fig. 3 as a vertical line bounded by $\sin^2\theta_w = 1$ and $v^2 = 0$. The intersection of our 95% contour with this line is essentially a fit to the Salam-Weinberg model. The limit on $\sin^2\theta_w$ is obtained as

$$\sin^2\theta_w < 0.63 \qquad 95\% \text{ c.l.}$$

III. MULTI GAUGE BOSON MODELS

Although the low Q^2 data from many sources have established the validity of the SU(2)xU(1) model of Salam and Weinberg, it has been pointed out by many authors [11,12,13] that it is possible to make a variety of gauge invariant extensions to SU(2)xU(1) which still retain the low Q^2 behaviour of SU(2)xU(1). Recent theoretical studies along these lines have been made by de Groot et al.[11] ($SU(2)xU(1)x\bar{U}(1)$), Barger et al.[12] (SU(2)xU(1)xSU(2)'), Elias et al.[13] ($SU(2)_L xSU(2)_R x U(1)_L xU(1)_R$), to quote three typical examples.

Our data have been compared with the models of de Groot et al. and Barger et al., since these two models represent what might be regarded as opposite ends of the theoretical spectrum. In both models two gauge bosons Z_1^o and Z_2^o with $m_1 <$ m(S-W) and $m_2 >$ m(S-W) are introduced.

The extension of the basic SU(2)xU(1) model leads to a modified form of the effective Langrangian :

$$\mathcal{L}_{eff} = \frac{\sqrt{2}G_F}{4} (j_3 - \sin^2\theta_w j_{em})^2 + Cj_{em}^2$$

The coefficient C represents the deviation from a single gauge boson situation. The choice of the gauge group extension (U(1) or SU(2)) determines the form of C as follows :

de Groot et al $\qquad C = \cos^4\theta_w \dfrac{(m_2^2-m_Z^2)\ (m_Z^2-m_1^2)}{m_1^2 m_2^2}$

Barger et al $\qquad C = \sin^4\theta_w \dfrac{(m_2^2-m_Z^2)\ (m_Z^2-m_1^2)}{m_1^2 m_2^2}$

The $SU(2)xU(1)x\bar{U}(1)$ model leads to a rather strong gauge boson coupling which would cause striking effects at PETRA energies if the mass of the lighter gauge boson were significantly lower than the 89 GeV/c^2 of the standard model, whereas in the SU(2)xU(1)xSU(2)' model the gauge bosons are an order of magnitude less strongly coupled. Our lepton pair data was compared with the theoretical cross sections given by these two models. With 95% confidence, we find

$$C \leq 0.06 \ .$$

This value leads to contours for the two models in the m_1-m_2 domain as shown in Fig. 4, where the regions lying above the curves are excluded at the level of 95%.

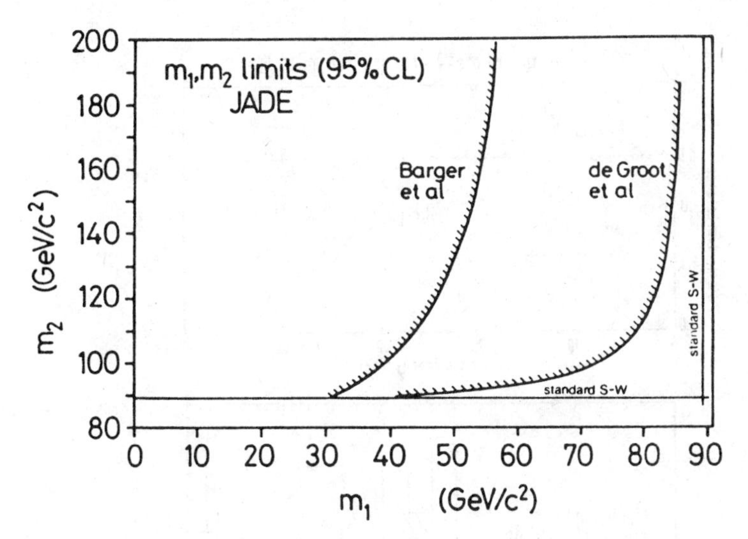

Fig. 4. The 95% c.l. contours for a fit to the gauge boson models of de Groot et al [11] and Barger et al [12].

$e^+e^- \to$ MULTIHADRONS

Part 1 of this chapter describes a jet analysis at a c.m. energy of about 30 GeV. We show in a model independent way that the observed planar events possess three-jet structure. Comparison with gluon bremsstrahlung yields a value for the strong coupling constant α_s. In part 2 a search for new flavour production at the highest PETRA energy of about 35 GeV is reported. Finally the neutral energy fraction is determined at c.m. energies of 12 GeV, 30 GeV and 35 GeV.

The selection criteria, efficiency calculation and radiative corrections are described elsewhere [14]. For the selected events the total visible energy $E_{vis} = \sum_i E_i$ and the longitudinal momentum balance $B_L = \sum_i p_i \cos\theta_i / E_{vis}$ were calculated, where the sums are taken over charged particles and shower energy clusters. Fig. 5a shows a correlation plot for B_L versus E_{vis} for data at $\sqrt{s}$ = 27.7 GeV. A cut of $|B_L| < 0.4$ reduces the background from beam-gas and two-photon processes. Fig. 5b shows the visible energy distribution after this cut together with the expected background from the two-photon processes. A further cut of $E_{vis} > E_{beam}$ eliminated the remaining beam-gas background and reduced the level of two-photon events to about 1%. This amount was verified by Monte-Carlo simulation.

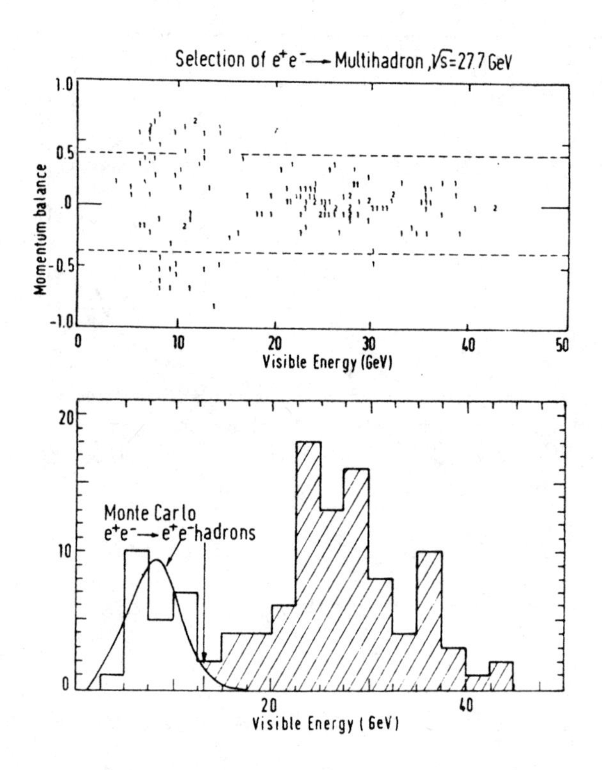

Fig. 5. a) Scatter plot of longitudinal momentum balance versus visible energy

b) Projection onto the visible energy axis for |momentum balance| < 0.4 .

I. JET-ANALYSIS AT $\sqrt{s} \sim 30$ GeV

The results described here are based on the combined sample of 89, 198 and 49 events obtained at $\sqrt{s}$ = 27.7 GeV, 30.0 GeV and 31.6 GeV, respectively, as well as 754 events recorded in an energy scan in the region 29.9 GeV $\leq \sqrt{s} \leq$ 31.5 GeV.

Ia. Event topology and evidence for three-jet structure

In lowest order QCD hadron production in e^+e^--annihilation proceeds through the reaction $e^+e^- \to q\bar{q} \to$ hadrons. Perturbative QCD predicts also higher order processes of the type $e^+e^- \to q\bar{q}g \to$ hadrons. This gluon bremsstrahlung will, depending on the energy of the gluon, broaden one jet or produce three-jet events with the axes of the three jets in a plane.

For the study of the event shape the normalized sphericity tensor [15] was constructed and diagonalized for each event :

$$T_{\alpha\beta} = \sum_i p_{i\alpha} \, p_{i\beta} \Big/ \sum_i p_i^2$$

where $p_{i\alpha}$ is the α-component (α = x,y,z) of the momentum of the i-th particle. The sum runs over all charged and neutral particles. The resulting eigenvalues Q_1, Q_2, Q_3 ($Q_1 < Q_2 < Q_3$) correspond to the lengths of the orthogonal axes ($\vec{n}_1$, $\vec{n}_2$, $\vec{n}_3$) of the momentum ellipsoid as indicated in Fig. 6 and they satisfy the constraint $Q_1+Q_2+Q_3 = 1$.

As can be seen from Fig. 6, $Q_2 - Q_1$ can be regarded as the planarity of the event, whereas $Q_2 + Q_1$ is a measure for the transverse momentum with respect to the highest momentum axis $\vec{n}_3$ and proportional to the sphericity S.

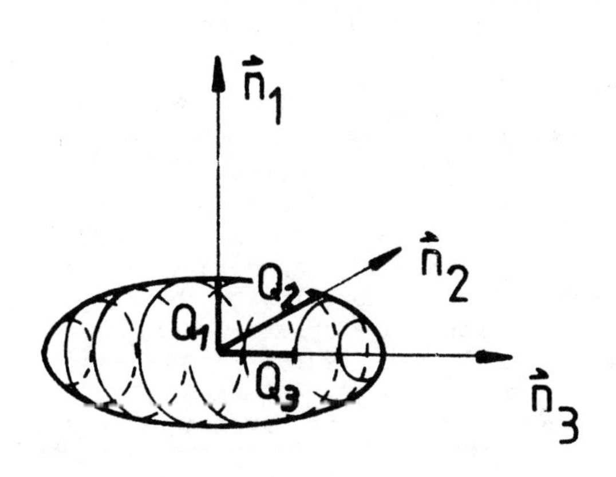

Fig. 6 Sphericity tensor

Each event is represented by a point in a triangle plot as shown in Fig. 7, where the two independent variables are chosen as Q_1 and $(Q_3 - Q_2)/\sqrt{3}$. Two-jet events will be located in the right-hand corner, spherical events in the left-hand upper corner and disk-like events in the left-hand lower corner.

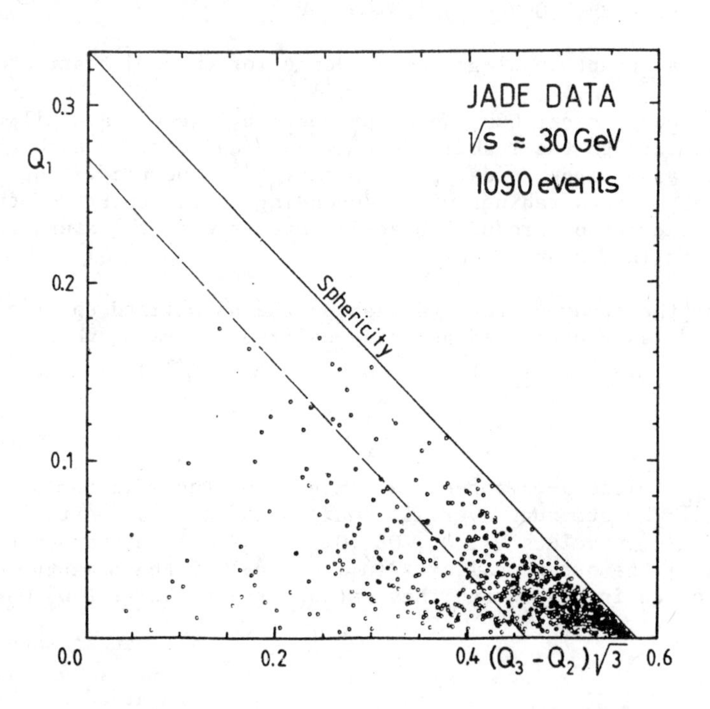

Fig. 7 The distribution of the eigenvalues Q_1, Q_2, Q_3 of the sphericity tensor T described in the text. The dashed line indicates planarity $(Q_2 - Q_1) = 0.1$.

The planarity axis runs perpendicular to the sphericity axis. Events with planarity $Q_2 - Q_1 > 0.1$ are therefore located below the dashed line indicated in Fig. 7. 181 events are observed with $Q_2 - Q_1 > 0.1$, whereas 57 events would be expected on the basis of the simple $q\bar{q}$-model [16]. The existence of planar events strongly suggests a three-body primary process such as gluon bremsstrahlung $e^+e^- \to q\bar{q}g$.

This was tested in the following way. For each of the 181 events with $Q_2 - Q_1 > 0.1$ the quantity thrust [17] $T = \max\left(\sum_i |p_{\parallel i}| / \sum_i |p_i|\right)$ was calculated *. A plane normal to the thrust axis separated each

* The limiting values of T are $T_{min} = 0.5$ (spherical event) and $T_{max} = 1.0$ (extreme two-jet event).

event into a "slim" and a "fat" jet by definition $\Sigma|p_T|_{slim} < \Sigma|p_T|_{fat}$. If the events possess a three-jet structure, the fat jet should consist of two jets. This should be visible in the rest frame of the fat jet where the two sub-jets will appear collinear [18]. Therefore all momentum four vectors of the particles (neutral and charged) in the fat jet were Lorentz transformed into the rest frame of that jet and in this frame the thrust T^* of these particles was calculated. As shown in Fig. 8a, the observed T^* distribution peaks at high T^*. If the gluon does not fragment too differently from quarks, this distribution should look like the thrust distribution of two-jet events produced at a c.m. energy equal to the average invariant mass of the fat jet.

In Fig. 8b the invariant mass distributions of the fat jet and of the events at $\sqrt{s} = 12$ GeV taken with the JADE detector are shown. Both distributions are similar indicating that the data at $\sqrt{s} = 12$ GeV can be used for comparison. The thrust distribution for these events is shown in Fig. 8c. The agreement between this distribution and the T^* distribution of the fat jet at $\sqrt{s} \approx 30$ GeV alone (also shown in Fig. 8c) is good. Therefore in its own rest frame the fat jet shows a two-jet structure similar to that of the two-jet events produced at $\sqrt{s}$ equal to the average invariant mass of the fat jet. This means the whole event at $\sqrt{s} \approx 30$ GeV possesses a three-jet structure as expected from gluon bremsstrahlung. It should be emphasized that the above conclusion was reached in a completely model independent way. An example for a three jet event compatible with hard gluon bremsstrahlung is shown in Fig. 9.

As a check the same analysis was repeated without the planarity cut $Q_2 - Q_1 > 0.1$ (dashed curve in Fig. 8d) and for the slim jet alone (full line in Fig. 8d). Both thrust distributions clearly peak at lower T^* (less jet-like structure).

Ib. Determination of α_s

In the process $e^+e^- \rightarrow q\bar{q}g$ the coupling at the quark-gluon vertex is determined by the strong coupling constant α_s. A quantitative evaluation of the data should lead to a determination of α_s. For that purpose the data at $\sqrt{s} \approx 30$ GeV were compared with a jet model [19] which includes gluon bremsstrahlung $q\bar{q}g$ in addition to $q\bar{q}$.

Fig. 10 shows the planarity distribution of the observed events. Comparing the high planarity tail with the $q\bar{q}g$ model prediction a value $\alpha_s = 0.17 \pm 0.02$ is obtained. The corresponding curve is given as a full line in Fig. 10. The error quoted is statistical only. The systematic error is 0.05 due to uncertainties in the Monte-Carlo input, e.g. ratio of pseudoscalar to vector meson generation or ansatz of the fragmentation functions. With the so derived value of $\alpha_s = 0.17$, 176 events are expected for the region $Q_2 - Q_1 > 0.1$ in the Q-plot of Fig. 7, whereas 181 events are seen.

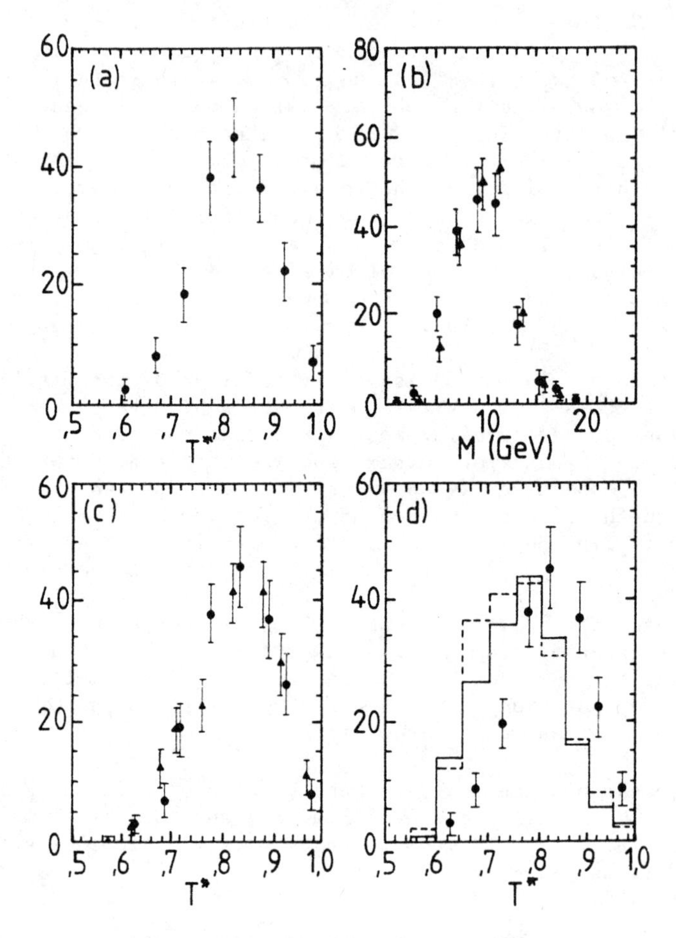

Fig. 8 The three-jet nature of the planar ($Q_2-Q_1 > 0.1$) events at $\sqrt{s} \approx 30$ GeV.

a) The observed distribution of T^* (the thrust of a jet in its rest system) for the fat jet in planar events.

b) The observed invariant mass distribution of the fat jet (dots) in planar events at $\sqrt{s} \approx 30$ GeV compared with the invariant mass distribution of the events at $\sqrt{s} = 12$ GeV (triangles).

c) The observed distribution of T^* for the fat jet in planar events at $\sqrt{s} \approx 30$ GeV (dots) compared with the thrust distribution for events at $\sqrt{s} = 12$ GeV (triangles).

d) T^*-distribution for the fat jet (dots) and for the slim jet (full line) in planar events together with the T^*-distribution for the fat jet (dotted line) in all events.

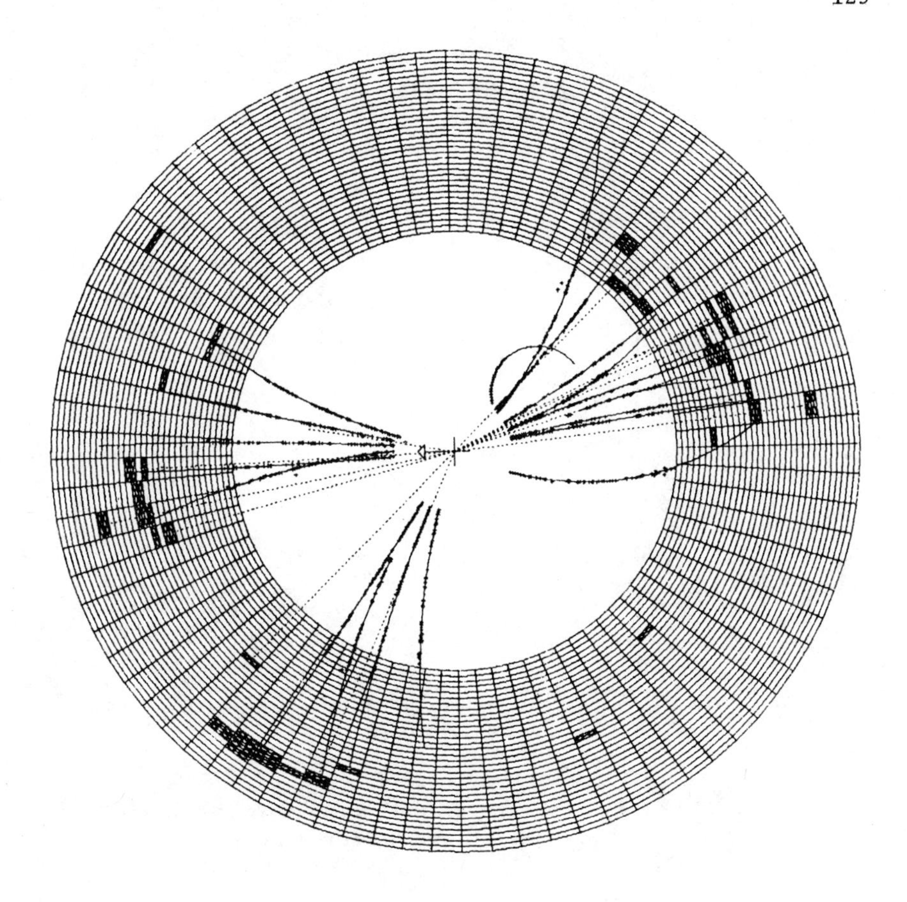

Fig. 9 Three-jet event, perspective view along beam line.

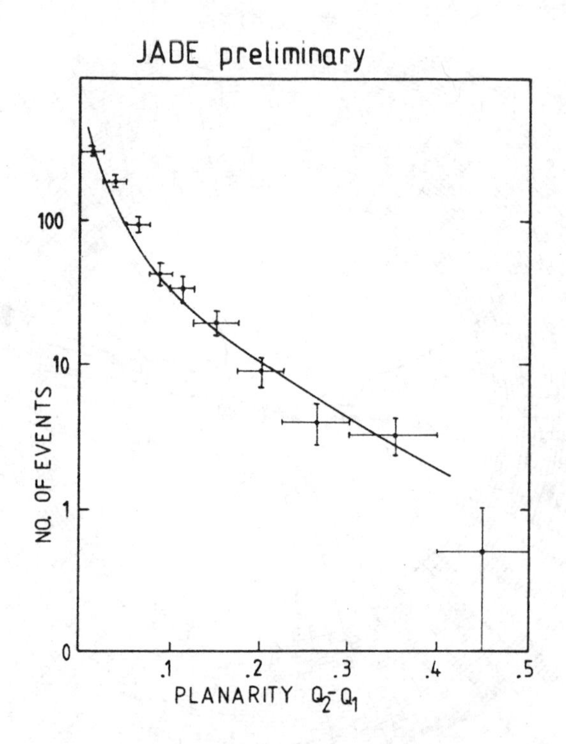

Fig. 10 The planarity distribution compared with the prediction of the $q\bar{q}g$ model.

II. SEARCH FOR NEW FLAVOUR PRODUCTION AT $\sqrt{s} \approx 35$ GeV

One method for detecting new flavour production is the measurement of the quantity R, the total hadronic cross section σ_{had} normalized to the muon pair cross section $\sigma_{\mu\mu}$:

$$R = \frac{\sigma_{had}}{\sigma_{\mu\mu}} = 3 \cdot \sum_i Q_i^2 \, (1 + \alpha_s/\pi)$$

Thus an increase in R is expected of $\Delta R = 0.3$ and $\Delta R = 1.3$ far above the thresholds for production of quarks with charge Q = 1/3 and Q = 2/3, respectively.

Our measurements of R at $\sqrt{s}$ between 12 GeV and 35.8 GeV are plotted in Fig. 11. No indication of a step in R of 1.3 units is observed as expected for the production of top mesons. The production of quarks with Q = 1/3 however cannot be excluded.

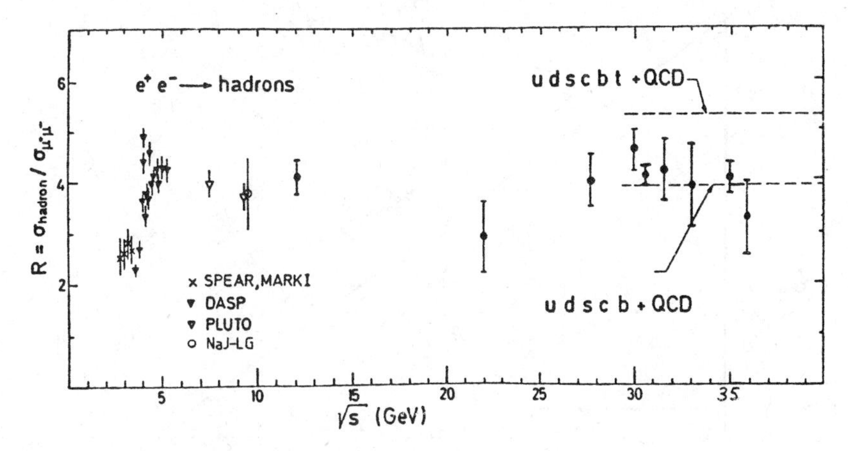

Fig. 11 Ratio R of hadronic to muon pair cross section as function of $\sqrt{s}$. The predictions with and without a t quark are indicated.

A more sensitive method for new flavour detection is the study of the topological distribution of the final state hadrons. Heavy particles with new flavour are expected to decay predominantly into multihadron final states, which near threshold would yield a relatively wide angled distribution in contrast to the narrow jet-like distributions from known low-mass flavours.

Diagonalisation of the sphericity tensor as described in a previous chapter leads to the Q-plot shown in Fig. 12. The event sample shown consists of 26, 214 and 26 events at $\sqrt{s}$ = 33.0 GeV, 35.1 GeV and 35.8 GeV, respectively. Spherical events are expected in the upper left corner. 3 events are observed in the indicated region with S > 0.55 and Q_1 > 0.075, whereas 20 events are expected from $q\bar{q}$-model calculations with the flavours u, d, s, c, b and t with a mass m_t = 14 GeV/c^2. From $q\bar{q}$ calculation without top production but including bremsstrahlung $q\bar{q}g$ one expects 1.5 events. The data are inconsistent with the production of a new flavour with charge Q = 2/3 up to $\sqrt{s} \approx$ 35 GeV.

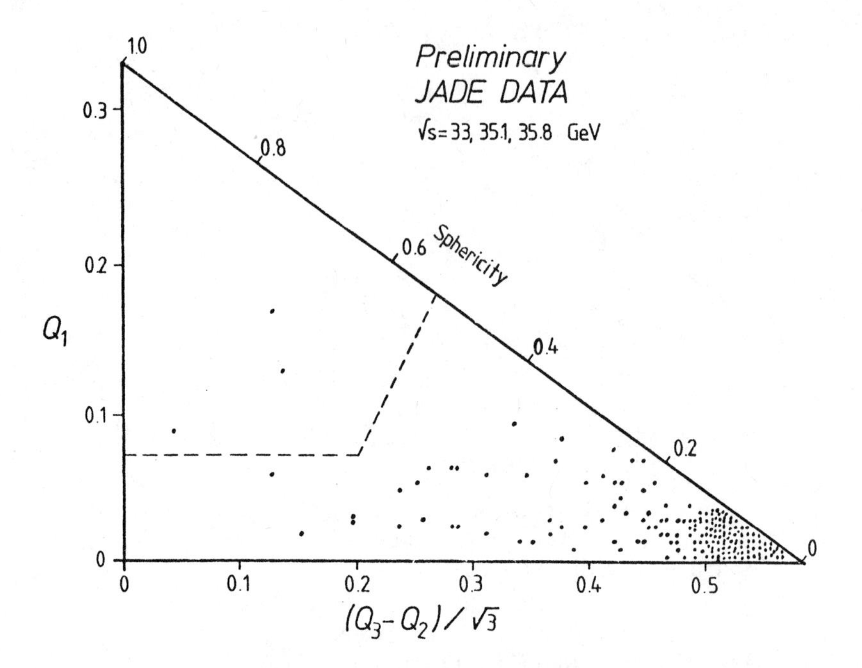

Fig. 12 The distribution of the eigenvalues Q_1, Q_2, Q_3 of the sphericity tensor T described in the text. The cuts $Q_1 > 0.075$ and sphericity > 0.55 are indicated.

III. NEUTRAL ENERGY FRACTION

The JADE detector with its large solid angle covered by lead glass counters is capable of measuring the photon and the neutral energy fractions.

The photon energy fraction has been determined from the total energy recorded in the lead glass arrays by subtracting the energy deposited by charged particles. This is done on a statistical basis by using published data on charged pions of various momenta detected in lead glass counters [20]. The measured data have been corrected for acceptance losses, photon conversions, interacting neutrons and long lived kaons.

The neutral energy fraction, i.e. the fractional energy carried away by photons, long lived neutral hadrons and neutrinos, has been obtained from the charged tracks only, using energy conservation.

$$\frac{E_{neut}}{\sqrt{s}} = 1 - \frac{E_{char}}{\sqrt{s}}$$

Corrections have been applied for photon conversion and hadronic interactions in the beam pipe, the acceptance losses and neutral kaons and Λ-particles decaying in the detector.

The observed fractional energies are shown in Fig. 13. They are approximately constant between 12 GeV and 35 GeV centre of mass energy.

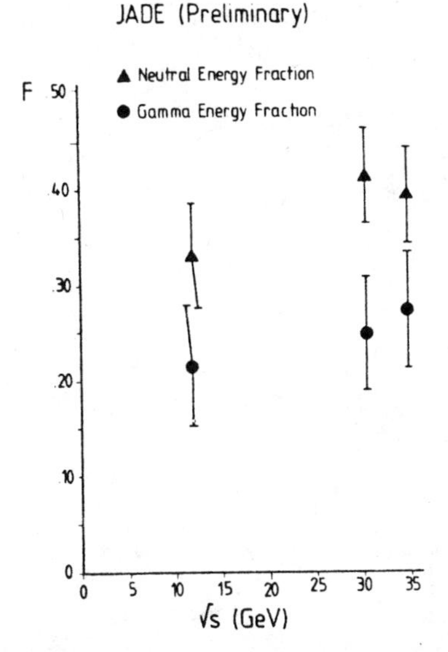

Fig. 13 Neutral energy fraction and gamma ray energy fraction as a function of $\sqrt{s}$.

SEARCH FOR FREE QUARKS AND HEAVY STABLE PARTICLES

Free quarks may be produced either exclusively ($e^+e^- \to q\bar{q}$, collinear two prong events) or inclusively, i.e. together with ordinary hadrons inside a multihadronic event. We have performed a search for quarks in both types of reactions at c.m. energies between 30 GeV and 35 GeV. Another object of interest are long lived particles with integer charge : If the bottom meson (B) were produced in multihadronic events with sufficiently long lifetime [21] it would be recognized as a heavy stable particle.

Particle identification has been done in the jet-chamber by calculating the energy loss dE/dx in the chamber gas from the sum of the pulseheights measured at both ends of the wire. Up to 48 samplings per track are obtained. To get a reliable estimate for the mean energy loss, independent of Landau fluctuations, the 60% lowest pulseheights are averaged and used for particle identification (method of truncated mean). Before averaging, each individual sampling has to be corrected for different gas and electronic amplification, path length of the track in the drift space, drop of pulseheight with increasing drifttime due to gas contamination and a saturation effect for tracks perpendicular to the wire. A more detailed description of these corrections can be found in Ref. 3. For electrons from Bhabha scattering a resolution of 14% FWHM is obtained (Fig. 14). For tracks in jet-like events a degradation in resolution (22% FWHM) is observed which comes from crossing and overlapping tracks yielding fewer useful samplings per track.

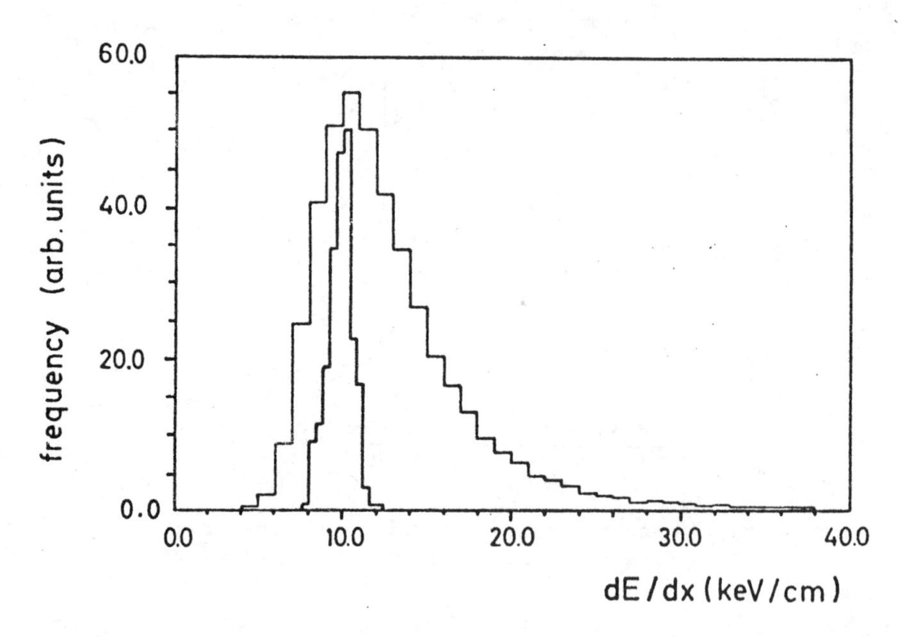

Fig. 14 Landau distribution and truncated mean (60% lowest pulseheights) for electrons from Bhabha scattering.

In Fig. 15 a scatter-plot, energy loss versus momentum, for particles from beam-gas reactions is shown. Contributions from pions, protons and electrons are visible. From the figure a relativistic rise of $(48 \pm 5)\%$ is obtained in good agreement with previous measurements [22].

Fig. 16 shows the observed energy loss in the chamber versus apparent momentum (assuming charge 1) for multihadronic events at $\sqrt{s}$ between 30 and 35 GeV. Also shown are the lines for the known stable particles π, K, p (solid lines) and a hypothetical particle with a mass of 5 GeV and charge 2/3 and 1, respectively (dotted lines). Since dE/dx as a function of momentum depends in good approximation on the variable $\beta\gamma = p/m$, curves for other particle masses can be obtained by parallel shifts. The relatively high number of protons is explained by pions reacting in the beam pipe or pressure

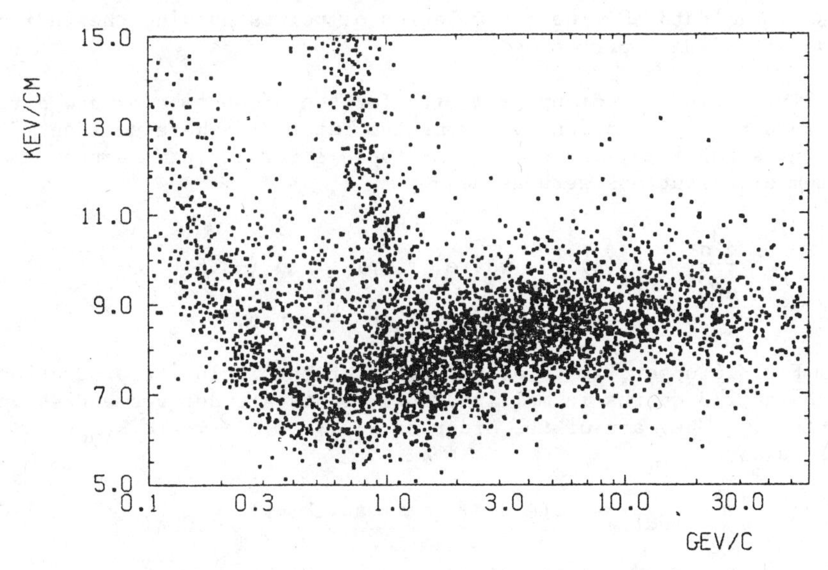

Fig. 15 Energy loss as function of momentum for particles from beam-gas reactions with E_{beam} = 15 GeV. Note the suppressed zero.

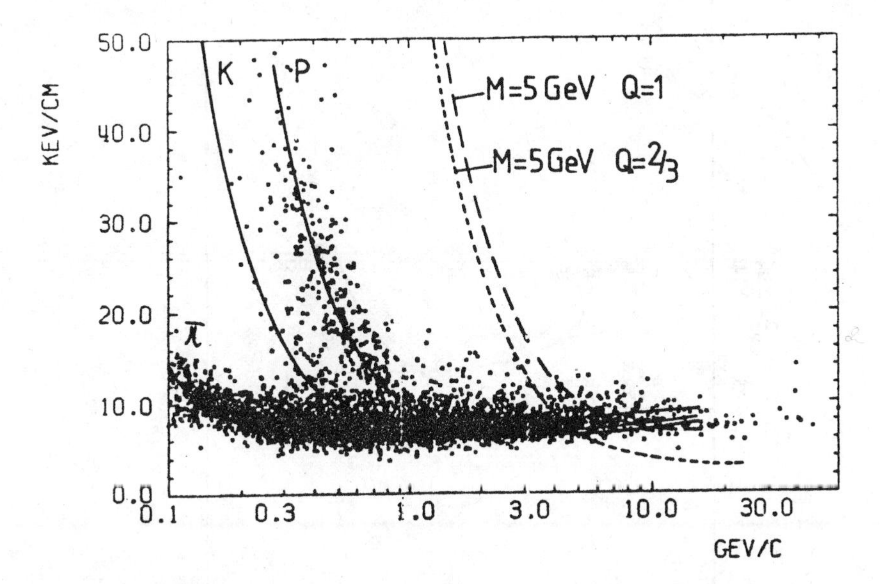

Fig. 16 Observed energy loss as function of apparent momentum for multihadron events at $\sqrt{s}$ between 30 GeV and 35 GeV. The lines for the known stable particles π, K, p (solid lines) and a hypothetical particle with a mass of 5 GeV and charge 2/3 and 1 (dotted lines) are also shown.

vessel. The data show no accumulation of points outside the indicated bands of ordinary particles.

The derivation of upper limits for the production of new particles requires integration over momentum intervals where no identification is possible. To correct for these intervals, two extreme momentum distributions were assumed :

$$E\frac{d^3\sigma}{dp^3} \sim f_i(p) \qquad \text{where} \quad f_1(p) \sim \exp(-3.5\,E)$$

$$\text{and} \quad f_2(p) = \text{const.}$$

From Fig. 16 upper limits of 90% c.l. for the inclusive production of charge 2/3 quarks in multihadronic events are derived and shown in Fig. 17. They are of the order of $R_{incl} \sim 10^{-2}$ where R_{incl} is defined as

$$R_{incl} = \sigma(e^+e^- \to q\bar{q} + \text{anything}) \,/\, \sigma_{\mu\mu} \,.$$

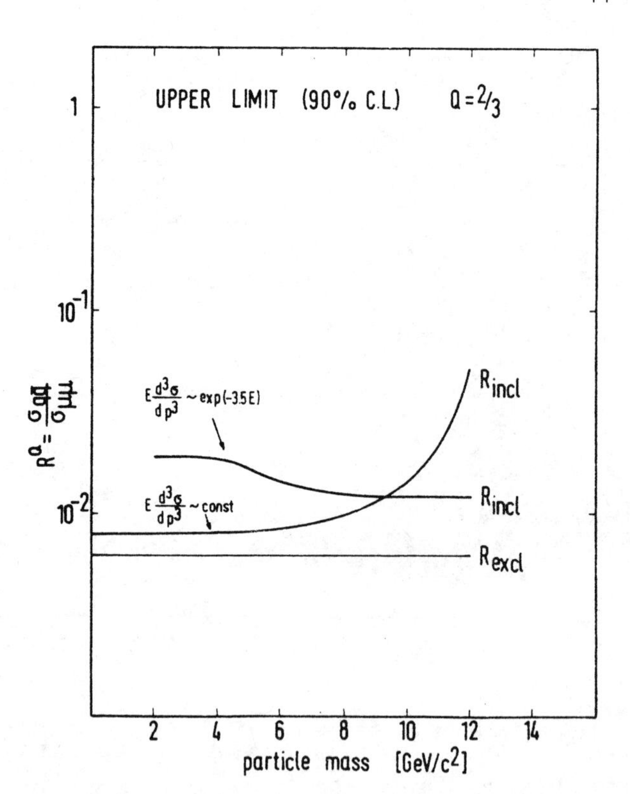

Fig. 17 Upper limits for inclusive and exclusive production of quarks with charge 2/3. See text for the definition of the quantity R.

The data sample taken with the "two-prong trigger" and selected according to the criteria described in a previous chapter allows to derive upper limits for the exclusive production of free quark pairs with charge 2/3. These limits are also shown in Fig. 17, where R_{excl} is defined in analogy to R_{incl} as

$$R_{excl} = \sigma(e^+e^- \rightarrow q\bar{q}) \;/\; \sigma_{\mu\mu}$$

To obtain an upper limit on the lifetime τ of the B-meson (charge 1) we assume a cross section for B-meson pair production of R = 3/9 and again two different momentum distributions $f_1(p) \sim \exp(-3.5 \cdot E)$ and $\frac{E}{4\pi p^2} f_2(p) = \text{const.}$

The results are given in Fig. 18. The 90% c.l. upper limit for the lifetime of B-mesons is of the order of $2 \cdot 10^{-9}$s. The previous limit [23] is $5 \cdot 10^{-8}$s for M = 5 GeV/c^2. We have improved this by an order of magnitude.

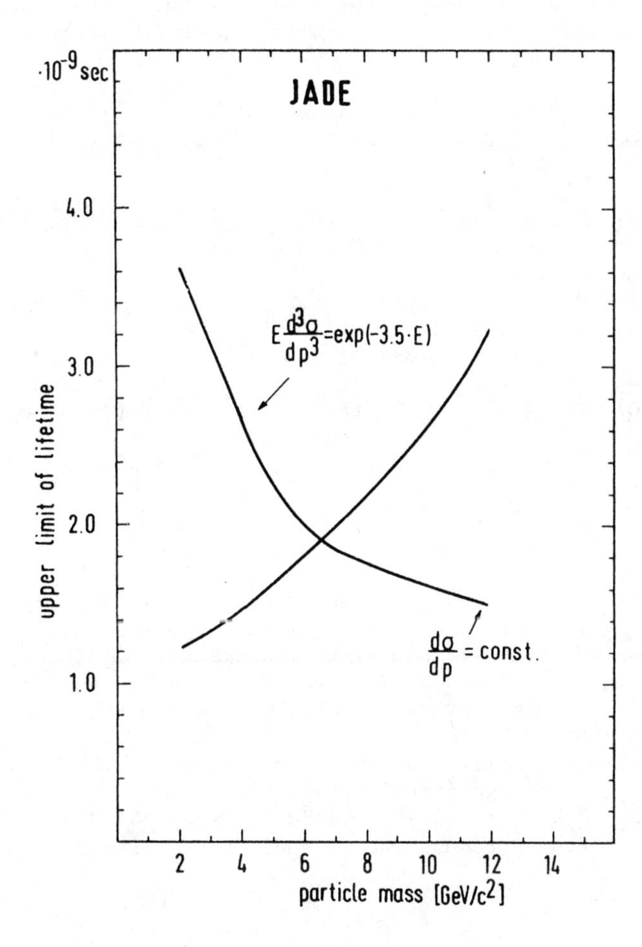

Fig. 18

Upper limit for the lifetime of B-mesons for two different momentum distributions.

CONCLUSIONS

1) The validity of QED has been tested and lower limits on hypothetical cut-off parameters are found to be up to 100 GeV.

2) Limits on the neutral current parameters have been derived in a model independent way at Q^2 of about 1000 GeV^2.

3) For multi gauge boson models limits are placed on low mass neutral weak bosons.

4) Planar events are observed at a level far above the statistical fluctuations of the two-jet process. It has been shown in a model independent way that these events possess a three-jet structure as expected from gluon bremsstrahlung $e^+e^- \to q\bar{q}g$.

5) The strong coupling constant has been determined to be $\alpha_s = 0.17 \pm 0.02$ (statistical) ± 0.05 (systematic).

6) R values are measured to be about 4 at $\sqrt{s}$ between 12 GeV and 35.8 GeV compatible with the production of quarks with only the known flavours.

7) No evidence is obtained for events with spherical hadron distribution. Open top production is unlikely at $\sqrt{s}$ below 35 GeV.

8) The neutral energy fraction has been measured and found to be constant at $\sqrt{s}$ between 12 GeV and 35 GeV.

9) No evidence is seen for the inclusive or exclusive production of quarks with charge 2/3. Upper limits are derived which are of the order of $0.01 \cdot \sigma_{\mu\mu}$.

10) Upper limits for the lifetime of B-mesons of the order of $2 \cdot 10^{-9}$s are obtained.

REFERENCES

1. J. Heintze, Nucl. Instr. and Meth. 156, 227 (1978)
W. Farr et al., Nucl. Instr. and Meth. 156, 283 (1978)
H. Drumm et al., IEEE Trans.Nucl. Sci. 26.1, 81 (1979)
2. W. Farr and J. Heintze, Nucl. Instr. and Meth. 156, 301 (1978)
3. H. Drumm et al., Proc. Int. Wire Chamber Conf., Vienna 1980 and DESY 80/38
4. JADE collaboration, W. Bartel et al., DESY 80/14 to be published in Phys.Lett.
5. F.A. Berends et al., Nucl. Phys. B61, 414 (1973)
F.A. Berends et al., Nucl. Phys. B68, 541 (1974)

6. F.A. Berends and G.J. Komen, Phys.Lett. 63B, 432 (1976)
7. S.D. Drell, Ann.Phys. 4,75 (1958)
T.D. Lee and G.C. Wick, Nucl. Phys. B9, 209 (1969)
8. A. Litke, Harvard University, Thesis (unpublished) (1970)
9. R. Budny, Phys. Lett. 55B, 227 (1975)
10. F.A. Berends et al., Nucl. Phys. B57, 381 (1973)
11. E.H. de Groot et al., Phys.Lett. 90B, 470 (1980) and University of Bielefeld preprint BI-TP-79/39 (1979)
12. V. Barger et al., University of Wisconsin and Hawaii preprints UW-COO-881-126 (1980) and UW-COO-851-133 (1980)
13. H. Georgi and S. Weinberg, Phys. Rev. D17, 275 (1978)
Q.Shafi and Ch. Wetterich, Phys.Lett. 73B, 65 (1978)
V. Elias et al., Phys.Lett. 73B, 451 (1978)
V. Barger and R.J.N. Phillips, Phys.Rev. D18, 775 (1978)
14. JADE collaboration, W. Bartel et al., Phys.Lett. 88B, 171 (1979)
15. J.D. Bjorken and S.J. Brodsky, Phys.Rev. D1, 1416 (1970)
16. R.D. Field and R.R. Feynman, Nucl. Phys. B136, 1 (1978)
17. S. Brandt et al., Phys. Lett. 12, 57 (1964)
E. Fabri, Phys.Rev.Lett. 39, 1587 (1977)
A. de Rujula et al., Nucl. Phys. B138, 387 (1978)
18. J. Ellis and I. Karliner, Nucl. Phys. B148, 141 (1979)
19. P. Hoyer et al., DESY 79/21 (1979)
20. D.P. Barber et al., Nucl. Instr. and Meth. 145, 453 (1977)
21. F.N. Cahn, Phys.Rev.Lett. 40, 80 (1978)
H. Fritzsch, Phys.Lett. 78B, 611 (1978)
S.L. Glashow, Harvard Univ. Preprint
E.W. Lee, S. Weinberg, Phys.Rev.Lett. 38, 1237 (1977)
22. A.H. Walenta et al., Nucl. Instr. and Meth. 161, 45 (1979)
23. D. Cutts et al., Phys.Rev.Lett. 41, 363 (1978)
R. Vidal et al., Phys.Lett. 77B, 344 (1978)

RECENT RESULTS OF MARK J: PHYSICS WITH HIGH ENERGY ELECTRON-POSITRON COLLIDING BEAMS

(the AACHEN, DESY, MIT, NIKHEF, PEKING Collaboration)

presented by

M. Chen

ABSTRACT

We present data that establishes the validity of Q.E.D. to a distance $< 2 \times 10^{-16}$ cm. Relative cross sections and event distributions show that there is no new charge 2/3 quark pair production up to $\sqrt{s} = 35.8$ GeV. We have discovered 3-jet events; the rate of their production and their distributions agree with Q.C.D. predictions. We have measured the strong interaction coupling constant α_s.

ISSN:0094-243X/80/620140-53$1.50

1) THE MARK J EXPERIMENT

1.1 Physics Objectives

The MARK J detector[1], which identifies and measures the energy and direction of muons, electrons, charged and neutral hadrons with close to uniform efficiency and with $\sim 4\pi$ acceptance, is capable of fulfilling a broad range of physics objectives. Some of the prime physics goals of the experiment are:

1) To study the various QED processes shown in Figure 1 and to study the universality of the known charged leptons in their electromagnetic interactions. At PETRA the available c.m. energy is $\sqrt{s} = 37$ GeV (q^2 up to 1400 GeV^2). Since first order QED processes exhibit a $^1/s$ cross section dependence the MARK J can probe the validity of QED with an order of magnitude greater sensitivity than that previously available in earlier colliding beam experiments performed at storage rings at SLAC, DESY, and the CEA in the range of $q^2 \lesssim 50$ GeV^2.

2) To search for new quark flavors by studying the energy and angular distributions of inclusive muon production in hadronic events (Figure 2a).

3) Using the distributions of μe and μh final states shown in Figure 2b to search for the existence of new charged leptons heavier than the tau.

4) To measure the total hadronic cross section (Figure 3) and thereby the structure and energy dependence of the total cross section, in order to search for new thresholds in the hadronic final state continuum, and to search directly for more J-like particles which appear as sharp resonances.

5) To study the topology of hadronic events by measuring the direction and energy of charged and neutral particles. In particular, at PETRA energies, the fragmentation of hard gluons emitted in association with quark-antiquark pairs leads to the creation of additional gluons and quarks, resulting in the production of multi-jet events. Study of the properties of these jets enables us to make a direct comparison with the predictions of QCD[2]. The rate of 3-jet events relative to 2-jet events enables us to measure directly the strong interaction coupling constant α_s.

6) To measure the charge asymmetry expected from the interference of weak and electromagnetic interactions in the production of $\mu^+\mu^-$ pairs. As shown in Figure 4, diagrams in which a virtual photon is exchanged or in which a Z^0 vector boson is exchanged both contribute to $\mu^+\mu^-$ production. The interference can be understood in terms of a variety of models based on the weak interaction Lagrangian

$$L_{int} = i\,\bar{\mu}\,\gamma^\tau\,(g_V - g_A\,\gamma^5)\,\mu Z_\tau.$$

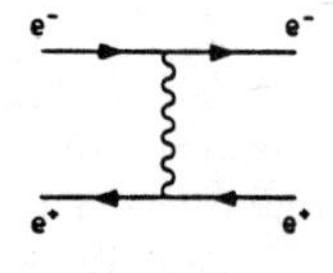

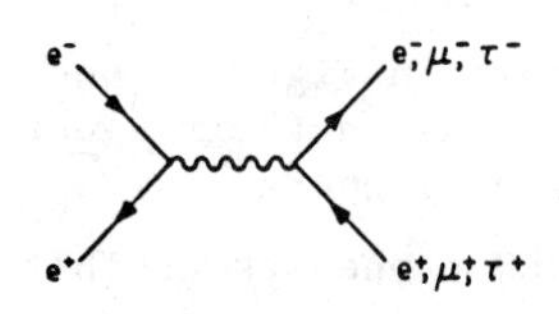

Fig. 1. Electron, muon and tau pair production in lowest order.

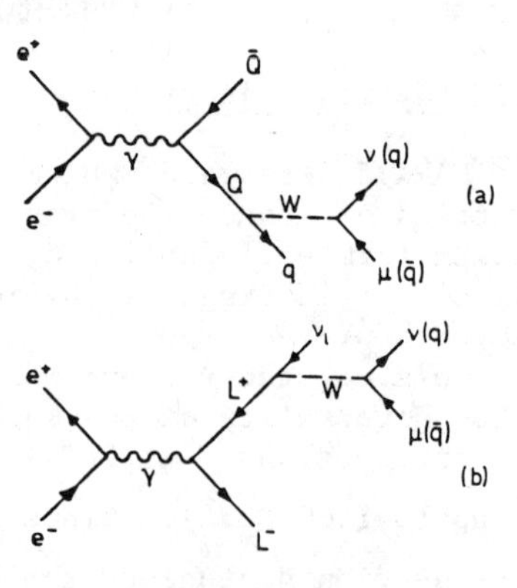

Fig. 2. a) Diagram for production and decay of heavy quarks in e^+e^- annihilation. b) Diagram for production and decay of heavy leptons in e^+e^- annihilation.

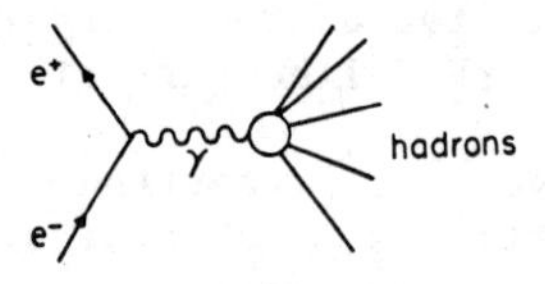

Fig. 3. The reaction $e^+e^- \to$ hadrons in lowest order.

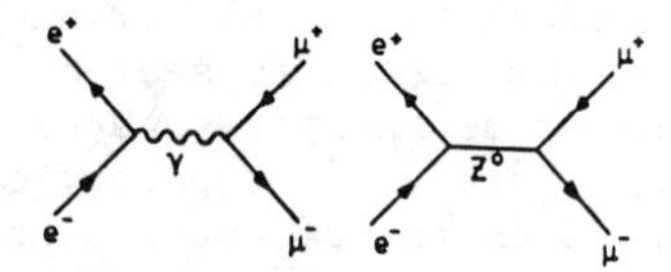

Fig. 4. First order electromagnetic and weak processes contributing to the reaction $e^+e^- \to \mu^+\mu^-$.

In the simple V-A model for example, one assumes $g_V = g_A = g$, where $g^2/M_Z^2 = G\sqrt{2}$, and where G is the Fermi coupling constant. In the now standard Glashow-Weinberg-Salam (GWS) model the couplings are expressed in terms of the single parameter θ_W, the Weinberg angle:

$$g_V = 1/4\ g \cos\theta_W\ (3\tan^2\theta_W - 1), \text{ and } g_A = 1/4\ g \sec\theta_W\ .$$

In order to distinguish between theoretical hypotheses, we can use the forward-backward charge asymmetry $A \equiv \sigma_- - \sigma_+ / \sigma_- + \sigma_+$, where σ_- (σ_+) corresponds to the μ^- (μ^+) appearing in the forward hemisphere. At $\sqrt{s}$ = 30 GeV, with a total time-integrated luminosity of 10^{38} cm^{-2}, one obtains $\sim 10^4$ events in a 4π detector, leading to a 10 standard deviation asymmetry effect in the V-A model and a 5 standard deviation effect in the GWS model[6].

Because the expected asymmetry is small and because higher order QED processes also produce sizable charge asymmetry at small angles, the measurement of asymmetry requires attention in reducing and understanding systematic errors in the detector design.

One notes that before the direct observation of the Z^o, the precise determination of the charge asymmetry arising from weak-electromagnetic interference is the most important verification of the idea of the unified electromagnetic and weak theory.

2.2 The Detector

The MARK J detector is shown in Figures 5 and 6. It is designed to distinguish charged hadrons, electrons, muons, neutral hadrons and photons and to measure their directions and energies. It covers a solid angle of $\phi = 2\pi$ and $\theta = 12^o$ to 168° (θ is the polar and ϕ is the azimuthal angle). The detector, which consists of five magnetized iron toroids built around a non-magnetic inner detector complemented by end caps, was designed to be insensitive to the effects of synchrotron radiation. Particles leaving the interaction region first pass through a five millimeter thick aluminium beam pipe, with an outer diameter of 190 mm. The aperture of the beam pipe is large enoughso that the synchrotron radiation produced in the final PETRA bending magnets and quadrupoles will pass unobstructed through the entire detector. Two thick copper absorbers are located symmetrically around the interaction region, at a distance of 1 meter, to trap synchrotron radiation reflected back towards the interaction region by collimators just in front of the last PETRA quadrupoles. The detector layer structure is best understood by referring to Figure 7.

During the first nine months of operation, a ring of 2 x 16 lucite Cerenkov counters each covering an azimuthal sector of 22.5^o and a polar-angle region from $9^o < \theta < 171^o$ surrounded the beam pipe.

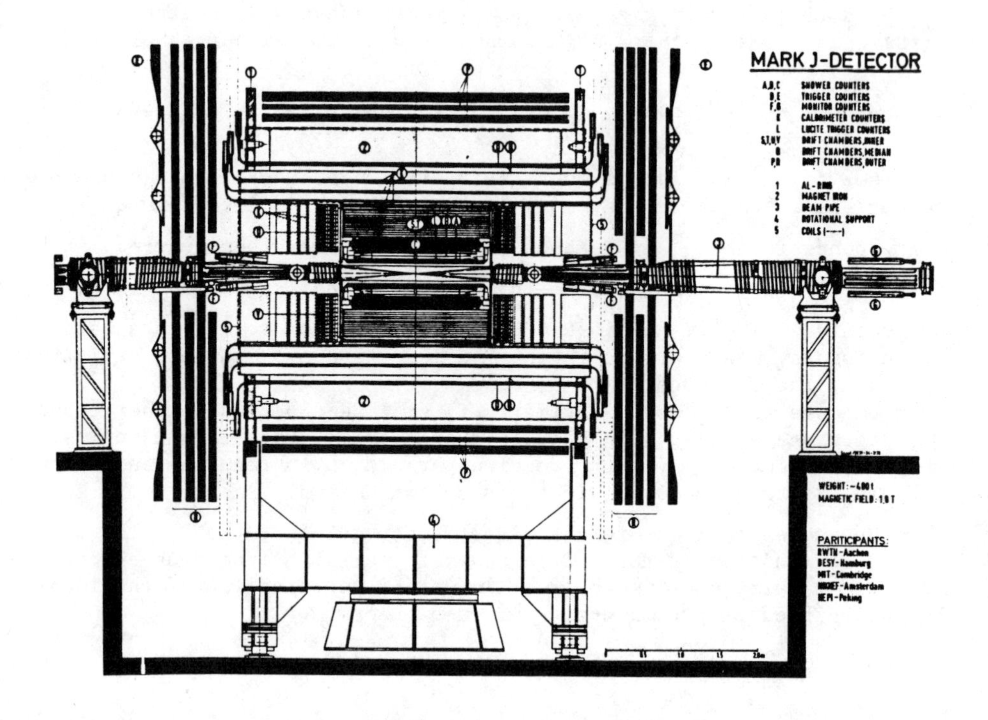

Fig. 5. The MARK J detector in a side view.

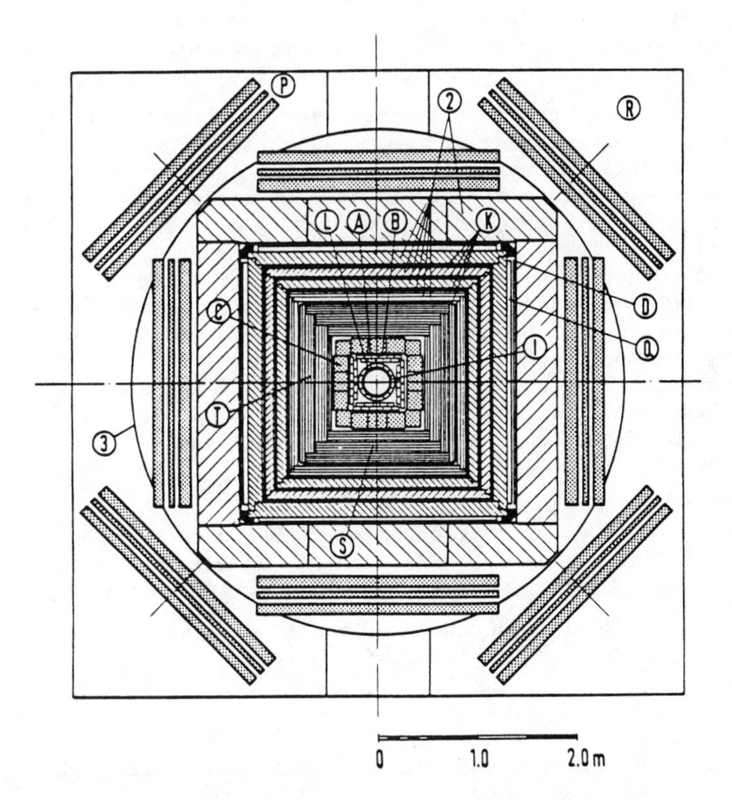

Fig. 6. The MARK J detector in end view. Beam pipe (1), drift tubes (DT), shower counters (A, B, C), inner drift chambers (S, T), calorimeter counters (K), outer drift chambers (Q, P, R), and magnetized iron (2).

These counters are divided at $\theta = 90^o$ to permit a crude determination of the momentum balance between the forward and backward hemispheres. The counters are insensitive to the effects of synchrotron radiation and can be used to separate charged from neutral particles.

In the latter part of 1979 the lucite counters were replaced by a four-layer inner track detector of 992 drift tubes. Each tube is 300 mm long and 10 mm wide and has a spatial resolution of 300 microns. The tubes, which are arranged perpendicular to the beam line, distinguish charged from neutral particles in the angular range $30^o < \theta < 150^o$ and reconstruct the position of the event vertex along the beam line to an accuracy of two millimeters. The distribution of event vertices obtained using the drift tubes is shown in Figure 8. The observed r.m.s. width of 1.27 cm is compatible with that expected from the known bunch length of the machine.

Particles then pass through 18 radiation lengths of shower counters used to identify and measure the energy of electrons, photons, charged and neutral hadrons. This inner calorimeter is divided into three layers of shower counters (labelled A, B and C in Figure 7). Each counter is constructed of 5.0 mm thick pieces of scintillator alternated with lead plates of equal thickness. The A and B counters are each 3 radiation lengths thick, while the C shower counter is a total of 12 radiation lengths thick (measured normal to the surface of the counter).

The twenty A shower counters are each 2 m long and cover the angular region of $\theta = 12^o$ to 168°. The 24 B counters are constructed identically to the A counters and cover an angular region from $\theta = 16^o$ to 164°.

Since every shower counter is viewed by one phototube at each end, the longitudinal (z) position of particle trajectories can be determined by comparing the relative pulse heights from each end of the counter. Timing information provides another measure of the longitudinal position. The trajectory location determined by this method was found to be in excellent agreement with the data from the drift tubes (see Section 1.2a).

Twelve planes of drift chambers (labelled S and T) measure the angles of particles penetrating the inner electromagnetic calorimeter. Each of the sense wires is connected to its own amplifier and time digitizer. Both end cap regions are covered by an additional ten planes of drift chambers (labelled U and V) of similar construction. These chambers are protected from beam backgrounds from the interaction region by the shower counters A, B and C. The energy sampling elements of the calorimeter K, shown in Figure 7, are 192 scintillation counters arranged in four layers. The main body of the calorimeter is composed of the magnetized iron plates which are also used to momentum-analyze muons. These plates range in thickness from 2.5 to 15 cm. Hadrons penetrating the inner shower counter layers, and secondary particles produced by hadronic showers initiated in the inner layers, deposit most of their remaining energy in the calorimeter K. The energy sampled by the K counters is thus used to

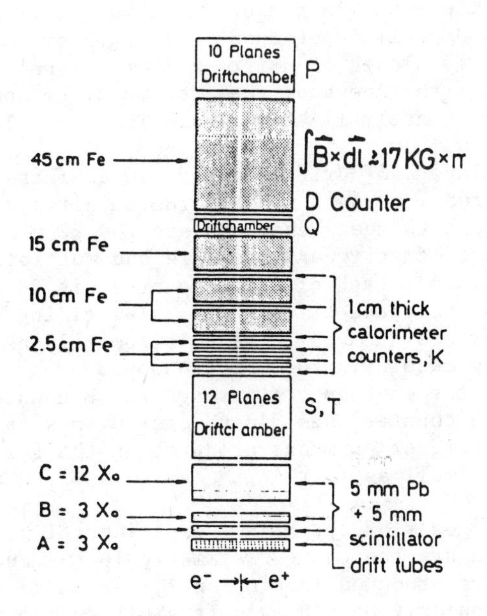

Fig. 7. The MARK J detector showing the outer drift chambers.

help distinguish hadrons from electrons, and to help identify minimum-ionizing particles.

Muons are identified by their ability to penetrate the iron of the hadron calorimeter. The low-momentum cut-off is about 1.3 GeV/c at normal incidence. The initial muon trajectory is measured in the S and T (U and V) chambers and in the drift tubes.

The bend angle and position of muons exiting from the calorimeter are measured in 10 planes of drift chambers, labelled R and P in Figure 7. The total thickness of the iron is 87 cm and it has a bending power of approximately 17 kG-meters. The typical bend angle for a 15 GeV muon is 30 mrad.

An additional 2 layers of drift chambers (Q chambers) are situated amidst the iron layers to measure the muon tracks in the bending plane. Adjacent to these chambers are the 32 muon trigger counters marked (D) used to trigger on single and multiple muon events and to reject cosmic rays. Each of these counters is 30 cm wide and 450 cm long and has a phototube at each end. The timing difference between real dimuon events and cosmic rays is about 10 ns. These counters have a timing resolution of about 400 ps.

Covering each of the end cap regions are the E counter hodoscopes. Each of these counters has dimensions 80 cm x 450 cm x 1 cm and they are used to trigger on muons produced in the forward and backward directions as well as to reject cosmic rays and beam-gas background.

One of the prime goals of the MARK J experimental program (see Section 1.1) is to measure the charge asymmetry in the angular distribution of muon pairs produced in e^+e^- annihilation to an accuracy of $\sim$ 1%. This goal can only be achieved if small systematic effects due to variations in chamber efficiency and counter gains, and slight asymmetries in the construction of the magnet and the positions of particle detectors in space, do not influence the overall charge asymmetry measurement. In order to isolate and subsequently eliminate the effects of these systematic errors in the measurement, the supporting structure is designed so that the entire detector can be rotated azimuthally about the beam line by $\pm 90^o$ and 180° about a vertical axis. The rotation about the vertical axis maps θ into $180^o - \theta$, and is therefore most useful in checking the measurement of the front-back charge asymmetry. The azimuthal rotation, which is used to check for beam polarization, can also be used to aid in the charge asymmetry measurement in the presence of polarized beams.

For the data in this report detectors E and R were not used.

The luminosity monitor consists of two arrays of twenty-eight lead glass blocks[3] (labelled G in Figure 7), each with dimensions of 8 cm x 8 cm x 70 cm located 5.8 m from the interaction point. They are designed to measure Bhabha events at small scattering angles ($\sim$ 30 mrad). Scintillators (F) in front of the lead glass define the acceptance and the lead glass counters measure the energy of the electron pairs.

The trigger is arranged in two stages. The first stage is a

fast loose trigger generated from the counter hit information with the following requirements:

i) For electron pairs we require at least 0.5 GeV total energy deposited in opposite quadrants of the A and B counters.

ii) For muon pairs we require at least two A and two B counters in coincidence with a pair of D counters which are coplanar within 50°.

iii) For single muon events we require at least two A, two B, two C and one D counters to be triggered.

iv) For hadrons we require at least four A and three B counters; each triggered quadrant must be in coincidence with the opposite quadrant.

All triggers are required to be in coincidence with the beam crossing signal.

After the fast trigger, a second stage imposes two more selections depending on event type. For electron pairs and hadron events the total energy deposited in the inner calorimeters A, B and C is determined by measuring the pulse area of linearly added signals. We require at least 13% of the total C.M.S. energy for hadrons and at least 10% for electron pair events. For muon pairs, single muon and hadron events, a microprocessor applies a loose track requirement demanding at least three pairs of wires to be hit in the S or T chambers.

2) PHYSICS RESULTS

2.1 Test of Quantum Electrodynamics and of Universality for Charged Leptons

There have been many experiments testing quantum-electrodynamics (QED) with electrons, muons and photons at electron-positron storage rings. Notable experiments[4] were done by Alles-Borelli et al., Newman et al., Augustin et al., O'Neill et al., and by our group at PETRA[5] up to a center of mass energy of 17 GeV. For a comprehensive review of QED work, see Brodsky and Drell[6]. Much has been learned about the properties of the heavy lepton tau since the original search began at ADONE on $e^{+} + e^{-} \rightarrow \mu e + \ldots$ [7]. The discovery of the τ lepton at SLAC[8] and its subsequent confirmation at DESY[9] has inspired further studies. We know it is a spin 1/2 particle which decays weakly[10] and whose properties are very similar to the muon.

In this experiment we study the reactions $e^{+} + e^{-} \rightarrow \ell^{+} + \ell^{-}$ for all the known charged leptons ($\ell = e, \mu, \tau$) by measuring the dependence of the cross section on center of mass energy or scattering angle over a wide range of PETRA energies. These measurements enable us to compare the data with predictions of quantum electrodynamics, to test the universality of these leptons at very small distances, and to set a limit on the charge radius of these particles. Up to the present time the reactions:

$$e^+ + e^- \rightarrow e^+ + e^- \text{ (Bhabha scattering)} \quad (2)$$

$$e^+ + e^+ \rightarrow \mu^+ + \mu^- \quad (3)$$

$$\begin{array}{l} \quad\quad\quad\quad\ \ \ulcorner\!\!\longrightarrow \text{(h's or e)} + \nu\text{'s} \\ e^+ + e^- \rightarrow \tau + \bar{\tau} \\ \quad\quad\quad\quad\quad\ \llcorner\!\!\longrightarrow \mu + \nu\text{'s} \end{array} \quad (4)$$

have been measured at the center of mass energies $\sqrt{s}$ = 12, 13, 17, 22, 27.4, 30, 31.6, 35, 35.8, and 36.6 GeV.

a) Bhabha Scattering

The Bhabha events are identified by requiring two back-to-back showers in the A, B and C counters which are collinear to within 20° in ϕ and θ and with a measured total shower energy greater than 1/3 of the incident beam energy. Photons emitted close to either electron are included in the electron momentum. From the measurement of the acollinearity angle $\Delta\theta$, and the acoplanarity angle $\Delta\phi$, we observe that most of the events are in the region $\Delta\theta < 4^o$, $\Delta\phi < 4^o$. Because there are few events near the 20° cuts in $\Delta\theta$ we conclude that the background to Bhabha scattering is negligible.

To eliminate most background from hadron jets, the energy in the K counters was required to be less than 7% of the total energy. Because the QED test is most sensitive to background in the large angle region, all events having θ larger than 60° were scanned on graphic displays which showed the distribution of counter hits. On the basis of a Monte Carlo study of hadron events, we conclude that the background from this source is less than 1% of the events. As mentioned above, the acceptance for $e^+e^- \rightarrow e^+e^-$ was computed using Monte Carlo techniques and is defined by the geometry of the first shower counter array A. Both energy and acceptance losses in the corners were found to be small.

The first order QED photon propagator produces an s^{-1} dependence in the $e^+e^- \rightarrow e^+e^-$ cross section. Thus when radiative corrections have been taken into account in the data, the quantity s $d\sigma/d\cos\theta$ vs $\cos\theta$ should be independent of s. This distribution is plotted for the data at $\sqrt{s}$ = 13, 17 and 27.4 GeV in Figure 9. Excellent agreement with QED predictions is seen. To express this agreement analytically, we compare our data with the QED cross section in the following form (since charge is not distinguished here)[11]:

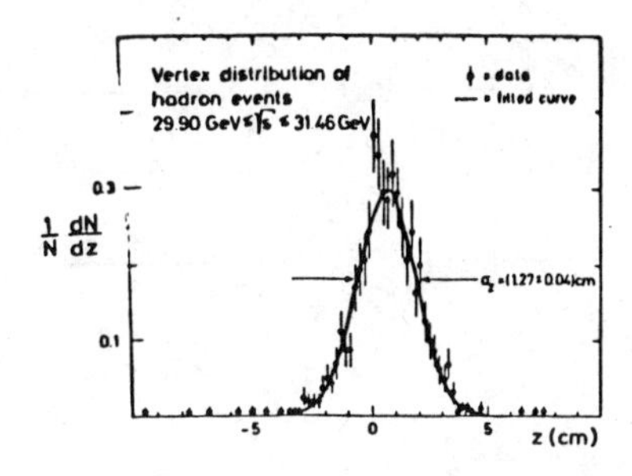

Fig. 8. Distribution of event vertices along the beam direction reconstructed using drift tube tracks.

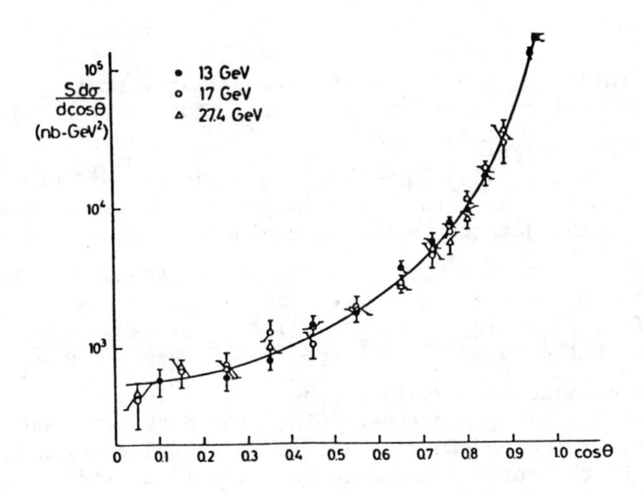

Fig. 9. The differential cross sections $d\sigma/d\cos\theta$ for $e^+e^- \rightarrow e^+e^-$ at s of 13, 17, and 27.4 GeV.

$$\frac{d\sigma}{d\Omega} = \frac{\alpha^2}{2s}\left\{\frac{q'^4 + s^2}{q^4} F_s^{\,2} + \frac{2q'^4}{q^2 s} \mathrm{Re}(F_s F_T^*) + \frac{q'^4 + q^4}{s^2} F_T^{\,2}\right.$$

$$\left. + \frac{q^4 + s^2}{q'^4} F'^{\,2}_s + \frac{2q^4}{q'^2 s} \mathrm{Re}(F'_s F_T^*) + \frac{q'^4 + q^4}{s^2} F_T^{\,2}\right\}\left\{1 + C(\theta)\right\} ,$$

where

$$F_s = 1 \mp q^2/(q^2 - \Lambda_{s\pm}^{\,2})$$

is the form factor of the spacelike photon, $F'_s = 1 \mp q^2/(q^2 - \Lambda_{s\pm}^{\,2})$,

$$F_T = 1 \mp s/(s - \Lambda_{T\pm}^{\,2})$$

is the form factor of the timelike photon, $q^2 = -s \cos^2(\theta/2)$, $q'^2 = -s \sin^2(\theta/2)$, Λ is the cutoff parameter in the modified photon-propagator model[12] and $C(\theta)$ is the radiative correction term as a function of θ.

The radiative correction to the e^+e^- elastic scattering process was calculated using the program of Berends for these particular event selection criteria[13].

In order to establish lower limits on the cut-off parameters a Monte Carlo program was used to generate e^+e^- pairs which were then traced through the detector with the inclusion of measured θ, ϕ resolutions. A χ^2 fit to all of the 13, 17 and 27.4 GeV data is then made using the Monte Carlo generated angular distribution. The normalization is treated in two ways: (1) the total number of Monte Carlo events in the region $0.90 < \cos\theta < 0.98$ was set equal to the total number of measured events in the same region, (2) the minimum-χ^2 for the entire data sample determined the normalization. The two methods agree with each other to within 3% and give essentially the same result in the cut-off parameter Λ. The lower limits of Λ at 95% confidence level under various assumptions are shown in Table I. The JADE and PLUTO groups have also analyzed their QED reactions and have obtained similar conclusions with regard to the validity of QED at small distances (see Section 2.4).

b) Muon and Tau Pair Production[14]

The MARK J detector is designed to distinguish muons from electrons and hadrons and to distinguish back-to-back muon pairs from cosmic ray muons. Muon identification is also aided by the short decay path allowed to hadrons before reaching the shower counters. In addition to the cuts described in Section 1.3b, single muons are

TABLE I

Cut-off Parameters in GeV for photon form factors from Bhabha scattering.

Λ	$1 - \frac{q^2}{q^2 - \Lambda_+^2}$	$1 + \frac{q^2}{q^2 - \Lambda_-^2}$
Λ_S	91	152
Λ_T	58	64
$\Lambda_S = \Lambda_T$	97	157

identified as particles which:

i) are reconstructed in the inner drift chambers to come from the interaction region;

ii) leave minimum ionizing pulse heights in the A, B, C, K, and D counters, a total of seven layers;

iii) leave a track in the outer drift chambers (P) and thus fall into an angular range $45^o < \theta_\mu < 135^o$.

In addition back-bo-back muon pairs from reaction (3) are distinguished from cosmic rays by the requirement that:

i) the D counter timing signals are coincident with one another (and not relatively off time as in the case for cosmic rays traversing the detector);

ii) the muons should be collinear and coplanar, and they should pass through the intersection region.

A Monte Carlo study shows that the $\mu^+\mu^-$ acceptance, which is dominated by the geometrical acceptance of the P drift chambers, is 41% ± 3% independent of beam energy.

Tau leptons from reaction (4) are identified by detecting μ-hadron and μ-electron final states. The cross section is determined using the known branching ratio of $\tau \to \mu\ \nu\bar{\nu}$ (16%) and $\tau \to$ (e, lepton + hadron, or multi-hadrons) + ν (84%)[10]. The muons, hadrons and electrons are identified as described previously. The total deposited energy of hadrons (or electrons) is required to be greater than 2 GeV.

The major background to reaction (4) is the two-photon process:

$$e^+ + e^- \to e^+ + e^- + \mu^+ + \mu^- \qquad (7)$$

This process becomes important at high energies since the total cross section grows as $\ln(\sqrt{s}/m_e)^2 \ln(\sqrt{s}/m_\mu)$ with increasing beam energy in contrast to reaction (3), the rate for which falls as 1/s. The observed cross section for the two-photon process is suppressed by a factor $\sim 10^3$ relative to the total cross section by momentum cuts on the muons and by energy and angle cuts on the electrons, but rates of the accepted events remain significant.

In Figure 10 we show the calculated cross section in our detector when the observed particles are

(a) two μ's only,

(b) only one μ and one e and,

(c) two μ's and one e.

The cross section for each of these configurations were computed using the Monte Carlo integration program of Vermaseren[12]. The computations were compared to the cross sections measured for reaction (7) using the following cuts:

$$168^o \geq \theta_e \geq 12^o \qquad E_e \geq 2\ \text{GeV}$$

$$135^o \geq \theta_{\mu_1} \geq 45^o \qquad P_{t_{\mu_1}} \geq 1.5 \text{ GeV (first muon)}$$

$$147^o \geq \theta_{\mu_2} \geq 33^o \qquad P_{t_{\mu_2}} \geq 0.8 \text{ GeV (second muon)}$$

The measured cross sections, also shown in Figure 10, agree well with the calculations in all cases.

The cross section for case (a) above is much larger than that for the process

$$e^+ + e^- \rightarrow \tau + \bar{\tau} \rightarrow \mu^+ + \mu^- + 4\nu ,$$

and the $\mu^+\mu^-$ events from the two processes are hard to distinguish. We therefore exclude case (a) from the sample of $\tau\bar{\tau}$ candidates. In case (b) the electron from the two-photon process is strongly peaked in the small angle region. Typically, $d\sigma_{\mu e}/d(\cos\theta_e)$ decreases by two orders of magnitude from $\cos\theta_e = 0.98$ to 0.80. Furthermore, the observed muons and electrons tend to be coplanar because of conservation of transverse momentum. By requiring $30^o < \theta_e < 150^o$ for the $\tau\bar{\tau}$ sample, and by requiring that the final state μe by collinear within 30°, we were able to reduce the two-photon contribution to a negligible level ($< 10^{-3}$ picobarns). The rate for case (c) is small and can readily be separated from the τ events.

The fact that the μ-hadron events produced by reaction (4) in our energy region are almost collinear can also be used to distinguish reaction (4) from μ-hadron events produced by the semi-leptonic decay of particles with c (charm) or b (bottom) quantum number. In the latter cases the muon is accompanied by hadrons emitted close to the muon direction.

The measured muon momentum and hadron energy for the $\tau\bar{\tau}$ candidates is in agreement with calculations based on the known decay properties of the τ lepton[15].

The acceptance is calculated using a Monte Carlo method to generate $\tau\bar{\tau}$ production from reaction (4) including radiative corrections. We obtain a detection efficiency of $\sim$10% for τ pairs at various energies, when requiring one decay muon to be detected in association with a single electron, or one or more hadrons.

The resultant $e^+e^- \rightarrow \mu^+\mu^-$ and $\tau\bar{\tau}$ cross sections as a function of s are plotted in Figures 10 and 11 together with the QED prediction. We see that from $q^2 = s = 169$ to $q^2 = 1300$ GeV^2 the data agree well with the predictions of QED for the production of a pair of point-like particles. In particular, Figure 12 represents the first evidence that the τ lepton is a point-like particle over a large range of q^2, and demonstrates that it belongs in the same family as the electron and muon. To parametrize the maximum permissible size (radius) of the particles, we use a form factor:

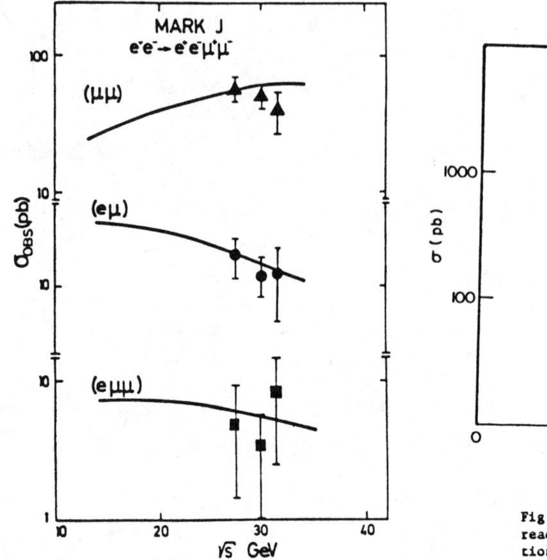

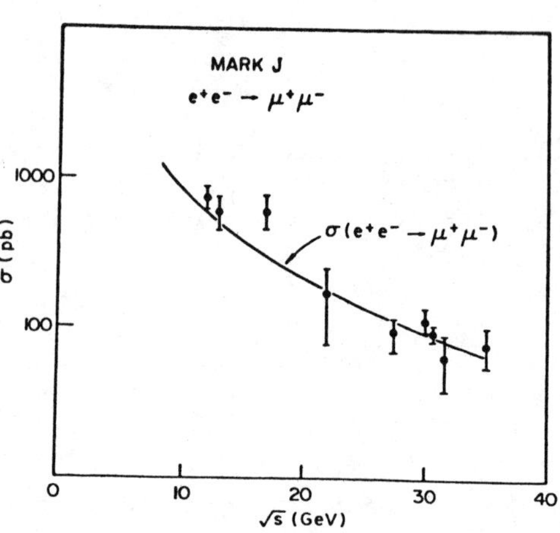

Fig. 11. Observed cross section for the reaction $e^+e^- \to \mu^+\mu^-$ compared to predictions of QED.

Fig. 10. The observed cross sections of $e^+ + e^- \to e^+ + e^- + \mu^+ + \mu^-$ in our detector as functions of $\sqrt{s}$ when the observed particles are: a) two μ's b) one μ and one e c) two μ's and one e. The solid lines are Monte Carlo calculations of the yield from two photon diagrams and the points are the measurements.

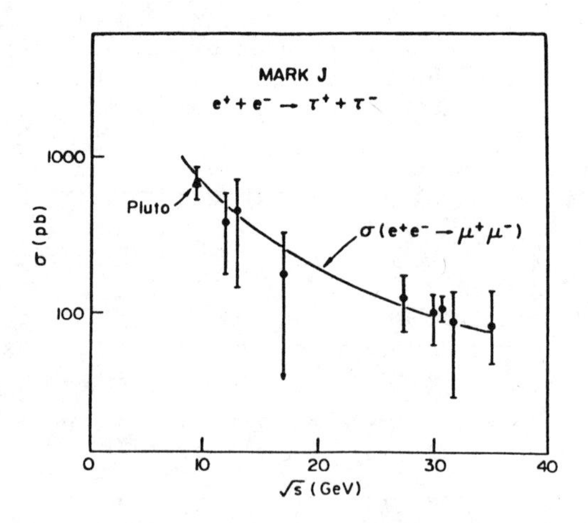

Fig. 12. Observed cross section for the reaction $e^+e^- \to \tau^+\tau^-$ compared to predictions of QED.

$$F_{\ell} = 1 \mp \frac{q^2}{q^2 - \Lambda^2_{\ell\pm}} \quad (\ell = e, \mu, \tau) .$$

By comparing our data with the cross sections including this form factor we find lower limits on the cut off parameters, at the 95% confidence level, summarized in Table II.

Thus, from Heisenberg's Uncertainty Principle, all the known charged leptons are point-like particles in their electromagnetic interactions, with characteristic radii $< 2 \times 10^{-16}$ cm.

2.2 Hadronic Final States

a) Hadron Identification

The final selection for hadrons is made by scanning the events on an interactive graphics system after a preselection which includes energy and momentum balance cuts. In this scanning, hits observed in the drift tubes and in the drift chambers allow an easy determination of the event vertex. Thus beam gas events which do not come from the interaction region are readily recognized. The charged multiplicity observed in the drift tubes and the shower envelope in the counters is used to reject events of electromagnetic origin. For events which have low multiplicity both in the tubes and the counters, we require that a minimum fraction of the total energy is deposited in the hadron calorimeter (K in Figure 7) to further discriminate against purely electromagnetic final states.

The energy spectrum of the events passing the cuts is shown in Figure 13. For the analysis of hadronic final states we employed an additional energy cut of $E_{vis} > 0.5\sqrt{s}$ for measurement of total cross section (Section 2.2b), for most of the thrust analysis (Section 2.3a) and the study of inclusive muons in hadron events (Section 2.3c). This additional cut also reduces the contamination from beam-gas events, from two-photon processes[15] and $e^+e^- \rightarrow \tau^+\tau^-$ events which yield hadrons in the final state[13].

For reasons discussed in Section 2.3d, we employed an even more restrictive cut of $E_{vis} > 0.7\sqrt{s}$ for study of gluon effects in thrust distributions (Section 2.3a), for jet analysis using Fox-Wolfram moments (Section 2.3b), for the discovery of 3-jet events (Section 2.3d) and determination of the strong coupling constant α_s (Section 2.3e). Both of these cuts are indicated by arrows in Figure 13.

b) Total Hadronic Cross Section[16,17,18]

The total cross section for $e^+e^- \rightarrow$ hadrons was measured over a wide range of center of mass energies from 12 to 35.8 GeV, including results obtianed by extended periods of running at a fixed beam

TABLE II

Cut-off parameters in GeV for e^+e^-, $\mu^+\mu^-$ and $\tau^+\tau^-$ production using leptonic form factors.

ℓ	electron	muon	tau
Λ_-	142	150	157
Λ_+	94	126	76

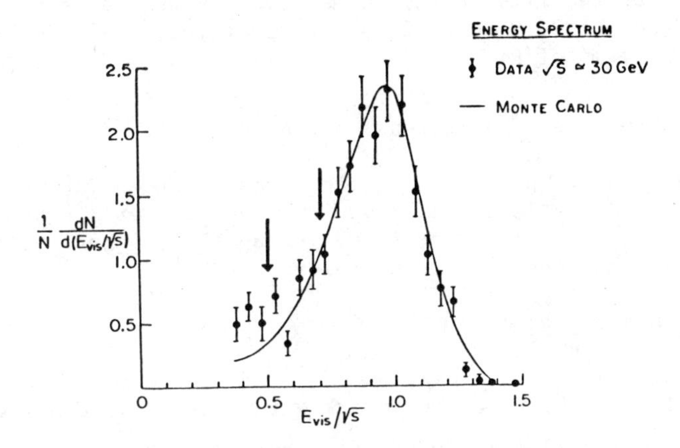

Fig. 13. Energy distribution of hadron events satisfying the cuts described in Section 2.2a.

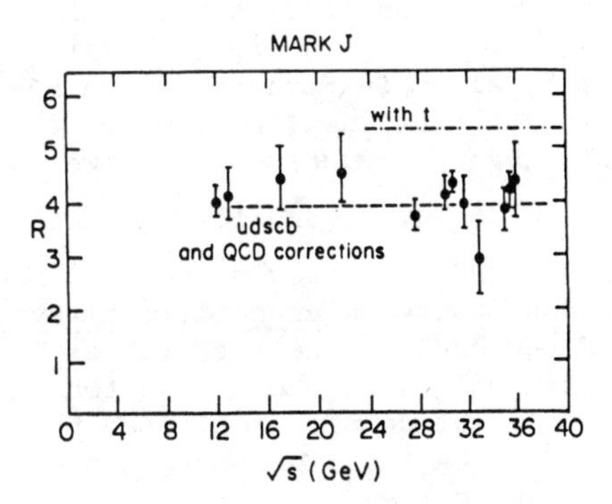

Fig. 14. The total relative hadronic cross section $R = (e^+e^- \to \text{hadrons})/\sigma(e^+e^- \to \mu^+\mu^-)$ at all energies covered by this experiment. Note that the point at 27.5 GeV corresponds to data taken over the range of 27.4 to 27.7 GeV, and the point at 30.7 GeV represents the energy scan from 29.9 to 31.5 GeV.

energy and by two fine energy scans covering the range of 29.92 to 31.46 GeV and of 35.0 to 35.26 GeV respectively. The results are expressed in terms of R:

$$R = \sigma(e^+e^- \to \text{hadrons})/\sigma(e^+e^- \to \mu^+\mu^-).$$

In the naive quark-parton model, the cross section for the hadron production process is simply given by the sum over flavors of the point-like $q\bar{q}$ pair cross sections. Using this picture with spin 1/2 massless quarks, and with three colors gives

$$R = 3 \sum_q e_q^2 ,$$

where e_q is the charge of the quark with flavor q. Considering the five known quark flavors (u, d, s, c and b), and correcting the naive model for gluon emission as predicted by QCD, we expect $R \simeq 4$, over the entire PETRA energy range, with only a slight decrease in R with increasing beam energy.

The MARK J results for R are shown in Figure 14 along with the QCD predictions. The data consist of a total of 2300 hadron events. The results are summarized in Table III. Figure 14 shows that the data agree with the QCD predictions for five quark flavors (represented by the lower dashed line), and that there is no sign up to 35.8 GeV of a threshold in the hadron continuum corresponding to a new heavy quark of charge 2/3 such as the top quark (the upper line).

The experimental R-values in Figure 14 have been corrected for initial-state radiative corrections ($\Delta R \simeq -0.3$) for contamination of the sample by hadronic events produced by the two photon process $e^+e^- \to e^+e^- +$ hadrons ($\Delta R \lesssim -0.1$), and for the contribution $e^+e^- \to \tau^+\tau^- \to$ hadrons + leptons (ΔR typically = -0.3). In addition to the statistical errors shown, there is an additional systematic uncertainty due to model dependence of the acceptance of 10%. The accuracy in evaluating the contributions of the two photon and tau-pair contributions to the hadronic event sample is limited by the lack of detailed experimental data on the high energy, high multiplicity states arising from these sources.

In addition to the $t\bar{t}$ contribution to the hadronic continuum, the toponium system should form one or more bound states, with the number of such states depending on the shape of the binding potential[16]. Interpretation of the vector mesons ρ, ω, ϕ, J, ψ', Υ, Υ', and Υ'', as nonrelativistic $q\bar{q}$ bound states, or "quarkonia" leads to the prediction that the gap between the lowest bound state and the continuum is probably $\sim$ 1 GeV, and very likely $\lesssim$ 2 GeV.

In order to check for the existence of $t\bar{t}$ bound states lying below 35.8 GeV, the two energy scans mentioned earlier were performed in 20 MeV center of mass energy steps (matching the r.m.s. energy spread of PETRA). The results of the scan are shown in Figures 15a

TABLE III

Results of R measurement. The errors are statistical only.

$\sqrt{s}$ [GeV]	$R \pm \Delta$ (statistical)
12.0	4.03 ± 0.28
13.0	4.1 ± 0.5
17.0	4.4 ± 0.6
22.0	4.7 ± 0.7
27.57 *)	3.8 ± 0.3
30.0	4.2 ± 0.3
29.90-31.46	4.33 ± 0.17
31.6	4.0 ± 0.5
33.0	2.9 ± 0.6
35.0	3.8 ± 0.4
35.0-35.28	4.2 ± 0.3
35.8	4.4 ± 0.7

*) = combination of 27.4 and 27.7 GeV data

and 15b, along with the results obtained earlier at 30.0 and 31.6 GeV and at 35 and 35.8 GeV respectively. The figures show that the data are entirely consistent with the predictions of QCD for u, d, s, c and b quarks, that is, with a constant value of R over the whole range. No single point lies $\geq 3\sigma$ above the QCD line, and there is no indication of an upward slope with increasing energy signaling the onset of a new contribution to the continuum. The values of R averaged over the individual energy range of the scan are 4.33 ± 0.17 for $30 < E_{cm} < 31.6$ GeV and 4.2 ± 0.3 for $35 < E_{cm} < 35.8$ GeV.

In order to set a quantitative upper bound on the possible production of a narrow resonance, the data in Figure 15 were fitted by a constant plus a gaussian:

$$R = R_o + R_V \, e^{\left[-(\sqrt{s}-M)^2/2\Delta_W{}^2\right]}$$

where R_o represents the non-resonant continuum, M is the mass of the resonance, Δ_W is the r.m.s. machine energy width, and R_V is the peak value of the resonant contribution. The largest value of R_V consistent with the data was determined by trying fits with M, the center of the gaussian, fixed at all the center of mass energies at which data were taken. The largest value of R_V was obtained at 31.32 GeV corresponding to an upper limit on the resonance strength

$$\Sigma_V \equiv \int \sigma_V(w)\,dw, \text{ where } w = \sqrt{s}\ ,$$

of 33 MeV-nb at the 90% confidence level. Using the relation between the resonance strength, the width into e^+e^- (Γ_{ee}), the hadronic width (Γ_h), the total width (Γ), and the hadronic branching ratio ($B_h \equiv \Gamma_h/\Gamma$):

$$\Sigma_V \equiv \int \frac{3\pi}{M^2} \frac{\Gamma ee \Gamma h}{(w - M)^2 + \Gamma^2/4} \, dw = \frac{6\pi^2}{M^2} B_h \Gamma_{ee}$$

we obtained

$$B_h \Gamma_{ee} < 1.3 \text{ KeV (90\% C.L.) for } 30 < E_{cm} < 31.6 \text{ GeV}$$

Similarly,

$$B_h \Gamma_{ee} < 1.2 \text{ KeV (90\% C.L.) for } 35 < E_{cm} < 35.28 \text{ GeV.}$$

This upper limit excludes the production of a vector particle consisting of a $q\bar{q}$ bound state where the quark has charge 2/3. On the

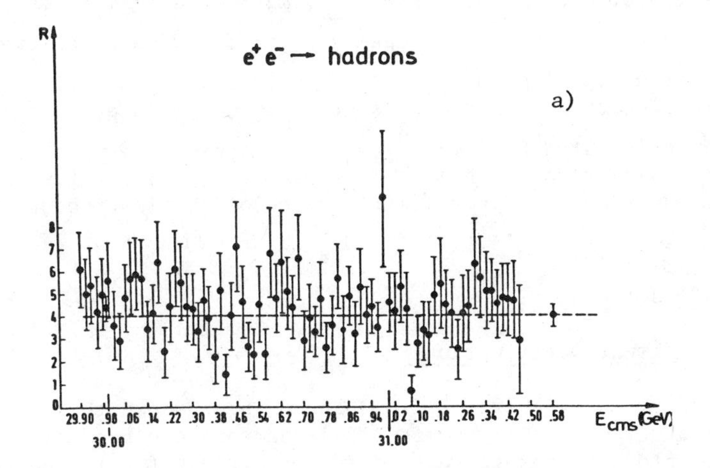

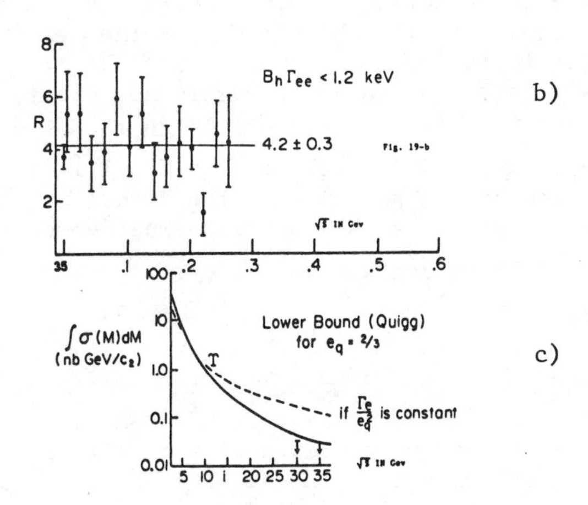

Fig. 13. a) R values measured during the energy scan between 29.9 and 31.5 GeV. The line represents the predictions of QCD. b) R values measured during the energy scan between 35.0 and 35.26 GeV. The line represents the predictions of QCD. c) Theoretical predictions of the production cross section of the ground state toponium in e^+e^- annihilation; slashed curve is calculated assuming Γ_e/e_q^2 is constant; solid line is the rigorous lower bound as calculated by Quigg in reference 21; indicates the experimental upper limit.

basis of the experimental fact that Γ_{ee}/e_q^2 is approximately constant for the vector meson ground states ρ, ω, ϕ, J and Γ, as is predicted by duality arguments[16], one expects $\Gamma_{ee} \sim 5$ KeV for the lowest mass meson in the toponium family. A rigorous lower bound of the production cross section of the ground state toponium in e^+e^- annihilation has been calculated by Quigg[16] as a function of the mass of the toponium. The theoretical predictions are shown in Figure 15c. We see the experimental upper limits are below the lower bound both at $E_{cm} = 31$ and $E_{cm} = 35.1$ GeV.

2.3 Jet Analysis

a) Thrust Distributions

Data at lower energies from SPEAR[19] have shown that the final state hadrons from the process $e^+e^- \rightarrow$ hadrons are predominantly collimated into two back-to-back "jets" in agreement with the expectations of simple models in which the time-like photon materializes initially into a quark antiquark pair. It is thus necessary to develop kinematic quantities which describe the jet-like nature of the MARK J hadronic events.

A jet analysis of the hadronic events has been devised using the spatial distribution of the energy deposited in the detector. For each counter hit, a vector $\vec{E}^i$ (the energy flow) is constructed, whose direction is given by the position of the signal in the counter, and whose magnitude is given by the corresponding deposited energy. A parameter thrust (T) is defined as[18]

$$T = \max \Sigma E^i_{/\!/} / E_{vis}$$

where $E^i_{/\!/}$ is the parallel component of E^i along a given axis, and the maximum is found by varying the direction of this axis, and where the sums are taken over all counter hits. The resultant direction is thus the direction along which the projected energy flow is maximized.

For events in which the spatial distribution of energy is isotropic, T is expected to approach 0.5. This would be the situation if, for example, the virtual photon materializes into two very heavy quarks, each with a mass close to the beam energy. Such quarks would be produced almost at rest. On the other hand, pairs of light quarks would move at high speed and the Lorentz boost of their hadronic fragmentation products would result in the hadrons being produced in narrow jets collimated around the initial quark directions. Higher beam energies would result in narrower jets, so that T should approach the value 1. Thus, thrust measurements can be used as a sensitive method to detect the presence of a new threshold due to new heavy quarks.

These expectations are illustrated qualitatively in Figure 16 in which the thrust distribution expected just below and just above the threshold for production of a new heavy quark is shown. Production of a new heavy quark would also result in lowering the average T as the energy is raised and the threshold is passed.

The normalized thrust distributions 1/N dN/dT for 13, 17, 22, and the combination of 27.4 and 27.7 GeV data (labelled 27 GeV combined) are shown in Figures 17a-d along with the Monte Carlo predictions of a quark-parton model with u, d, s, c, and b quarks and no gluon emission. (The individual distributions at 27.4 and 27.7 GeV are in agreement with each other.) As expected for production of final states with two jets of particles, the distributions become narrower and shift towards high thrust with increasing energy.

Figure 18 shows the normalized thrust distributions for combined data between 29.9 and 31.6 GeV and for data between 35 and 35.8 GeV where the measured energy of the selected events is at least 70% of $\sqrt{s}$. The curves show the Monte Carlo predictions with and without inclusion of gluons. As can be seen, these data lend support to the necessity of including gluons in the Monte Carlo program. The data is inconsistent with a Monte Carlo calculation which includes a charge 2/3 t quark produced as described previously. We thus conclude that there is no evidence for production of a new heavy quark with charge q = 2/3 e.

Figure 19 shows the average thrust <T> plotted at the six energies. The solid curves are from Monte Carlo calculations which include u, d, s, c and b quarks with gluon emission. The energy dependences of the data are smooth and show none of the steps which would have appeared at new quark thresholds.

b) Jet Analysis Using Fox-Wolfram Moments

Another method of jet characterization has been proposed by Fox and Wolfram. The simplest observables characterizing the three-dimensional shape of the energy distribution in a hadronic event are the Fox-Wolfram moments[20].

$$H_\ell = \sum_{i,j} \frac{\vec{E}_i \vec{E}_j}{E_{vis}^2} P_\ell(\cos\theta_{ij}), \text{ where } P_\ell \text{ is } \ell\text{th Legendre polynomial,}$$

and

$$\pi_1 = \sum_{i,j,k} \frac{\vec{E}_i \vec{E}_j \vec{E}_k}{(E_{vis})^3} (\hat{E}_i \cdot \hat{E}_j \times \hat{E}_k)^2$$

The H's parametrize the shape of the energy distribution by measuring the correlation between pairs of energy flow elements in terms of spherical harmonics. For cigar-shaped events expected from the

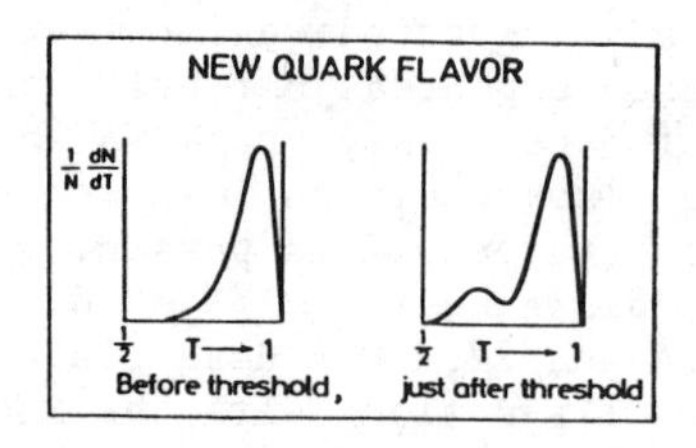

Fig. 16. A sketch showing the qualitative change in the thrust distribution expected when the energy is increased and the threshold for production of a new quark flavor is crossed.

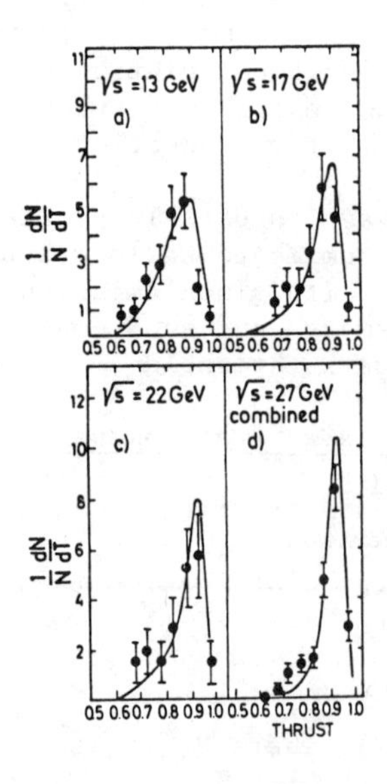

Fig. 17. Thrust distribution observed at $\sqrt{s}$ = a) 13, b) 17, c) 22, d) 27 GeV (see text). The solid line is the quark model prediction for u, d, s, c and b quarks with no gluon emission.

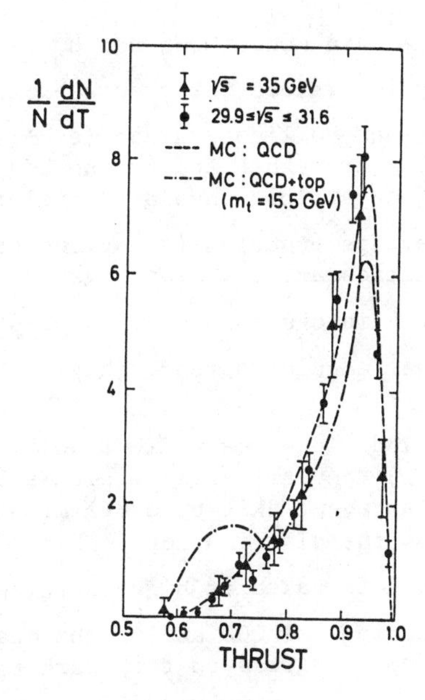

Fig. 18. The thrust distribution with a 70% energy cut for $\sqrt{s}$ = ~30 GeV. The curves are predictions for various models as described in the text.

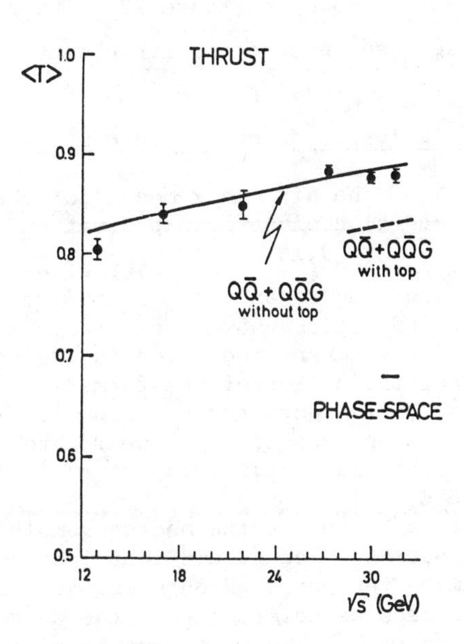

Fig. 19. Average value of thrust as a function of $\sqrt{s}$ together with the QCD prediction (solid line). The values expected from a phase space distribution and from a QCD model with a top quark are also shown.

process $e^+ + e^- \to q + \bar{q}$ the even moments H_2 and H_4 tend to be large, while they take relatively low values for events broadened on one side by the radiation of a hard non-collinear gluon ($e^+ + e^- \to q + \bar{q} + g$). Less significant differences are exhibited by the odd moments. The moment π_1, involving a three-particle angular correlation, is sensitive to the planarity of events expected from gluon bremsstrahlung.

Since all these quantities are invariant against a split-up of the particle energy vector E into separately measured components $\vec{E}'_n(\sum_n \vec{E}'_n = \vec{E}_i)$ having a small angular spread, they are well suited for detectors not identifying individual particle tracks. Since they are also rotationally invariant, they are easily calculable and do not involve the definition of an event axis by a maximization process.

Figures 20 and 21 show the distributions $1/N\ dN/dH_2$ and $1/N\ dN/dH_4$ of the combined data taken at high energies ($\sqrt{s} > 27$ GeV). They are compared to predictions of QCD and of the quark-parton model with $\langle P_t \rangle = 225$ MeV/c, where P_t refers to the quark transverse momentum in the fragmentation process. The experimental distributions are in excellent agreement with QCD and clearly rule out the simple quark model at this particular $\langle P_t \rangle$. The same is true for the differential distribution $1/N\ dN/d\pi_1$, shown in Figure 22. The predictions of the quark model for these shape parameters are, however, sensitive to the $\langle P_t \rangle$ chosen.

c) A Study of Inclusive Muons in Hadronic Events

In the framework of the six quark model for the weak decays of heavy quarks, (c, b and t) copious muon production is expected from the cascade decays $t \to b \to c$[9,10]. The onset of production of a new heavy lepton would also lead to an increase in muon production. Thus, in addition to indications based on thrust and R measurements, a measurement of inclusive muon production in hadronic final states should provide a clear indication of the formation of top quarks or new leptons. All the hadron data for $\sqrt{s}$ from 12 to 35.8 GeV has therefore been analyzed and scanned in a search for muons. The sample of events used in the inclusive muon survey is a subsample of that used to measure R.

The main sources of muons in the hadron sample are decay products of bottom and charm quarks. Background contributions to the muon signal, arising from hadron punch through and decays in flight of pions and kaons have been calculated using the Monte Carlo simulation to be $\sim$2% at these energies. The contribution of $\tau^+\tau^-$ events to the μ + hadron sample becomes negligible when the total energy cut and the energy balance cut are applied.

Table IV summarizes the results for the relative production rate

TABLE IV

Monte Carlo predictions and data for hadronic events which include muons.

E_{cm} GeV	Luminosity nb^{-1}	Number of Hadron Events	Number of Muon Events	% of Muon Events	Monte Carlo (no top) ±	Monte Carlo (with top) ±
12	97.7	239	2	0.8 ± 0.6	1.1 ± 0.3	
13	53	95	1	1.05 ± 1.0	1.25 ± 0.3	
17	60	67	2	3.0 ± 1.7	2.0 ± 0.3	
27.4	414	188	11	5.85 ± 1.8	3.3 ± 0.4	
30 to 31.6	2804	1147	43	3.75 ± 0.57	4.5 ± 0.5	7.8 ± 0.5
35 to 35.8	1885	640	22	3.5 ± 0.7	5.1 ± 0.4	7.8 ± 0.5

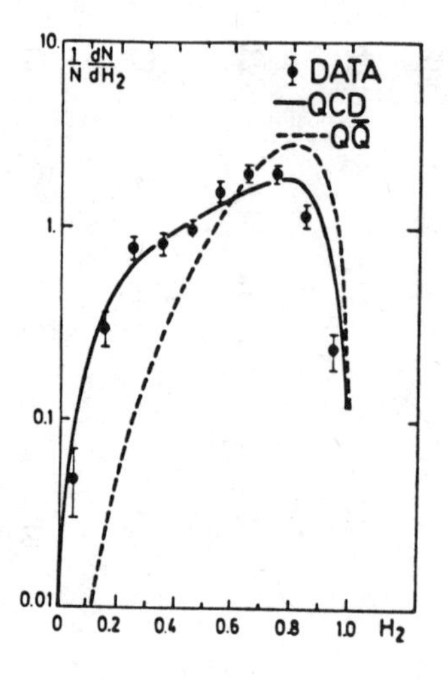

Fig. 20. The distribution of the combined high energy data ($\sqrt{s}$ > 27 GeV) in the Fox-Wolfram moment, H_2 compared to the prediction of QCD at α_s = 0.23 and the quark model with $<P_t>$ = 225 MeV.

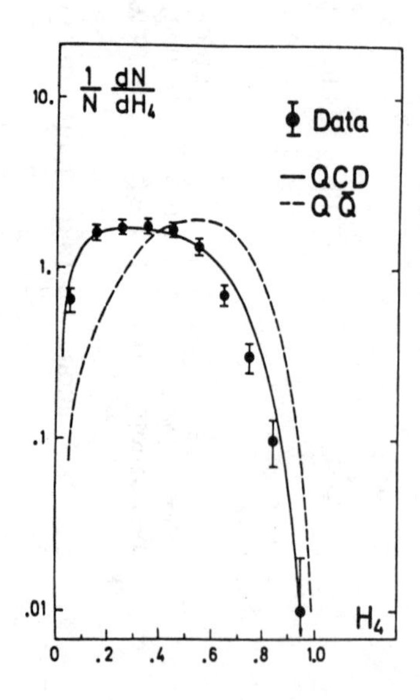

Fig. 21. The Fox-Wolfram moment H_4 distribution.

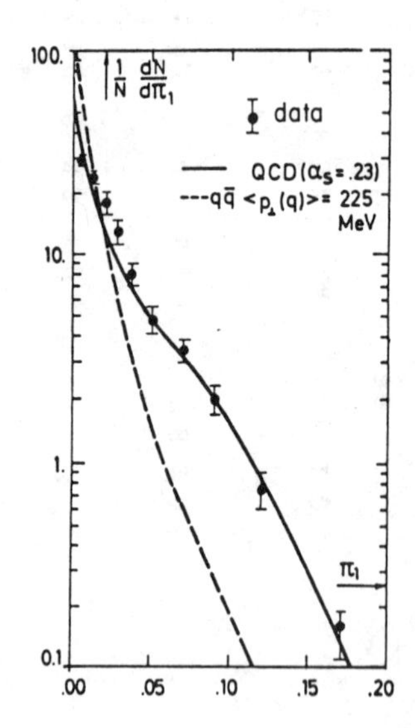

Fig. 22. The Fox-Wolfram moment π_1 distribution.

of hadronic events containing muons. The table demonstrates once again the absence of toponium up to 35.8 GeV. For $\sqrt{s} \geq 30$ GeV the observed rate of 3.75 ± 0.57% agrees with the Monte Carlo predictions for five quark flavors, but is approximately 5 standard deviations away from the prediction which includes the top quark.

The relative production rate of hadronic events containing muons as function of energy is shown in Figure 23a, together with the predictions based on the production of u, d, s, c and b quarks (slashed curve) and the production of top quarks (solid curve). We see the data agree with the five quark model and disagree with the six quark hypothesis. The predicted rate based on a theoretical model which assumes there are only five quarks in nature and the b quark decays only semi-leptonically is shown as -x- in Figure 23a. We see the data are lower than the prediction.

Figure 23b shows the thrust distribution of the hadronic events containing muons compared to a QCD calculation containing five quark flavors. There is very good agreement between the data and the Monte Carlo prediction. The scarcity of events at low thrust in the figure also rules out the existence of the top quark.

d) Discovery of Three Jet Events

In this section we review the detailed topological analysis which was used by the MARK J to unambiguously isolate the 3-jet events arising from the emission of hard non-collinear gluons[21]. Examination of the azimuthal distribution of energy around the thrust axis was used to obtain a sample of planar events. An analysis of the spatial distribution of the energy flow for the planar event sample established the underlying 3-jet structure in agreement with the QCD predictions for $e^+e^- \to q\bar{q}g$.

In the reaction $e^+ + e^- \to$ hadrons, the final states have many appearances: spherical, 2-jet like, 3-jet like, 4-jet like, etc. Events which fall into each of these visual categories can be produced by a variety of underlying processes:

$e^+ + e^- \to$ a phase space like distribution of hadrons

$e^+ + e^- \to q + \bar{q}$ (quark $\langle P_t \rangle = 200 - 500$ MeV)

$e^+ + e^- \to q + \bar{q} +$ gluon

and $e^+ + e^- \to q + \bar{q} + 2$ gluons

These alternatives make any conclusions which may be drawn from the jet-like appearance of individual events of little value in distinguishing between the models, nor can such an appearance provide information about the nature of the basic final-state constituents. Neutral particles carry a large fraction of the total energy. When the statistics are limited it is important to measure both charged and neutral particles. For a consistent analysis, one must collect a statistically significant number of events in a given kinematic

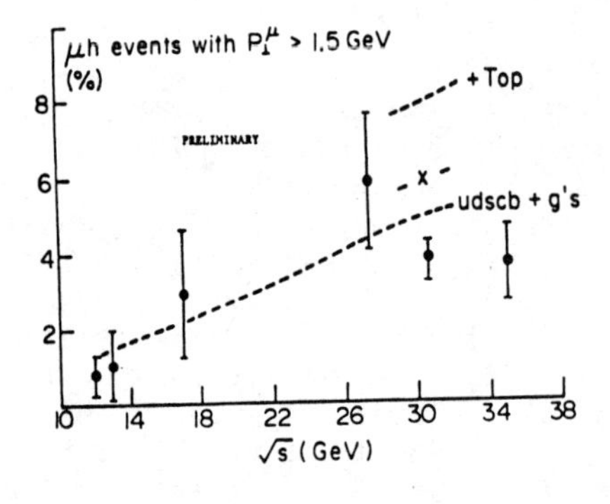

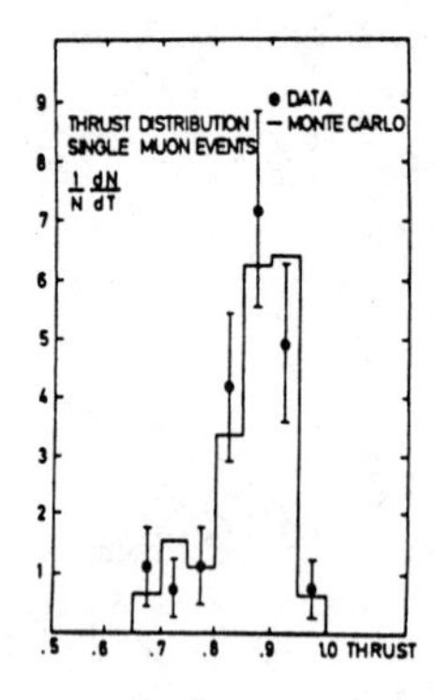

Fig. 23. a) The relative production rate of hadronic events containing muons as function of energy. The slashed curve is the prediction based on the production of five types of quarks. The solid curve includes the top quark as well. The -x- is calculated assuming that the b quark decays only semi-leptonically.
b) The thrust distribution of high energy hadron events ($\sqrt{s}$ > 27 GeV) containing at least one muon. The data (solid points) are compared to the prediction of QCD with five quark flavors.

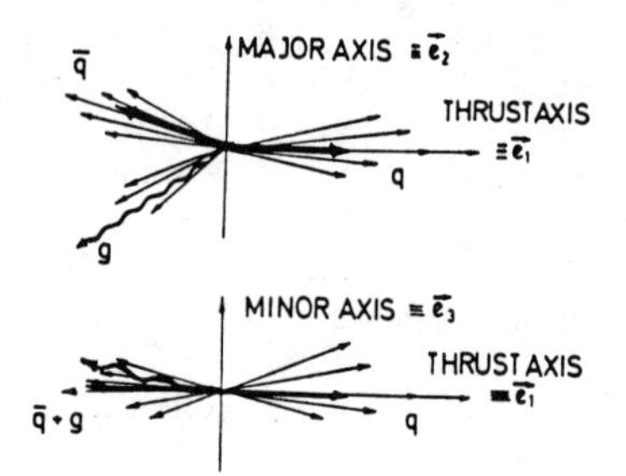

Fig. 24. A schematic view of the process $e^+e^- \to q\bar{q}g$, and the three resulting hadron jets showing the axes used to describe the event.

region and compare the number of events in the region with specific model predictions on a statistical basis. A meaningful comparison with models must take into account the uncertainties in the models such as the quark P_t distributions, fragmentation functions, etc. Before conclusions can be drawn, background contributions from other processes must be understood and kept small[22].

In order to exclude events where leading particles have escaped down the beam pipe, or where part of a broad jet is missed, we select only those events for which $E_{vis} > 0.7\sqrt{s}$. This cut also eliminates two-photon events and events where a hard photon is emitted in the initial state. The drift tubes enable us to separate more distinctly the distribution of charged particles from neutrals. Since neutral particles carry away a large portion of the total energy, they will not only affect the axes of the jets, but will also affect the identification of individual jets.

The jet analysis of hadronic events and the search for 3-jet events is based on a determination of the three dimensional spatial distribution of energy deposited in the detector. This method is quite different from the pioneering method used by the PLUTO and TASSO groups[23]. The characteristic features of hard non-collinear gluon emission in $e^+e^- \to q\bar{q}g$ are illustrated in Figure 24. Because of momentum conservation the momenta of the three particles have to be coplanar. For events where the gluon is sufficiently energetic, and at large angles with respect to both the quark and antiquark, the observed hadron jets also tend to be in a recognizable plane. This is shown in the upper part of the figure where a view down onto the event plane shows three distinct jets; distinct because the fragmentation products of the quarks and gluons have limited P_t with respect to the original directions of the partons. The lower part of the figure shows a view looking towards the edge of the event plane, which results in an apparent 2-jet structure. Figure 23 thus demonstrates that hard non-collinear gluon emission is characterized by planar events which may be used to reveal a 3-jet structure once the event plane is determined.

The spatial energy distribution is described in terms of three orthogonal axes called the thrust, major and minor axes. The axes and the projected energy flow along each axis T_{thrust}, F_{major} and F_{minor} are determined as follows:

(1) The thrust axis, $\vec{e}_1$, is defined as the direction along which the projected energy flow is maximized. The thrust, T_{thrust} and $\vec{e}_1$ are given by

$$T_{thrust} = \max \Sigma_i \frac{|\vec{E}^i \cdot \vec{e}_1|}{\Sigma_i \vec{E}^i}$$

where $\vec{E}^i$ is the energy flow detected by a counter as described above and $\Sigma_i \vec{E}^i$ is the total visible energy of the event (E_{vis})

(2) To investigate the energy distribution in the plane perpendicular to the thrust axis, a second direction, $\vec{e}_2$, is defined perpendicular to $\vec{e}_1$. It is the direction along which the projected energy flow in that plane is maximized. The quantity F_{major} and $\vec{e}_2$ are given by

$$F_{major} = \max \Sigma_i \frac{|\vec{E}^i \cdot \vec{e}_2|}{E_{vis}} \; ; \; \vec{e}_2 \, \vec{e}_1$$

(3) The third axis, $\vec{e}_3$ is orthogonal to both the thrust and the major axes. It is found that the absolute sum of the projected energy flow along this direction, called F_{minor}, is very close to the minimum of the projected energy flow along any axis, i.e.,

$$F_{minor} = \frac{\Sigma_i |\vec{E}^i \cdot \vec{e}_3|}{E_{vis}} \sim \min \frac{\Sigma_i |\vec{E}^i \cdot \vec{e}|}{E_{vis}}$$

If hadrons were produced according to phase-space or a $q\bar{q}$ two-jet distribution, then the energy distribution in the plane as defined by the major and minor axes would be isotropic, and the difference between F_{major} and F_{minor} would be small. Alternatively, if hadrons were produced via three-body intermediate states such as $q\bar{q}g$, and if each of the three bodies fragements into a jet of particles with $\langle P_t^h \rangle \sim 325$ MeV, the energy distribution of these events would be oblate (P_t^h refers to the final state hadrons). Following the suggestion of H. Georgi, the quantity oblateness, O, is defined as

$$O = F_{major} - F_{minor}$$

The oblateness is $\sim 2\, P_t^{gluon} \sqrt{s}$ for three-jet final states and is approximately zero for final states coming from a two-jet distribution.

Figure 25a shows the event distribution as a function of oblateness for the data at $\sqrt{s} = 17$ GeV where the gluon emission effect is expected to be small. The data indeed agree with both models, although the prediction with gluons is still preferred.

Figure 25b shows the event distribution as a function of oblateness for part of the data at $27.4 \leq \sqrt{s} \leq 31.6$ GeV as

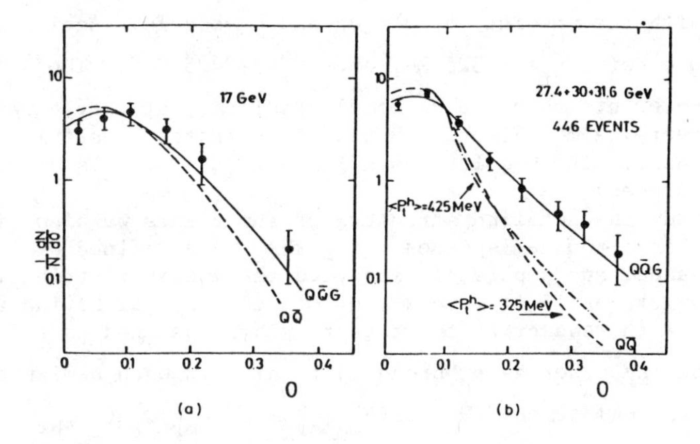

Fig. 25. Differential oblateness distribution at a) $\sqrt{s}$ = 27 GeV and b) at high energies (combined data) compared to the predictions of QCD (solid line) and quark model (dashed lines).

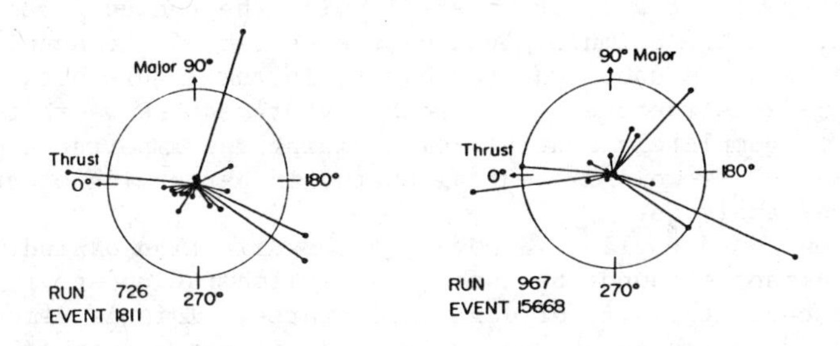

Fig. 26. Energy flow diagrams for two high energy hadronic events viewed in the major-thrust plane.

compared with the predictions of $q\bar{q}g$ and $q\bar{q}$ models. Again, in the $q\bar{q}$ model we use both $\langle P_t^h \rangle = 325$ MeV and $\langle P_t^h \rangle = 425$ MeV. The data have more oblate events than the $q\bar{q}$ model predicts, but they agree with the $q\bar{q}g$ model very well. Figure 25b also illustrates a useful feature of the oblateness: it is quite insensitive to the details of the fragmentation process.

To study the detailed structure of the events we also divided each event into two hemispheres using the plane defined by the major and minor axes, and separately analyzed the energy distribution in each hemisphere as if it were a single jet. The jet having the smaller P_t with respect to the thrust axis is defined as the "narrow" jet (n) and the other as a "broad" jet (b). In each hemisphere we calculate the oblateness, $O_n = 2(F^n_{major} - F^n_{minor})$, and $O_b = 2(F^b_{major} - F^b_{minor})$, and thrusts T_n and T_b.

One approach to analyzing the flat events for a possible 3-jet structure is illustrated by Figure 26. The figure shows the energy flow diagram for each of two high energy hadronic events, viewed in the event plane determined by the major and thrust axes. The energy flow diagram is a polar coordinate plot in which we summed the energy vectors E^i in 10° intervals. Each point in the plot represents the summed energy in an angular interval, with the radius given by the magnitude, and the azimuth given by the center of the angular interval in the event plane. The two events in the figure both show an apparent 3-jet structure. The second event also shows that one of the jets is completely neutral, emphasizing the importance of measuring charged as well as neutral energy in performing a consistent topological analysis.

As mentioned earlier, however, the examination of individual event appearances cannot be used to establish the underlying 3-jet structure characteristic of $q\bar{q}g$ final states. This is demonstrated by Figure 27, which shows two low thrust, planar events at 12 GeV center of mass energy. The events also show a distinct multi-jet structure. It should be noted that all the measured distributions at low energies (thrust, oblateness, etc.) are well described by a simple $q\bar{q}$ model, so that the suggestive event appearances are unrelated to gluon emission, but are dominated by fluctuations in the quark fragmentation process. The views of the events in the minor-thrust plane (looking at the edge of the event plane) also show that the events are planar.

At the 1979 International Symposium on Lepton and Photon Interactions at Fermilab[24], all DESY groups (JADE, MARK J, PLUTO and TASSO) reported evidence for hard gluon emission. The first statistically relevant results, establishing the 3-jet pattern from $q\bar{q}g$ of a sample of hadronic events were presented by the MARK J collaboration. These are shown in Figure 28 in which a sample of the events with low thrust and high oblateness, where the gluon-emission effect is expected to be relatively large, is selected for detailed

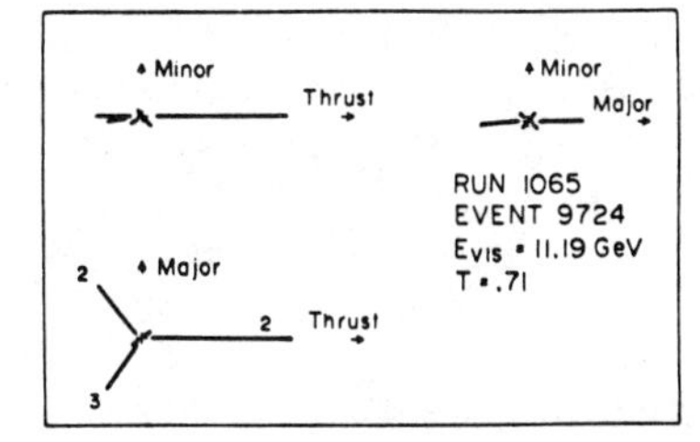

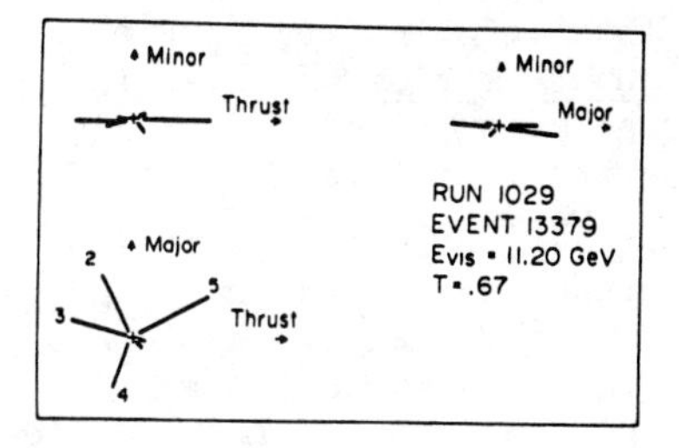

Fig. 27. Two events measured at $\sqrt{s}$ = 12 GeV with a) T = 0.71, b) T = 0.67. The lines show direction and magnitude of energy deposited in the calorimeter displayed in three projections. The events appear to have a multi-jet structure in the thrust-major plane. The view in the thrust-minor plane shows the events are flat. The top event is a 3-jet event. The bottom event is a planar multi-jet event. The numbers associated with the tracks in the Major-Thrust plane are the numbers of charged tracks in the jets.

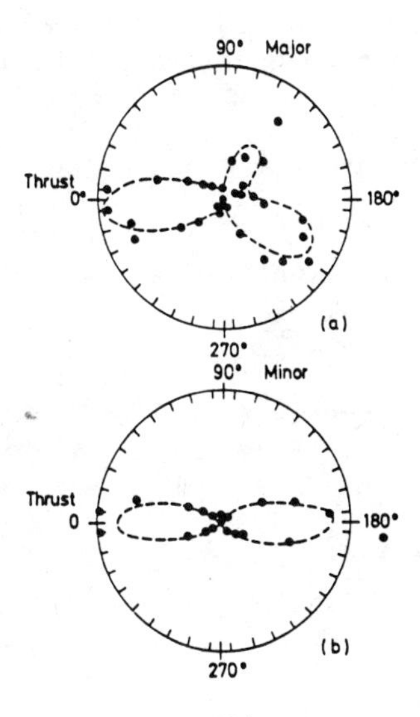

Fig. 28. a) Energy distribution in the plane defined by the thrust and major axes for all events with T < 0.8 and O_b > 0.1 at $\sqrt{s}$ = 27.4, 30 and 31.6 GeV obtained up to the time of the 1979 Fermilab Conference. The radial distance of the data points is proportional to the energy deposited in a 10° bin. The superimposed dashed line represents the distribution predicted by QCD. b) Measured and predicted energy distribution in the plane defined by the thrust and minor axes, which shows only 2-jets.

examination. The key feature of this figure is that it consists of the superposition of an entire event sample, and thus displays the average behavior of the energy flow for planar events at high energy. The event sample is composed of 40 events with $T < 0.8$ and $O_b > 0.1$ out of 446 hadronic events obtained up to the time of the Fermilab Conference in the energy range of 27-31.6 GeV. Each of the 40 events has been examined by physicists and found to be 3-jet in appearance like those shown in Figure 30. In superposing events the narrow jet of each event points in the direction of the thrust axis, and the other two jets are oriented in the other hemisphere according to their sizes.

The calculated Monte Carlo model predictions in the figure are compatible with the data with $\chi^2 = 67$ for 70 degrees of freedom. The accumulated energy distribution in the left part of the figure, showing the view in the plane defined by the thrust and minor axes, exhibits a flat distribution consistent with the model predictions.

These results can be contrasted to those obtained with a simple phase space model. When viewed in the major-thrust plane, phase-space shows three nearly identical lobes due to the method of selection used. However, at $\sqrt{s} = 30$ GeV these lobes are different in appearance from the jets shown in Figure 28. In general, one expects the three jets from $q\bar{q}g$ to become slimmer and easier to distinguish from the phase-space distribution as the center-of-mass energy increases. Using a χ^2 fit of the phase-space energy distribution to the data we found that $\chi^2 = 222$ for 70 degrees of freedom. Therefore, phase-space is inconsistent with the data. Furthermore, large contributions of phase-space distributions are ruled out by our thrust distributions as shown in Figures 17-19.

For the analyses performed in the earlier part of 1979, as shown in Figures 25 and 28, we used a Monte Carlo model implemented by P. Hoyer et al.[25]. In the analysis discussed below, the Monte Carlo of Ali et al.[2] was adopted. This model incorporates higher order QCD effects, the q^2 evolution of the quark and gluon fragmentation functions and the weak decays of heavy quarks. The QCD predictions presented in Figues 25 and 28 are not noticeably different for the two models.

Figures 29 and 30 show the event distribution as a function of O_n and O_b, compared to the predictions of the QCD model[26] and of two quark-parton models with quark $\langle P_t \rangle = 300$ MeV and 500 MeV respectively[27]. Figure 29 shows that the narrow jet distribution agrees with the various models indicating that it comes from a single quark jet. Figure 30, however, shows that the quark-parton models severely underestimate the number of events with $O_b > 0.3$ while the QCD model correctly predicts the observed distribution.

The T_n and T_b thrust distributions of the flat events in the

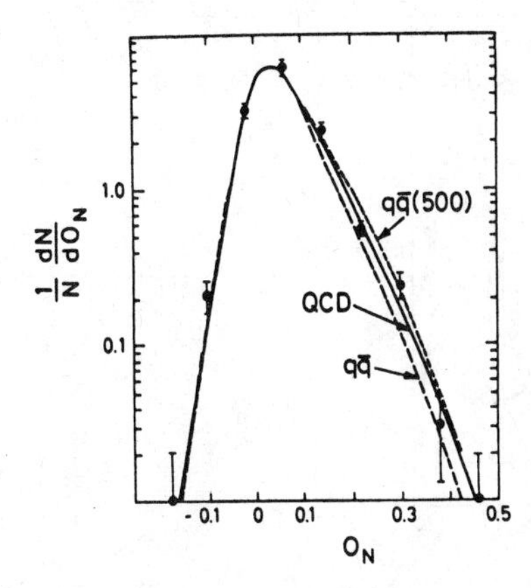

Fig. 29. The narrow jet oblateness distribution $1/N \ dN/dO_N$ for all hadron events with measured energy $E_{vis} \geq 0.7\sqrt{s}$. The data are compared to the predictions of the QCD model and to two quark-antiquark models with $<P_t> = 300$ MeV and $<P_t> = 500$ MeV respectively.

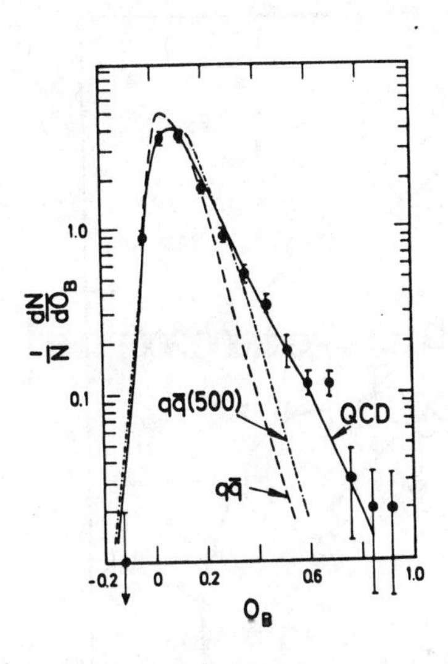

Fig. 30. The broad jet oblateness distribution $1/N \ dN/dO_b$ under the same condition as Figure 29.

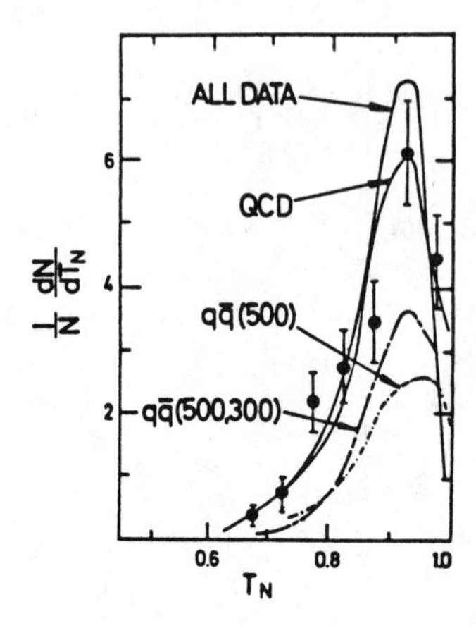

Fig. 31. The thrust distribution of the narrow jet events with $0_b \geq 0.3$. Also shown are the various model predictions including a flattened $q\bar{q}$ (500, 300) discussed in the text. Note the narrow jet thrust distribution is consistent with thrust distribution of all the hadron events labelled "ALL DATA." This curve is normalized to the data of Figure 18.

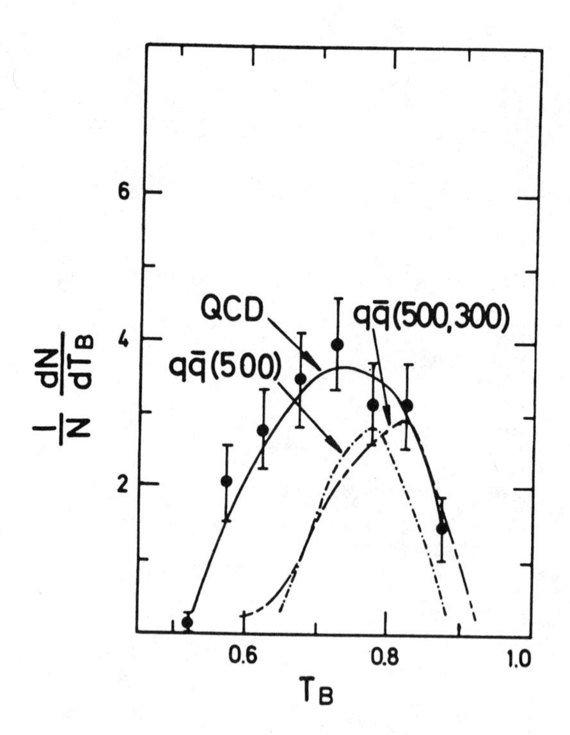

Fig. 32. The thrust distribution of the broad jets for events with $0_b > 0.3$. The curves are discussed in the text.

region $O_b > 0.3$ are shown in Figures 31 and 32 along with the QCD and high P_t $q\bar{q}$ models. The observed distribution is also compared to a "flattened" $q\bar{q}$ model in which the quarks have $<P_t> = 500$ MeV in the thrust-major plane and $<P_t> = 300$ MeV in the thrust-minor plane. The T_n distribution in Figure 31 is in good agreement with the QCD model predictions and has the same general shape as the thrust distribution for high energy $e^+e^- \rightarrow q\bar{q}$ reactions shown in Figures 17 and 18. As expected the T_b distribution, however, is much broader than that of T_n and agrees only with the QCD predictions.

The distributions in Figures 31 and 32 demonstrate that the relative yield of flat events, and the shape of these events as measured by T_n and T_b, can only be explained by the QCD model. The distributions T_n and T_b further exclude phase space (which peaks at lower thrust values) as well as qq models as possible explanations of the energy flow plots.

The development of the 3-jet structure with decreasing thrust and increasing oblateness, as predicted by QCD[28], is shown in a series of energy flow diagrams in Figures 33-35.

As seen in Figure 33 the events at high thrust values are dominated by a 2-jet structure characteristic of $e^+ + e^- \rightarrow q + \bar{q}$. In Figure 34, where the thrust is lower, we begin to see the appearance of the gluon jet; and in Figure 35 the 3-jet events are predominant. It is important to note that in all three cases the data agree with the QCD model prediction, showing the increased incidence of hard non-collinear gluon emission with decreasing thrust and increasing oblateness. The energy flow in the minor-thrust plane contains only two nearly identical lobes similar to the narrow jet in Figure 28b, in good agreement with QCD predictions. In Figure 36 we unfold the energy flow diagram of Figure 35 to see more clearly the comparison of the data with the predictions of QCD, $q\bar{q}(<P_t> = 500$ MeV), and a "mixed model" consisting of a combination of $q\bar{q}$ and phase-space contributions. All models in Figure 35 are normalized to have the same areas (as the data) before the individual cuts are imposed. The normalization for the mixed model was determined by adjusting the $q\bar{q}(<P_t> = 300$ MeV) and phase-space contributions to agree with the measured thrust distribution as illustrated in Figure 37. As seen in Figure 36, only QCD can describe the observed 3-jet structure. The conclusions obtained by the JADE, PLUTO and TASSO collaborations on their tests of QCD and studies of multi-jet events are in good agreement with us (see Section 2.4).

A very powerful method to eliminate phase-space as a source of the 3-jet event and in fact eventually to study the nature of the individual jets is shown in Figure 38. Here all the data with $27 < E < 35.8$ GeV are included. The thrust distributions of each of the jets, separated according to their position in the energy flow

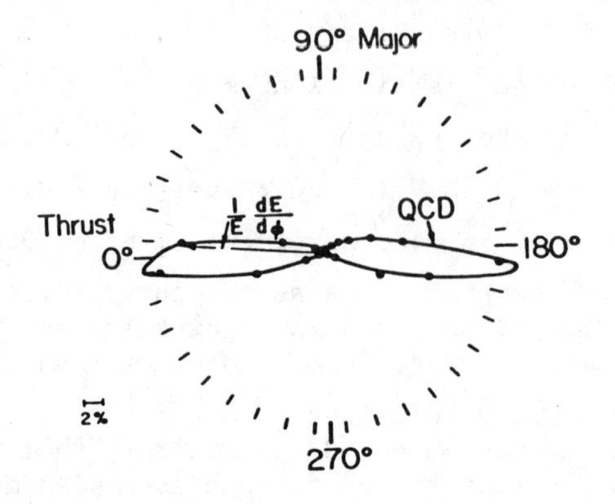

Fig. 33. Energy flow diagram in the thrust-major plane for high energy data (27–31.6 GeV) with T > 0.9. The solid line is the prediction of QCD.

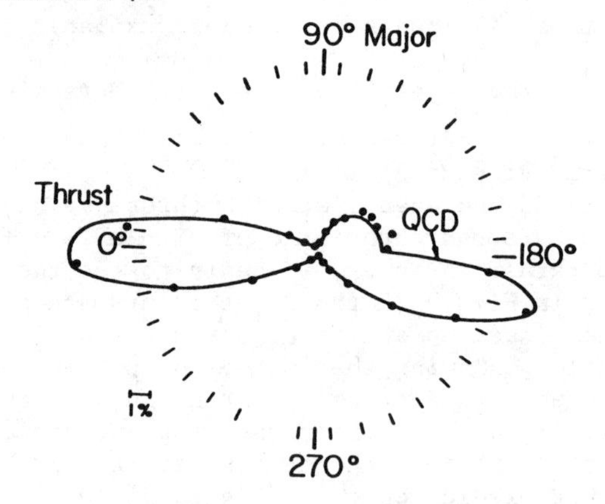

Fig. 34. Same as Figure 33 for events with $0.8 \leq T \leq 0.9$ and broad jet oblateness $O_b > 0.1$.

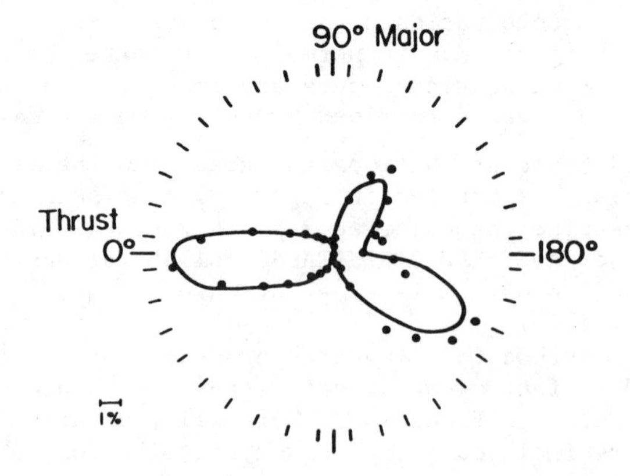

Fig. 35. Same as Figure 33 for events with $T < 0.8$ and $O_b > 0.1$.

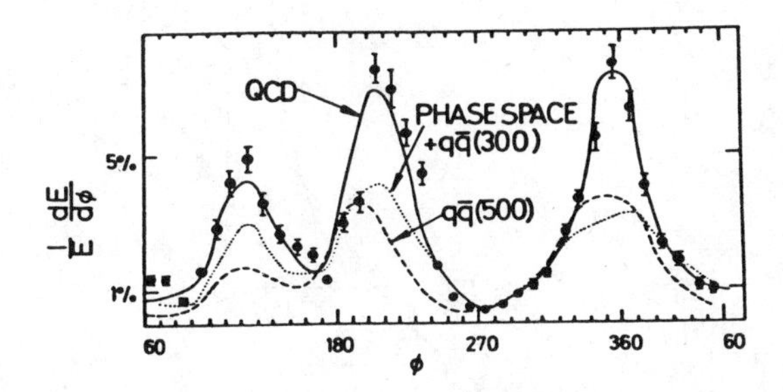

Fig. 36. The unfolded energy flow diagram of Figure 35 as compared to QCD, the quark model ($\langle P_t \rangle$ = 500 MeV) and a mixed $q\bar{q}$ and phase space model (see text).

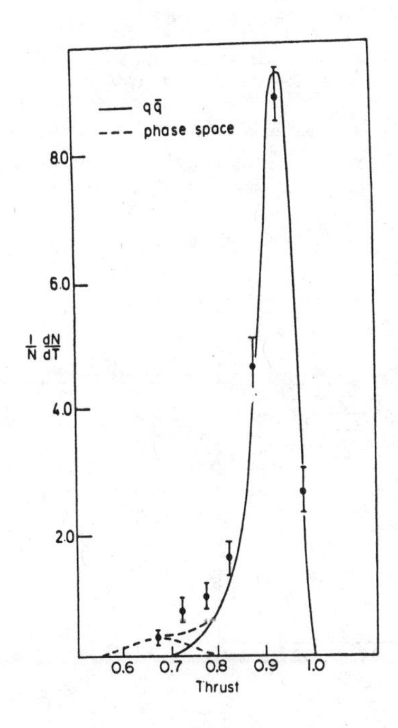

Fig. 37. Illustration of the method used in determining the maximum permissible admisture of phase-space to $q\bar{q}$ in the mixed model shown in Figure 36.

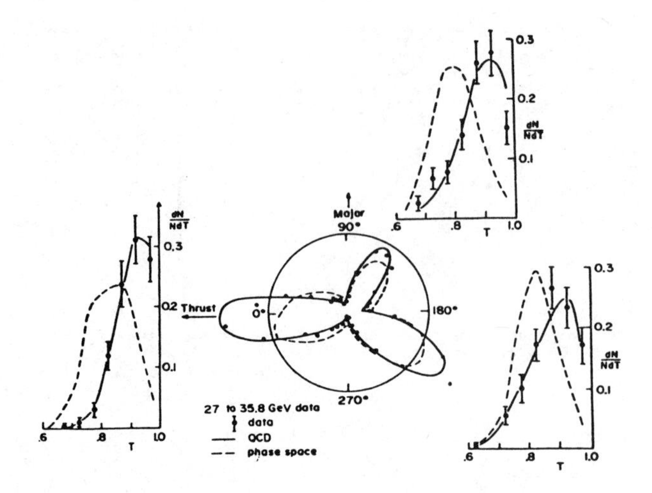

Fig. 38. Thrust distribution 1/N dN/dT for each individual jet in the 3-jet sample of Figure 35 which were selected by using $O_b > 0.1$ and thrust > 0.8. The corresponding distribution normalized to the same total number of 3-jet events are also shown for QCD (solid curve) and phase-space model (dashed curve).

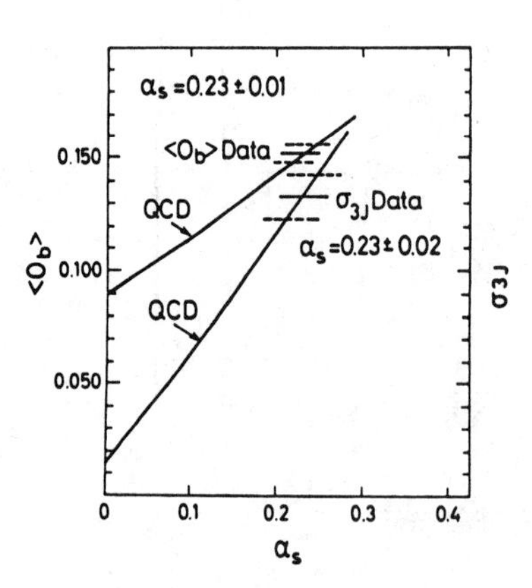

Fig. 39. The left graph: The average balue of oblateness $\langle O_b \rangle$ for all events with $E_{vis} \geq 0.7\sqrt{s}$ as a function of α_s, computed by varying α_s in the QCD model. The right graph: The fraction of hadronic events with $O_b > 0.3$ (σ_{3j}) as a function of α_s computed by varying α_s in the QCD model.

figure, are compared to the expectations of QCD and phase-space. The thrust distributions agree well with the QCD prediction but are in strong disagreement with phase space predictions.

e) Determination of the Strong Coupling Constant α_s

Recent experiments on scaling violations in lepton inelastic scattering[29], on high P_t events in dilepton production by hadrons[30], and multi-jet events in e^+e^- annihilations[18,21,31] all indicate that the results are explained naturally in the quantum chromodynamics (QCD) theory of the strong interactions of quarks and gluons[2]. The strong coupling constant $\alpha_s(q^2)$ between quarks and gluons has been measured indirectly in quarkonium bound states[32], and in deep inelastic experiments[29]. At PETRA, where the q^2 is much larger, computations are expected to be more reliable. In addition, high energy e^+e^- annihilations offer a more direct way of measuring α_s and testing perturbative QCD because it is expected to give rise to multijets which can be systematically identified.

The 3-jet events discussed in the previous section, which consist of $q\bar{q}g$ fragmentation products with relatively small backgrounds from fluctuations of phase-space-like processes or quark-antiquark intermediate states, allow us to make further comparisons of the event properties with the predictions of QCD. In particular the relative yield of 3-jet events and the shape distribution gives a way to measure directly α_s, the strong coupling constant.

We used several methods in determining the strong coupling constant α_s, including:

1) the average oblateness $\langle O_b \rangle$,

2) the fraction of events with $O_b > 0.3$,

3) the relative yield of events with $O_b - O_n > 0.3$ where O_n is constrained to be greater than zero.

For each quantity we allowed α_s to vary in the QCD model, and we then determined the range of α_s values for which the QCD model predictions agree with the data within errors. In particular, the samples obtained using criteria 2) and 3) consist predominantly of 3-jet events from $e^+e^- \rightarrow q\bar{q}g$, in which the gluon emitted is both very energetic and at a large angle with respect to both the quark and antiquark. This leads to an event sample where the number of events in the sample is a quasi-linear function of α_s, and in which the influence of non-perturbative effects which are not calculable in QCD is minimal. For criterion 2), for example, we observed 161 events, which matches the QCD model with $\alpha_s = 0.23$. The $e^+e^- \rightarrow q\bar{q}$ contribution is calculated to be 21 events. The predominance of $q\bar{q}g$ in a sample with $O_b > 0.3$ is maintained even if $\langle P_t \rangle$ is allowed to vary from 225 MeV to 500 MeV in the model. With $\langle P_t \rangle$ 500 MeV the $e^+e^- \rightarrow q\bar{q}$ contribution is calculated to be 58 events.

The methods described above yield a self-consistent set of α_s values, as illustrated in Figure 39. On the basis of the results of the three methods we obtain

$$\alpha_s = 0.23 \pm 0.02 \text{ (statistical error)} \pm 0.04 \text{ (systematic error)}$$

The large systematic error was mostly due to uncertainties in QCD calculations[2]. For method 2) the range of α_s due to variation in $\langle P_t \rangle$ from 225 to 400 MeV is $\pm$ 0.01 and the change in α_s due to different cuts in O_b from $O_b > 0.3$ to $O_b > 0.15$ or cuts in O_n from no cuts to $O_n < 0.1$, is -0.01. For method 2), changing the fragmentation function $zD(z)$ to $1-z$ for u, d and s quarks and $zD(z)$ to z for c and b quarks does not change the α_s value noticeably. Tables V and VI show in detail the change of α_s with respect to the O_b cuts and O_n cuts.

Our value of α_s is consistent with the values of the JADE group obtained with a different Monte Carlo program (see Section 2.4). The value of α_s[33,34] is in qualitative agreement with the values obtained in deep inelastic lepton nucleon scattering experiments[29], and in the analysis of the quarkonium states[32]. However, detailed comparison among these results cannot yet be made without accurate higher order QCD calculations.

2.4 Comparison With Other Experiments at PETRA

Our detector uses calorimetric techniques which are very different from other PETRA groups (JADE, PLUTO and TASSO) which use solenoidal magnetic field followed by particle identification devices. These differences in technique imply that the event selection criteria, Monte Carlo analysis programs, and the assignment of systematic errors are quite different. However, despite the different techniques used the results of all the PETRA groups are complementary and supportive of each other in their physics conclusions on the test of QED[35], on the measurement of R, the search for new flavors[31,36] and of the effects of hard gluon jets[31,37].

3) CONCLUSIONS

In the first one and one-half years of experimentation with a

TABLE V

Value of α_s with different O_b cuts, without cuts in O_n and $\langle P_t \rangle = 247$ MeV.

$O_b >$	α_s
0.15	0.22 ± 0.02
0.20	0.22 ± 0.02
0.25	0.22 ± 0.02
0.30	0.23 ± 0.02

TABLE VI

Value of α_s with different O_n cuts, with $O_b > 0.3$ and $<P_t> = 247$ MeV.

$O_n <$	α_s
No cuts	0.23 ± 0.02
0.24	0.23 ± 0.02
0.20	0.23 ± 0.02
0.16	0.23 ± 0.02
0.12	0.22 ± 0.03
0.08	0.22 ± 0.03

simple detector, we have obtained the following results:

1) We have established the validity of quantum electrodynamics to a distance $< 2 \times 10^{-16}$ cm. Quarks, electrons, muons and tau leptons are point-like with sizes smaller than 2×10^{-16} cm.
2) The relative cross sections and event distributions show that there is no new charge 2/3 quark pair production up to $\sqrt{s} = 35.8$ GeV.
3) The energy flow of events at high energies is in good agreement with quantum chromodynamics. The quantity of flat events and their distributions disagree with the simple quark antiquark model prediction.
4) We have discovered 3-jet events; the rate of their production and their distribution agree with the prediction of QCD.
5) We have measured the strong interaction coupling constant α_s.

There are two reasons which made it possible to obtain these results:

1) In e^+e^- collisions, the signal is clear and unique. Every event has a definite physical interpretation and can be analyzed in terms of QED or QCD. This is quite different from our previous experience with proton proton collisions where the signal of the virtual photon events is less than one part in 10^6 of the background.

2) PETRA was reliably constructed and available for use by experimentalists from the beginning.

REFERENCES

1. U. Becker et al., "A Simple Detector to Measure e^+e^- Reactions at High Energy," Proposal to PETRA Research Committee (March, 1976).
2. D. J. Gross and F. A. Wilczek, Phys. Rev. Lett. 30, 1343 (1973); H. D. Politzer, ibid. 30, 1346 (1973). J. Ellis et al., Nucl. Phys. B111, 253 (1976). T. deGrand et al., Phys. Lett. D16, 3251 (1977); G. Kramer et al., ibid. 79B, 249 (1978). A. DeRujula et al., Nucl. Phys. B138, 387 (1978). P. Hoyer et al., DESY Preprint 79/21 (unpublished). A. Ali et al., Phys. Lett. 82B, 285 (1979); DESY Report 79/54, submitted to Nucl. Phys. B.
3. P. D. Luckey et al., in Proceedings of the International Symposium on Electron and Photon Interactions at High Energies, Hamburg (Springer, Berlin, 1965) Vol. II, p. 397.
4. V. Alles-Borelli et al., Nuovo Cimento 7A, 345 (1972). H. Newman et al., Phys. Rev. Lett. 32, 483 (1974); J-E. Augustin et al., ibid. 34, 233 (1975); L. H. O'Neill et al., ibid. 37, 395 (1976).
5. D. P. Barber et al., Phys. Rev. Lett. 42, 1110 (1979).
6. S. J. Brodsky and S. D. Drell, Annu. Rev. Nucl. Sci. 20, 147 (1970).
7. ADONE Proposal INFN/AE-67/3, (March 1967), ADONE-Frascati (unpublished) and M. Bernardini et al., (Zichichi Group), Nuovo Cimento 17A, 383 (1973). S. Orito et al., Phys. Lett. 48B, 165 (1974).
8. M. Perl et al., Phys. Rev. Lett. 35, 1489 (1975); G. Feldman et al., ibid. 38, 117 (1977).
9. J. Burmester et al., Phys. Lett. 68B, 297 (1977); 68B, 301 (1977).
10. For a review of our present knowledge of the τ lepton, see Guenter Fluegge, Zeitschr. f. Physik C1, Particles and Fields, 121 (1979), and the references therein.
11. R. Hofstadter, in Proceedings of the 1975 International Symposium on Lepton and Photon Interactions at High Energies (Stanford Linear Accelerator Center, Stanford, California, 1975) p. 869.
12. S. D. Drell, Ann. Phys. (N.Y.) 4, 75 (1958). T. D. Lee and G. C. Wick, Phys. Rev. D2, 1033 (1970).
13. Y. S. Tsai, Phys. Rev. D4, 2821 (1971). H. B. Thacker and J. J. Sakurai, Phys. Lett. 36B, 103 (1971). K. Fukijawa and N. Kawamoto, Phys. Rev. D14, 59 (1976). See also Reference 28.
14. D. P. Barber et al., Phys. Rev. Lett. 43, 1915 (1979).
15. H. Terazawa, Rev. Mod. Phys. 45, 615 (1973). We used $\sigma(\gamma\gamma \to \text{multipion}) = 240\text{-}270/W^2$ nb. W is the energy of the two-photon system.
16. C. Quigg, Fermilab Conference 79/74-THY, September 1979. J. L. Rosner, C. Quigg, H. B. Thacker, Phys. Lett. 74B, 350 (1978). C. Quigg, contribution to the 1979 International Symposium on Lepton and Photon Interactions, Fermilab.

17. T. Applequist and H. Georgi, Phys. Rev. D8, 4000 (1973); A. Zee, ibid. D8, 4038 (1973).
18. D. P. Barber et al., Phys. Rev. Lett. 42, 1113 (1979); 43, 901 (1979). D. P. Barber et al., Phys. Lett. 85B, 463 (1979). For a theoretical discussion on the use of thrust variables see: E. Farhi, Phys. Rev. Lett. 39, 1587 (1977). S. Brandt et al., Phys. Lett. 12, 57 (1964). S. Brandt and H. Dahmen, Zeitschr. f. Phys. C1, 61 (1979).
19. R. Schwitters, in Proceedings of the 1975 International Symposium of Lepton and Photon Interactions at High Energies (Stanford Linear Accelerator Center, Stanford, Calif., 1975) p. 5. G. Hansen et al., Phys. Rev. Lett. 35, 1609 (1975). G. Hansen, Talk at the 13th Recontre de Moriond, Les Arcs, France (March 12-26, 1978), SLAC-PUB 2118 (1978).
20. G. C. Fox and S. Wolfram, Phys. Rev. Lett. 41, 1581 (1978). G. C. Fox and S. Wolfram, Nucl. Phys. B413, (1979). G. C. Fox and F. Wolfram, Phys. Lett. 82B, 134 (1979).
21. D. P. Barber et al., Phys. Rev. Lett. 43, 830 (1979).
22. Ch. Berger et al., Phys. Lett. 86B, 418 (1979), have presented a systematic study in the selection of 3-jet events from $q\bar{q}$ and 3-jet events from $q\bar{q}g$ (see Tables 1 and 2 of their paper).
23. Ch. Berger et al., Phys. Lett. 76B, 176 (1978). S. Brandt and H. Dahmen, Zeitschr. f. Phys. C1, 61 (1979). B. H. Wiik, in Proc. Intern. Neutrino Conf. (Bergen, Norway, June 1979). P. Soeding, European Physical Society Conference Report, July (1979). This report includes five events of 3-jets of charged particles. Each event has a measured energy $\sim 1/2$ of the total energy. Their prediction for $q\bar{q}g$ was 9 events with an unstated 3-jet background from $q\bar{q}$. Statistically significant results on planar events have been published (see Table 4) in R. Brandelik et al., DESY-Preprint 79/61 (1979).
24. H. Newman, paper presented at the 1979 Fermilab Conference (to be published in Proceedings). See Reference 26; See also S. Orito (the JADE Collaboration), ibid; Ch. Berger (the PLUTO Collaboration), ibid; G. Wolf (the TASSO Collaboration), ibid. For a summary of DESY work up to that time, see H. Schopper, DESY Report 79/79 (1979).
25. P. Hoyer et al., DESY Report 79/21 (1979).
26. The data in Figures 33 and 34 at $0 < O_b$ or $0 < O_n$ is due to statistical fluctuations, and the energy resolution $\Delta E/E \sim 20\%$ and possible higher order QCD effects.
27. In Figures 34-40 if we choose quark $\langle P_t \rangle < 300$ MeV we observe larger deviations between $q\bar{q}$ model and the data.
28. A. DeRujula et al., Nucl. Phys. B138, 387 (1978).
29. H. L. Anderson et al., Phys. Rev. Lett. 40, 1061 (1978). P. C. Bosetti et al., Nucl. Phys. B142, 1 (1978). J. G. H. deGroot et al., Zeitschr f. Phys. C1, 143 (1979). J. G. H. de Groot et al., Phys. Lett. 82B, 292 (1979); 82B, 456 (1979).

30. D. Antreasyan et al., (to be published).
31. W. Bartel et al., (the JADE Collaboration), Phys. Lett. 88B, 171 (1979); 89B, 136 (1979). Ch. Berger et al., (the PLUTO Collaboration), Phys. Lett. 81B, 410 (1979); 86B, 413 (1979); 86B, 418 (1979); 89B, 120 (1979). R. Brandelik et al., (the TASSO Collaboration), Phys. Lett. 83B, 261 (1979); 86B, 243 (1979); 88B, 199 (1979); 89B, 418 (1980).
32. M. Krammer and H. Krasemann, DESY Report 78/66 (1978).
33. D. P. Barber et al., Phys. Lett. 89B, 139 (1979).
34. It is not yet clear which q^2 should be used to extract Λ to compare quantitatively with other data.
35. S. Orito (the JADE Collaboration), invited talk 1979 Symposium at FNAL and DESY Report 79/64, 1979. Ch. Berger et al., (the PLUTO Collaboration), DESY Report 80/01.
36. Ch. Berger et al., (the PLUTO Collaboration), DESY Report 80/02. W. Bartel et al., (the JADE Collaboration), DESY Report 80/04.
37. W. Bartel et al., (the JADE Collaboration), DESY Report 79/80. Ch. Berger et al., (the PLUTO Collaboration), DESY Report 79/83.

RECENT RESULTS FROM THE TASSO DETECTOR AT PETRA

R. J. Barlow
Oxford University Nuclear Physics Laboratory, Oxford, England

ABSTRACT

The construction and performance of the detector are briefly described. Results are presented on QED reactions: bounds on cutoff parameters, weak-electromagnetic interference, and properties of the τ lepton. From hadronic events we have results on the search for the top quark, on inclusive distributions, on the identification of charged and neutral kaons and of baryons, and the yields and distributions of these particles. We have studied the properties of jets in 3 jet events, and have made a determination of the QCD coupling constant α_s of $0.17 \pm 0.02 \pm 0.03$ (at 30 GeV) independent of preconceived assumptions about the parameters of the jet fragmentation model.

INTRODUCTION

The TASSO detector has been running at PETRA since October 1978, and we now have data on the reaction $e^+e^- \to$ anything at CMS energies between 12.0 and 35.8 GeV. Results have appeared in various publications [1,2,3,4,5,6,7,8,9,10] to which the reader is referred for a complete description of the apparatus, selection procedures, etc.

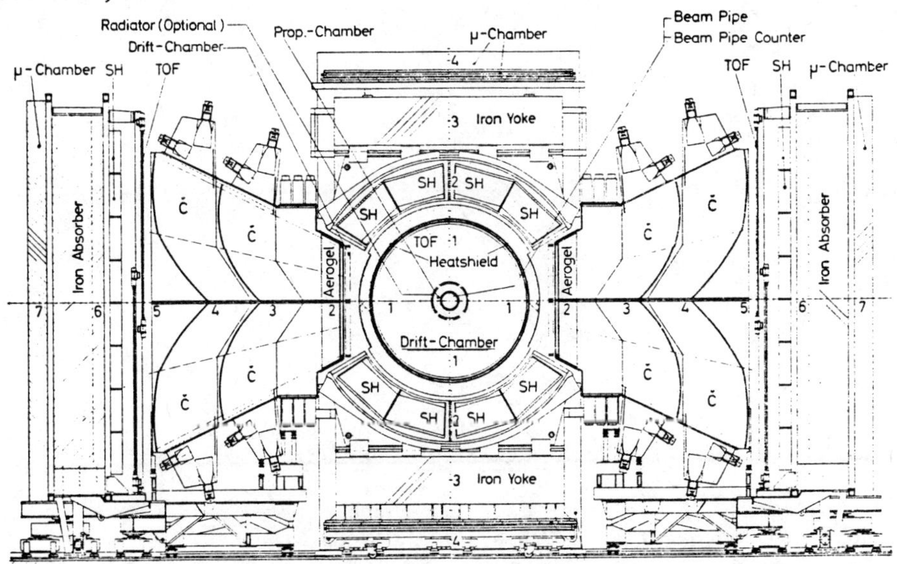

Fig. 1. The TASSO Detector

ISSN:0094-243X/80/620193-14$1.50

THE DETECTOR

Figure 1. shows a transverse cross-section through the TASSO detector. The central section, which covers ∿80% of the solid angle, is surrounded by a solenoid producing a 5 kG magnetic field. A particle produced at the interaction point passes, after leaving the beam pipe, first through a cylindrical proportional chamber, with 4 anode planes, each with 480 wires (along the beam axis) and 8 cathode planes, each of 120 strips (wound helically), then through a larger drift chamber with 15 planes of sense wires, 9 along the axis and 6 at $\sim\pm4^{\circ}$, for a stereo effect, with spatial resolution 220 μ and momentum resolution $\Delta p/p$ of 2%p (GeV). Finally it reaches a layer of scintillator, (48 segments), used for the trigger and for time-of-flight measurement (resolution $\sigma \simeq$ 400p sec) just inside the coil. Figure 2. shows an event in the central detector.

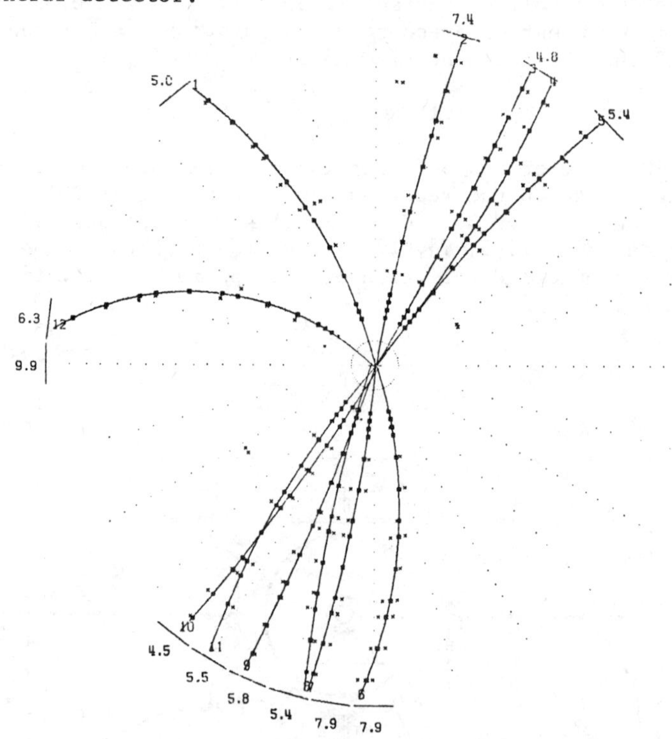

Fig. 2. An Event in the Central Detector

Outside the solenoid coil lie the electromagnetic shower counters. Above the coil, and at the ends, these are of Lead/Liquid Argon design, with a resolution of ∿10%/√E in energy and ∿5mm in position. Below the coil they are (at present) of a Lead/scintillator sandwich type, energy resolution ∿14%, and granularity ∿50cm.

To the sides are the two hadron identification arms that give the detector its name. First 3 Cerenkov counters (Aerogel, Freon 114, CO_2), then another time-of-flight system, similar to the first but with the advantage of a much longer flight path, behind which are shower counters of the lead/scintillator type just described. Finally after an iron filter in the arm, and also around the magnet yoke in the central detector, lie arrays of proportional tubes for muon identification.

Hardware processors attached to the proportional and drift chambers, combined with the inner TOF signals, give triggering information on whether a beam crossing produced tracks in the chambers; the actual trigger combines these outputs with shower counter etc. signals, and the PETRA strobe frequency of 250 KHz is reduced to a usable rate of 1-2 Hz for events read into the computer and written to tape. Dead time from all sources is 6% at 1 Hz.

To remove background cuts are made offline: typically, we require 4 or more reconstructed tracks passing close to the origin, a total charged energy of at least 8 GeV, and that not all tracks lie in the same hemisphere. For a typical PETRA filling, lasting between 2 and 4 hours, we might collect an integrated luminosity of $\sim 20 nb^{-1}$, and of the $\sim$15,000 events read in, $\sim$10 might be good hadronic events.

QED PROCESSES[7]

Cutoff parameters Λ_+ and Λ_- obtained from the 4 processes:

$$e^+e^- \to e^+e^-$$
$$e^+e^- \to \mu^+\mu^-$$
$$e^+e^- \to \tau^+\tau^-$$
$$e^+e^- \to \gamma\gamma$$

are shown in Table 1.

Reaction	Fitted Parameters (GeV^{-2})	lower limits (95 % c.l.)	
		Λ_+ (GeV)	Λ_- (GeV)
$e^+e^- \to e^+e^-$	$1/\Lambda_S^2 = (1.7 \pm 4.1) \cdot 10^{-5}$	108	138
	$1/\Lambda_T^2 = (7.9 \pm 12.6) \cdot 10^{-5}$	59	88
	$1/\Lambda^2 = (1.4 \pm 4.0) \cdot 10^{-5}$ ($\Lambda_S = \Lambda_T$)	112	139
$e^+e^- \to \mu^+\mu^-$	$1/\Lambda_T^2 = (4.2 \pm 6.9) \cdot 10^{-5}$	80	118
$e^+e^- \to \tau^+\tau^-$	$1/\Lambda_T^2 = (2.0 \pm 10.2) \cdot 10^{-5}$	73	82
$e^+e^- \to \gamma\gamma$	-	34	42

Table 1.

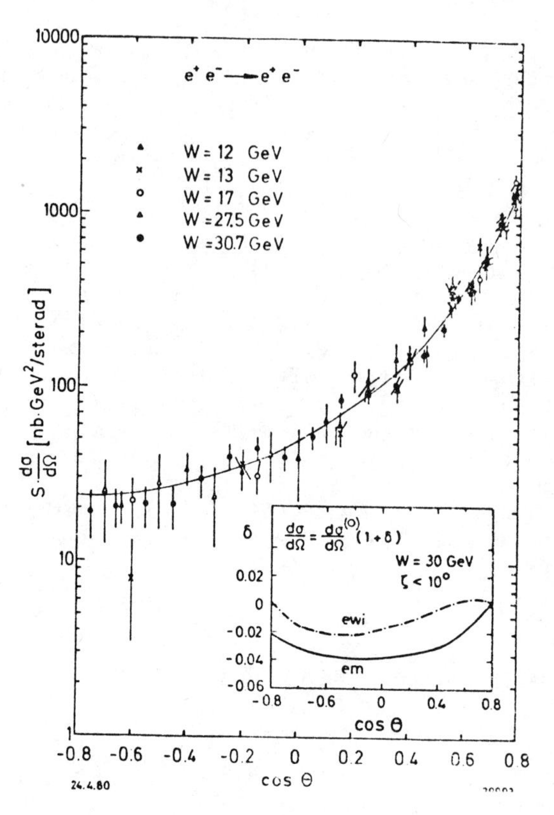

Fig. 3. Angular Distribution of Bhabha Scattering Events

Figure 3. shows the angular distribution for Bhabha scattering, and points at all energies can be seen to agree with the QED prediction. Radiative corrections have been applied to the data, and as these are much larger than the expected weak-electromagnetic interference effects (see inset) it is essential to be confident in the accuracy of the corrections. Our basis for such confidence appears in Figure 4., the distribution in the measured acolinearity angle between the outgoing electron and positron. This differs from zero slightly due to our finite experimental resolution, and there is also a tail of events with larger angles due to photon emission. Our calculation of the combined effect reproduces the data perfectly. When weak-electromagnetic interference is included in the formula for the angular distribution, and the χ^2 agreement between curve and data points studied as a function of the Weinberg angle, we obtain a limit (95% C.L.)

$$\sin^2\theta_W < 0.55$$

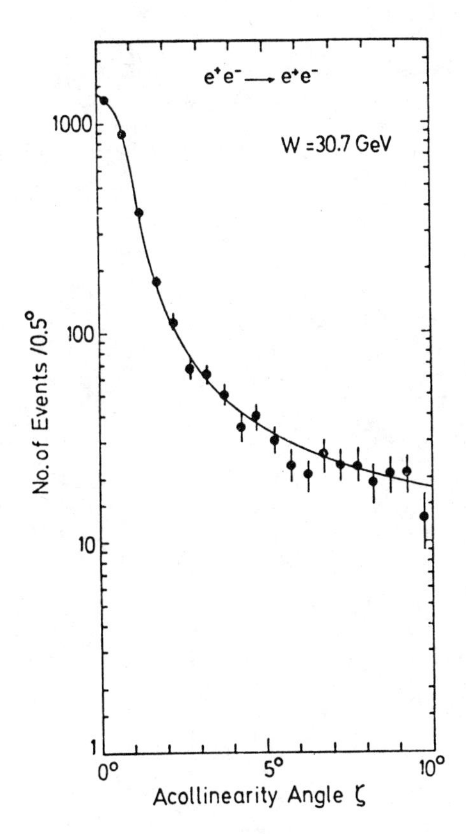

Fig. 4.

Another limit, $\sin^2\theta_W < 0.62$ is obtained from the μ pairs; this comes purely from the total cross section measurement (i.e. the limit on Λ_-). The forward backward asymmetry of -0.01 ± 0.12 is not yet sufficiently accurate for meaningful comparison with predictions.

Production of τ pairs has been studied extensively[6]. Advantages of such studies at these energies are (1) the low multiplicity 4 and 6 prong τ pair events stand out clearly from the (typically 12 prong) hadronic processes, and (2) the τ lifetime in the laboratory is enhanced by a very large gamma factor. On a sample of 70 events we establish a new lower limit for the τ lifetime (95% C.L.) of $1.4\ 10^{-12}$ seconds and thence the τ coupling to its neutrino is at least 0.46 times that of the electron to its neutrino. Figure 5. shows the decay distribution from the τ → 3 prong decays.

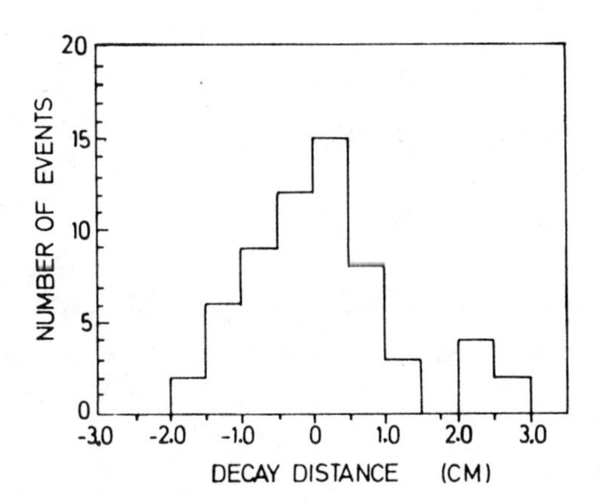

Fig. 5.

The search for the top quark, looking for a step in R and/or mean sphericity[4], or for the narrow onium statues[3] continues unsuccessful. We see no sign of any step in R up to a CMS energy of 35.8 GeV, and an energy scan at the top of the PETRA energy range is now in progress.

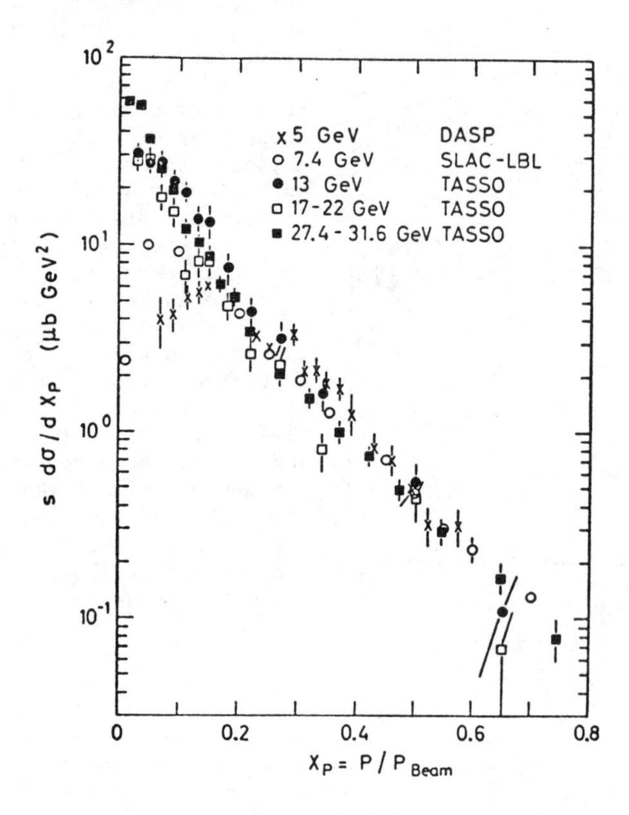

Fig. 6. × Distribution at all Energies

In the inclusive distributions an increase in the number of slow particles appears. In Figure 6, the distribution in the scaling variable ×, it can be seen that although the 13-30 GeV data scale down to low ×, the distribution is no longer described by a single exponential due to an excess of low-× tracks. It is this excess that produces the faster-than-logarithmic rise in the total charged multiplicity[5].

To identify K^o particles[9] by the decay $K_S^o \rightarrow \pi^+\pi^-$ we form the invariant mass of those $\pi^+\pi^-$ pairs satisfying the following criteria:

1. Both tracks pass more than 4mm from the origin (in the r-∅ plane).
2. The reconstructed total momentum of the pair points back to the origin (within 10°).
3. The V vertex lies within 45cm of the origin (so there is a good length of measured track).
4. Both tracks start at the V, not before it.
5. They do not form a γ.

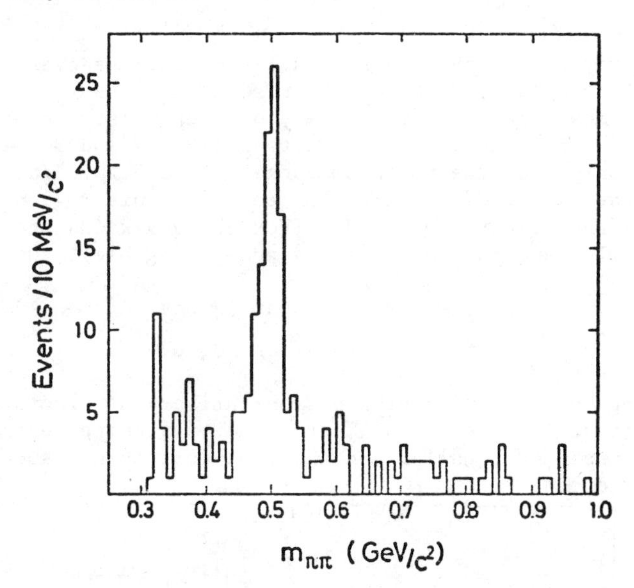

Fig. 7. K^o Mass Signal

The resulting invariant mass plot is shown in Figure 7, with a very satisfactory peak at the K^o mass. From 812 events we see 90 K^o particles, on a background of 26 extrapolating from adjacent mass bins (or 25, from subjecting the like sign dipion pairs to the same selection criteria). The selection efficiency is of the order of 25%, varying strongly with momentum - slow K^os are lost to cut (1), fast ones to cut (3). When the efficiency correction factor (computed from Monte Carlo) is applied to the observed $c\tau$ distribution, we obtain an exponential with the slope given by the known K^o lifetime, justifying the correctness of the calculation. In our presented results, the observed K^o distributions have been corrected for this inefficiency, other K_S^o decay modes, and the K_L^o fraction.

Charged kaons, protons and antiprotons are identified by the time-of-flight. The inner TOF system provides π-K separation up to 0.6 GeV/c, and (anti)proton separation up to 1.0 GeV/c; for the outer TOF these numbers are 1.1 and 2.2 GeV/c respectively. Corrections and losses - misidentified electrons, interactions in the beam pipe and coil, decays, track confusion etc., are not small, but are calculable.

The total number of neutral kaons per event is 1.4 ± 0.3 (statistical) ± 0.14 (systematic). Comparing this with the point-like μ pair cross section gives

$$R_{K^o} = \frac{\sigma(K^o X) + \sigma(\overline{K^o} X)}{\sigma(\mu^+\mu^-)} = 5.6 \pm 1.1 \pm 0.8$$

at a CMS energy of 30 GeV. This is a large increase over the value of 2.10 ± 0.14 at 7 GeV[11]. The rise is commensurate with the rises in the mean charged multiplicity between these 2 energies.

For (anti)protons we have results at 12 GeV and 30 GeV cms energy, but only over the momentum range 0.5 to 2.2 GeV/c; presumably the number of (anti)protons outside this region is small and can be safely neglected. (For charged kaons the energy range is smaller and the overspill is non negligible). We obtain:

$$R_{p,\bar{p}} \begin{array}{l} = 0.9 \pm 0.3 \quad \text{at 12 GeV} \\ = 1.5 \pm 0.4 \quad \text{at 30 GeV} \end{array}$$

corresponding to production of a baryon-antibaryon pair in 26 ± 9% of all events at 12 GeV, 39 ± 11% at 30 GeV (assuming neutrons and protons are produced equally, and that not more than 1 such pair is produced per event).

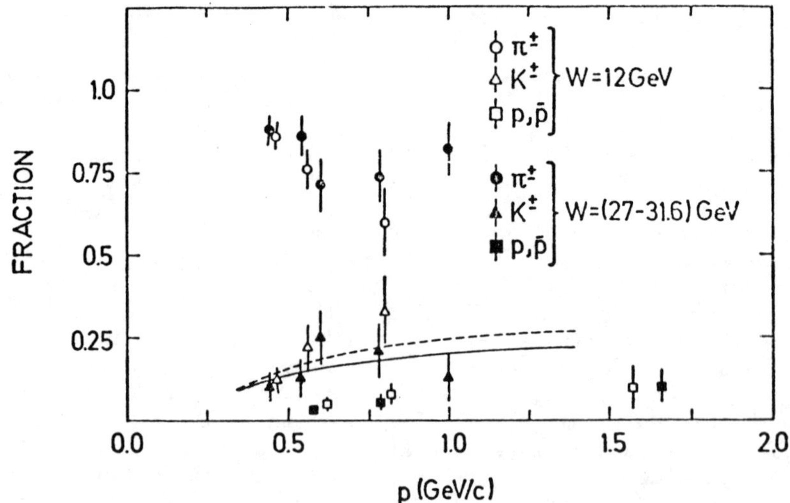

Fig. 8. Kaon and Proton Fractions

The fractions of charged Kaons and protons as a function of particle momentum are shown in Figure 8, and are seen to rise as the momentum increases. The $K^{\pm}$ fraction (and other features of Kaon production) are well described by the model of Field and Feyman[12], shown in the diagram. (The full line is the 30 GeV prediction, the dotted line that at 12 GeV).

The scaling cross section for charged and neutral kaons together is shown in Figure 9, along with some low energy K^{o} values[11], with which they scale very nicely. Figures 10A and 10B show the $\pi^{\pm}$, $K^{\pm}$ and $p,\bar{p}$ distributions at 12 and 30 GeV. All these curves show the same excess of slow particles previously mentioned, and the K and p behaviour appears to have the same form as that of the pions, with perhaps a slightly smaller overall magnitude.

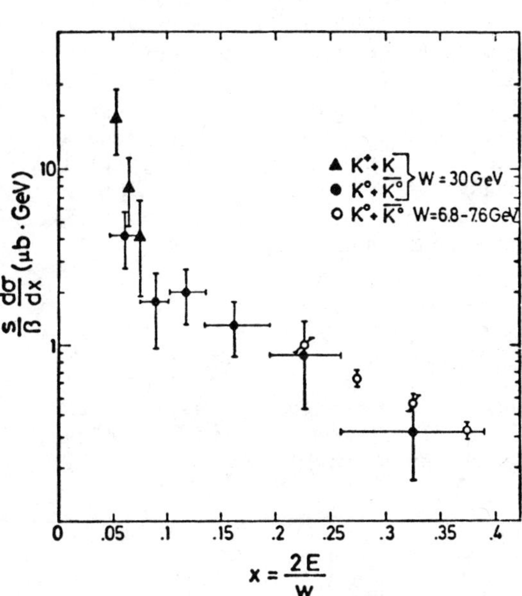

Fig. 9. × Distribution – K^{o} Particles

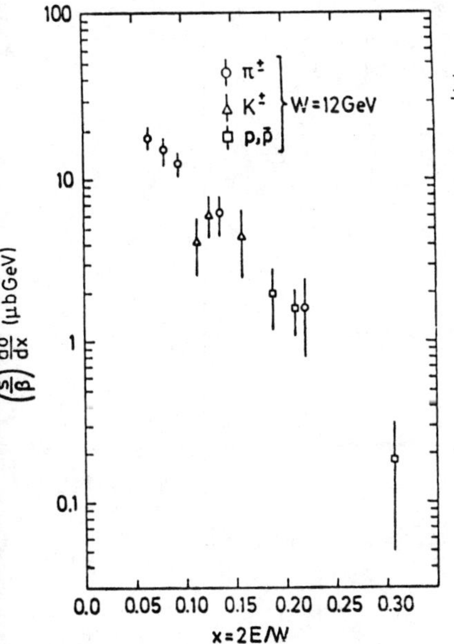

Fig. 10A × Distribution – Charged Particles (12 GeV)

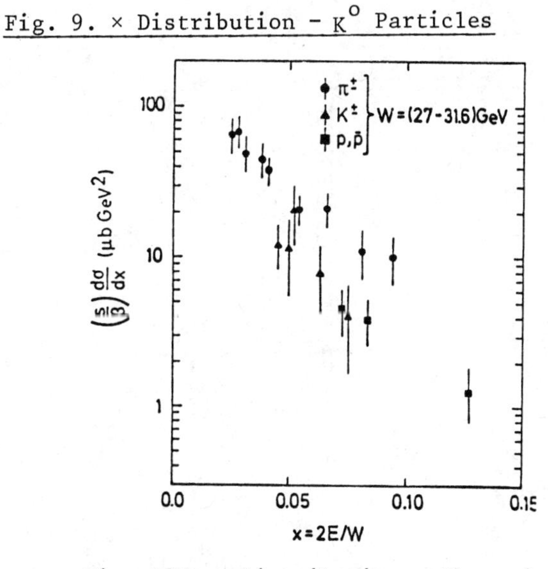

Fig. 10B × Distribution – Charged Particles (30 GeV)

PLANAR EVENTS AND QCD

The appearance at high energies of a class of events with high sphericity (low thrust), but of a structure not spherical but planar has been discussed extensively by previous speakers at this symposium, and in the literature[2], so the arguments will not be repeated here; we will take it as a starting point.

These planar events can usefully be analysed by the method of Wu and Zobernig[13], which used an analogue of sphericity for 3 jets in 2 dimensions; briefly one considers only the components of particle momenta in the event plane, and selects that division of the particles into 3 contiguous sectors, each with its own jet axis, such that the sum over particles of $q_\perp^2$ the squared momentum transverse to the appropriate jet axis, is minimised. This gives a reasonably reliable reconstruction of the original parton directions, and the quantity J_3

$$J_3 = \frac{2}{N\ \text{particles}^{-3}} \sum (q^{i\,2}_\perp/\sigma^2)$$

where σ is a typical $P_\perp$, say 300 MeV/c. It can be seen that $J_3 \sim 1$ if the planar event is truly 3 jetlike, whereas if it is isotropic (in the event plant) it will be larger. Figure 11 shows the J_3

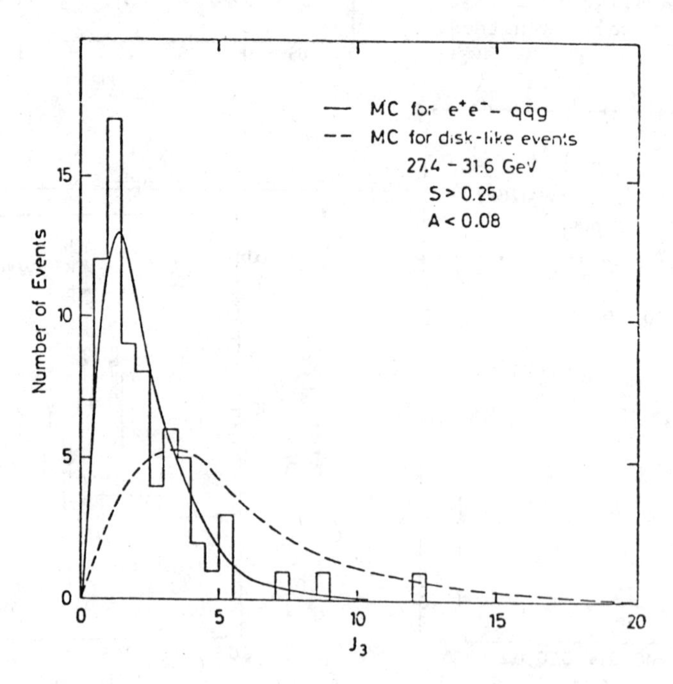

Fig. 11. Trijettiness

distribution for the planar event sample, and the data indeed peak around 1, and are totally unlike the isotropic Monte Carlo curve (dotted line). The solid line is a QCD prediction which will be discussed later - it is not relevant to the current argument that the high sphericity events are not merely planar, they are 3 jetlike.

We can now investigate the nature of the jets produced in 3 jet reactions. Figure 12A shows the P_T^2 distribution for particles

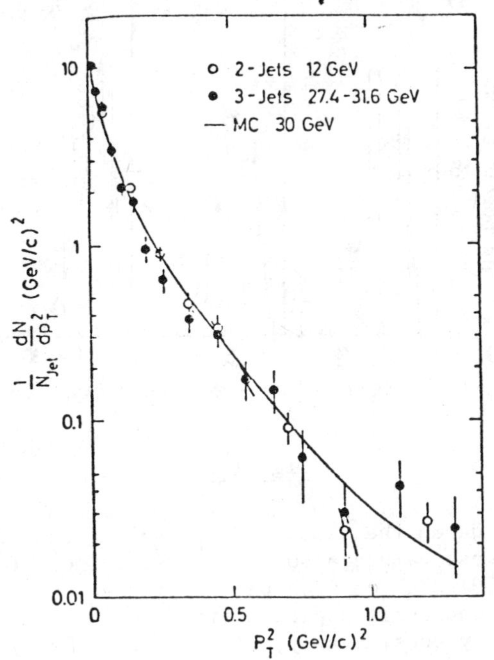

Fig. 12A. P_T^2 for Jets from 2 and 3 Jet Events

in 3 jet events (about their appropriate jet axis), with for comparison the 12GeV (strictly 2 jet) data, and there is no discernible difference. Figure 12B shows the charged multiplicities of the 3 jets, ordered by their energies, and the mean multiplicities of them all agree well with ½ the value of the event multiplicity at a beam energy equal to that of the jet concerned. At this stage, therefore, we can discern no difference between jets produced in 3 jet events and those produced in 2 jet events.

Up to this point the discussion has been completely model independent - now I shall mention QCD, and our determination of α_s[8]. We have used QCD Monte Carlo programs[14, 15] to generate events; the jet fragmentation depends on 3 parameters:

(1) σ the rms p_T of hadrons from a fragmenting parton.
(2) a_F parametrises the quark fragmentation.
(3) $P/(P + V)$ the probability that a pair of quarks in a meson will have spins antiparallel.

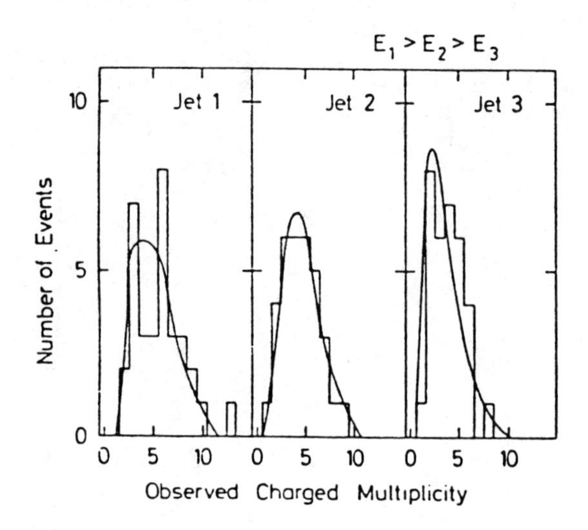

Fig. 12B.

We have considered the values of a_F ranging from 0.1 to 0.9, and for each of these values also considered values of $P/(P + V)$ between 0.1 and 0.9, and for each pair fitted α_s and σ, fitting the 2-D sphericity-aplanarity plot, using events with $S > 0.25$. The resulting σ vary considerably, but α_s is remarkably stable; we obtain an average value of 0.16 ± 0.04 (statistical), with an uncertainty of 0.017 from the variation due to different a_F and $P/(P + V)$ values.

Placing a slightly higher reliance on the model, one can use the low S data to fit a_F, and $P/(P + V)$; doing this we find the following values describe our data best:

$$\sigma = 0.32 \pm 0.04 \text{ GeV/c}$$

$$a_F = 0.57 \pm 0.20$$

$$P/(P + V) = 0.56 \pm 0.15$$

fitting α_s with these fixed gives a value 0.17 ± 0.02 (statistical) ± 0.03 (systematic). (N.B. Corrections due to hard photon emission are fully included).

Inserting these parameters into the model gives distributions in many quantities in good agreement with the data - the J_3, P_T^2,

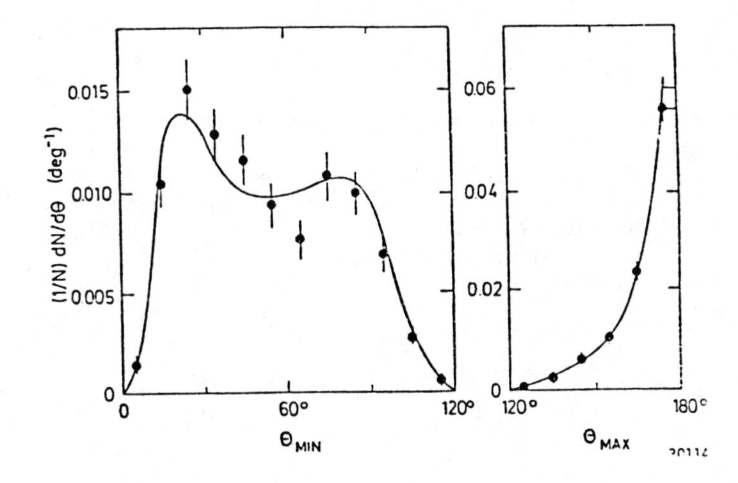

Fig. 13. Jet Angles

and N_{ch} plots have been presented already in Figures 11 and 12; Figure 13 shows the data and the model predictions for θ_{min}, the smallest of the 3 angles between the 3 jets, and θ_{max}, the largest.

CONCLUSIONS

QED continues pointlike down to distances of 10^{-16}cm. The angular distribution of Bhabha scatters produces a limit of $\mathrm{Sin}^2\theta_W < 0.55$. The τ lifetime is smaller than 1.4 10^{-12} seconds.

No sign of the t quark threshold is seen up to 35.8 GeV.

Kaon and baryon production is significant, with $\sim$3 kaons and $\sim\frac{1}{2}$ a baryon-antibaryon pair per event. Scaling cross sections have the same form as for pions.

QCD described our 3 jet data well, and we obtain a value of α_s of 0.17 ± 0.02 ± 0.03, independent of assumptions about the jet model parameters.

REFERENCES

Members of the TASSO collaboration are:

R. Brandelik, W. Braunschweig, K. Gather, V. Kadansky, K. Lübelsmeyer, P. Mättig, H. -U. Martyn, G. Peise, J. Rimkus, H. G. Sander, D. Schmitz, A. Schultz von Dratzig, D. Trines, W. Wallraff (AACHEN): H. Boerner, H. M. Fischer, H. Hartmann, E. Hilger, W. Hillen, G. Knop, P. Leu, B. Löhr, R. Wedemeyer, N. Wermes, M. Wollstadt (BONN): D. G. Cassel, D. Heyland, H. Hultschig, P. Joos, W. Koch, P. Koehler, U. Kötz, H. Kowalski, A. Ladage, D. Lüke, H. L. Lynch, G. Mikenberg, D. Notz, J. Pyrlik, R. Riethmüller, M. Schliwa, P. Söding, B. H. Wiik, G. Wolf (DESY): R. Fohrmann, M. Holder, G. Poelz, O. Römer, R. Rüsch, P. Schmüser (HAMBURG): D. M. Binnie, P. J. Dornan, N. A. Downie, D. A. Garbutt, W. G. Jones, S. L. Lloyd, D. Pandoulas, J. Sedgbeer, S. Yarker, C. Youngman (IMPERIAL COLLEGE, LONDON): R. J. Barlow, I. Brock, R. J. Cashmore, R. Devenish, P. Grossmann, J. Illingworth, M. Ogg, B. Roe, G. L. Salmon, T. Wyatt (OXFORD): K. W. Bell, B. Foster, J. C. Hart, J. Proudfoot, D. R. Quarrie, D. H. Saxon, P. L. Woodworth, (RUTHERFORD LABORATORY): Y. Eisenberg, U. Karshon, D. Revel, E. Ronat, A. Shapira (WEIZMANN INSTITUTE): J. Freeman, P. Lecomte, T. Meyer, Sau Lan Wu, G. Zobernig (WISCONSIN)

1. R. Brandelik et al. Physics Letters 83B 261 (1979)
2. R. Brandelik et al. Physics Letters 86B 243 (1979)
3. R. Brandelik et al. Physics Letters 88B 199 (1979)
4. R. Brandelik et al. Z. f. Physik C 4 87 (1980)
5. R. Brandelik et al. Physics Letters 89B 418 (1980)
6. R. Brandelik et al. DESY Report 80/12 (1980)
7. R. Brandelik et al. DESY Report 80/33 (1980)
8. R. Brandelik et al. DESY Report 80/40 (1980)
9. R. Brandelik et al. DESY Report 80/39 (1980)
10. R. Brandelik et al. "Charged Pion, Kaon, Proton and Anti-Proton Production in High Energy e^+e^- Annihilations" To appear shortly as a DESY report.
11. V. Lüth et al. Physics Letters 70B 120 (1977)
12. R. D. Field and R. P. Feynman Nuclear Physics B136 1 (1978)
13. S. L. Wu and G. Zobernig Z. f. Physik C 2 107 (1979)
14. P. Hoyer et al. Nuclear Physics B161 349 (1979)
15. A. Ali et al. DESY Report 79/86 (1979)

SELECTED RESULTS FROM THE MARK II AT SPEAR*

Daniel L. Scharre
Stanford Linear Accelerator Center,
Stanford University, Stanford, Ca. 94305

ABSTRACT

Recent results on radiative transitions from the $\psi(3095)$, charmed meson decay, and the Cabibbo-suppressed decay $\tau \to K^* \nu_\tau$ are reviewed. The results come primarily from the Mark II experiment at SPEAR, but preliminary results from the Crystal Ball experiment on ψ radiative transitions are also discussed.

I. INTRODUCTION

This talk is the first of two reviewing recent results from SPEAR. I will concentrate on results from the Mark II experiment[1] and have selected three topics to discuss: radiative transitions from the $\psi(3095)$, charmed meson decays, and the Cabibbo-suppressed decay $\tau \to K^* \nu_\tau$. Some preliminary results from the Crystal Ball experiment[1] on ψ radiative transitions will also be presented.

The discussion on radiative transitions will be restricted to transitions from the ψ to ordinary hadrons, where ordinary hadrons are defined to be those which to first order do not contain charmed quarks. The status of the $\eta_c(2980)$ and radiative transitions to this state will be discussed in the next talk.[2] I will begin with a brief discussion of inclusive photon production at the ψ. This leads naturally into a discussion of the four exclusive radiative transitions which we (i.e., the Mark II) observe. Three of these transitions, to the η, $\eta'(958)$, and $f(1270)$, have been previously observed. Our results are in reasonable, but not perfect, agreement with the previous branching fraction measurements. The fourth observed transition is to a state which we tentatively identify as the $E(1420)$. This transition has not been previously observed.

The discussion on charmed meson decays will cover new results on exclusive decay modes of the D and general inclusive properties of

*Work supported by the Department of Energy, contract DE-AC03-76SF00515.

ISSN:0094-243X/80/620207-46$1.50

the final states produced in D decays. Measurements of the mean charged particle multiplicity, the kaon fraction, and the electron fraction in D decays will be discussed. From the electron fractions, we have determined the inclusive D^o and D^+ semileptonic branching fractions, which allows a determination of the ratio of the D^o and D^+ total lifetimes.

Finally, we have measured the branching fraction for the Cabibbo-suppressed decay $\tau \to K^* \nu_\tau$. The measured branching fraction is consistent with expectations based on the previously measured, Cabibbo-favored decay $\tau \to \rho \nu_\tau$.

II. EXPERIMENTAL APPARATUS

The Mark II magnetic detector was designed to be a second-generation, general-purpose detector for doing physics at SPEAR (and a first-generation detector for doing physics at PEP). The design was based on the earlier SLAC-LBL magnetic detector[3] which was responsible for much of the pioneering work in e^+e^- annihilation physics done at SPEAR. The Mark II detector has significant advantages over the previous detector in its charged particle momentum resolution, time-of-flight resolution, neutral particle detection, lepton identification, and trigger design.

A schematic of the Mark II detector is shown in Fig. 1 viewed perpendicular to the beam direction. A particle traveling radially outwards from the interaction region passes through a 16-layer cylindrical drift chamber, a layer of scintillation counters which provide time-of-flight (TOF) information, a 1-radiation length aluminum coil (which provides a solenoidal field of approximately 4.2 kG in the central detector), a set of lead-liquid argon (LA) shower counter modules, and finally a set of proportional counters for muon identification. In addition to the muon counters which are shown above and below the detector in Fig. 1, there are additional muon proportional counter layers on both sides of the detector. Also, there are two additional shower counter modules (one LA module and one module consisting of two multi-wire proportional chamber planes) which cover the endcap regions.

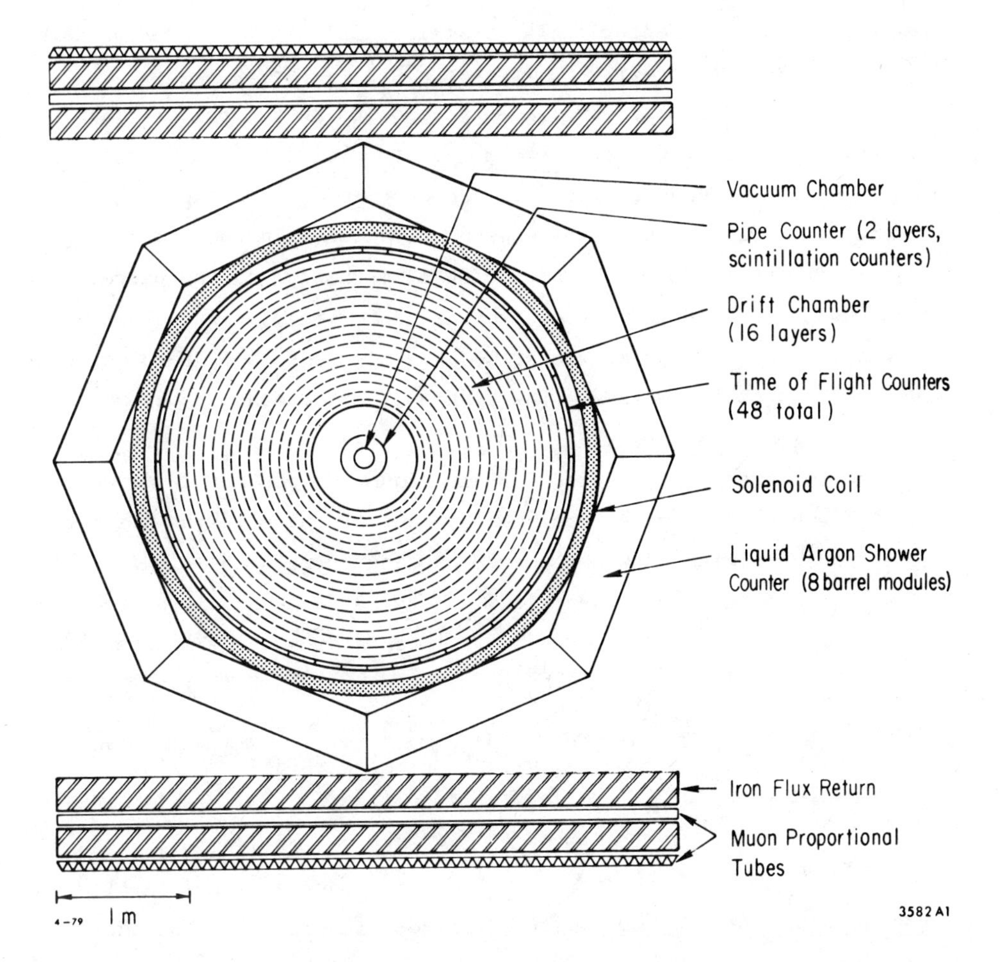

Fig. 1. Schematic of the Mark II magnetic detector viewed perpendicular to the beam direction. Not shown are the muon proportional counters on both sides of the detector and the endcap shower counters.

Charged tracks are reconstructed from hits in the 16 cylindrical drift chamber layers[4] of radii 0.37 m to 1.51 m which provide solid angle coverage over 85% of 4π sr. The azimuthal coordinates of charged tracks are measured to an rms accuracy of approximately 200 μm at each layer. The polar coordinates are determined from the 10 stereo layers oriented at ±3° to the beam axis. The momentum resolution can be expressed as $\delta p/p = [(0.0145)^2 + (0.005\ p)^2]^{\frac{1}{2}}$, where p is the momentum in GeV/c. The first term is the contribution from multiple Coulomb scattering and the second is the contribution from the measurement error.[5] The tracking efficiency is greater than 95% for tracks with $p > 100$ MeV/c over 75% of 4π sr.

Photons are identified by energy deposits in the shower counter modules. The shower counter system[6] consists of eight LA barrel modules which surround the central detector and two endcap modules. The endcap modules have not been used extensively in the analysis, and I will restrict the remainder of the discussion to the barrel module system. Each module consists of approximately 14 radiation lengths of lead and argon, with readout strips parallel, perpendicular, and at 45° to the beam axis. The system of eight modules covers approximately 64% of 4π sr.

The rms energy resolution for electrons and photons detected in the LA has been determined from measurement of pulse height distributions from particles of known energy passing through the liquid argon. Figure 2 shows the energy deposited in the LA for a sample of Bhabha events at 4.16 GeV center-of-mass energy ($E_{c.m.}$). The curve is a Gaussian resolution function, with $\delta E/E = 0.12\ E^{-\frac{1}{2}}$ (E in GeV), convoluted with the radiative tail. This expression provides a reasonable representation of the energy resolution for both electrons and photons of energies down to a few hundred MeV.

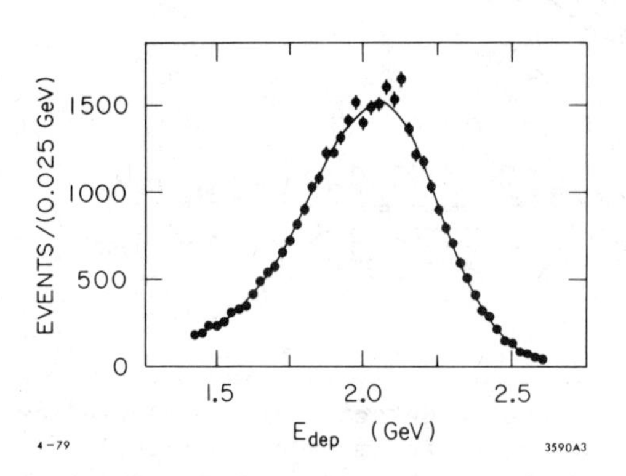

Fig. 2. Energy deposited in the LA for a sample of Bhabha events at $E_{c.m.} = 4.16$ GeV. Curve is described in text.

The rms angular resolution obtained for electromagnetic showers is typically 4 mrad, both in azimuth and dip angle, for high energy particles. At lower energies, the angular resolution can be a factor of two worse.

Figure 3 shows the detection efficiency for γ's and π^o's[7] in the LA as a function of energy. The measured values for the γ efficiency were obtained from 2-constraint (2C) fits to 2-prong and 4-prong events at the ψ according to the hypotheses $\psi \to \pi^+\pi^-\gamma(\gamma)$ and $\psi \to \pi^+\pi^-\pi^+\pi^-\gamma(\gamma)$,

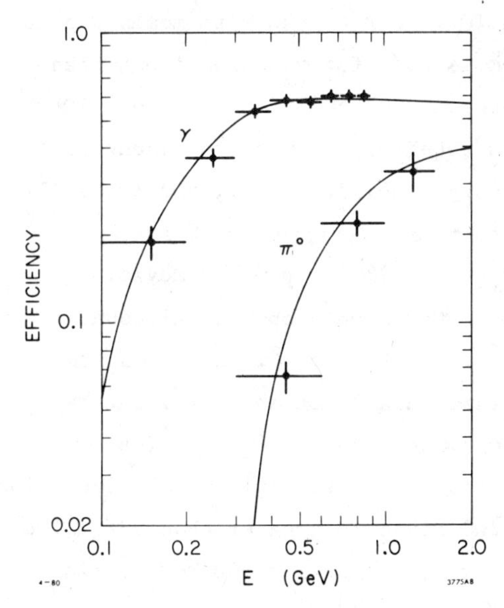

Fig. 3. Detection efficiency for γ's and π^o's in the LA as a function of energy. Data points are measured values and curves are Monte Carlo estimates.

where a particle in parenthesis is meant to imply an unobserved particle. The π^o mass constraint was imposed on the $\gamma(\gamma)$ system. The detection efficiency was calculated from the fraction of events in which the missing γ was observed and tracked in the LA. Corrections were made to correct for the geometrical bias imposed by the requirement that all charged particles be observed in the detector. A similar procedure was utilized to measure the π^o efficiency from 1C fits according to the hypotheses $\psi \to \pi^+\pi^-(\pi^o)$ and $\psi \to \pi^+\pi^-\pi^+\pi^-(\pi^o)$. The π^o detection efficiency was calculated from the fraction of events in which the missing π^o was observed, with similar corrections for geometrical bias. However, these π^o efficiency measurements must be considered as lower limits since the decays $\psi \to \pi^+\pi^-\gamma$ and $\psi \to \pi^+\pi^-\pi^+\pi^-\gamma$ will successfully fit the corresponding hypothesis in which the γ is replaced by a π^o, but no π^o will be observed. The solid curves represent the results of Monte Carlo determinations of the γ and π^o efficiencies employing the EGS electromagnetic shower development code.[8]

Particle identification for hadrons is accomplished with 48 TOF scintillation counters which surround the drift chamber and cover approximately 75% of 4π sr. The rms time resolution is 0.30 ns for hadrons. The average flight path of 1.85 m provides a separation of pions from kaons up to momenta of 1.35 GeV/c and kaons from protons up to 2.0 GeV/c at the 1-σ level.

Identification of electrons utilizes both the LA system and the TOF system. For tracks with momenta less than 300 MeV/c, the electron identification is based solely on TOF. For tracks with momenta greater than 500 MeV/c, only the LA system is used for electron identification. For tracks with momenta between 300 and 500 MeV/c, both TOF and LA pulse height information is required for electron identification. The hadron misidentification probability (i.e., the probability that a hadron will be misidentified as an electron) is 7% for $p < 500$ MeV/c, 4% for $p = 600$ MeV/c, and 2% for $p = 800$ MeV/c.

Muons are identified by hits in the muon proportional counter system. This system consists of layers of proportional tubes (two below and on each side of the detector and three above it) interleaved with steel and covers approximately 55% of 4π sr. The detection efficiency for muons within the solid angle of the detection system with momenta greater than 700 MeV/c is greater than 98%. The hadron misidentification probability is 4% for $p = 800$ MeV/c, 11% for $p = 900$ MeV/c and 2% for $p > 1$ GeV/c.

The detector is triggered with a two-stage hardware trigger,[9] selecting (with efficiency greater than 99%) all interactions emitting two or more charged tracks, each with transverse momentum greater than 100 MeV/c, within the solid angle covered by the drift chamber, one of which must be within the central region of the drift chamber which covers 67% of 4π sr. The luminosity is measured with independent shower counters detecting Bhabha scattering at 22 mrad, and checked against wide-angle Bhabha events observed in the central detector.

It is customary in a talk such as this to show a computer reconstruction of a "typical" event to impress other experimenters with how well the detector works. Figure 4 shows a resconstruction of an event corresponding to the $\psi'(3684)$ decay[10]

$$\psi' \to \pi^+\pi^-\psi \quad , \quad \psi \to \mu^+\mu^- \quad .$$

One can see that there is no problem tracking the charged particles through the drift chamber. The flight time of each track is determined by a TOF counter (represented by a rectangle at the end of each reconstructed drift chamber track). Finally, unambiguous identifications of the muons are made in the muon proportional counter system.

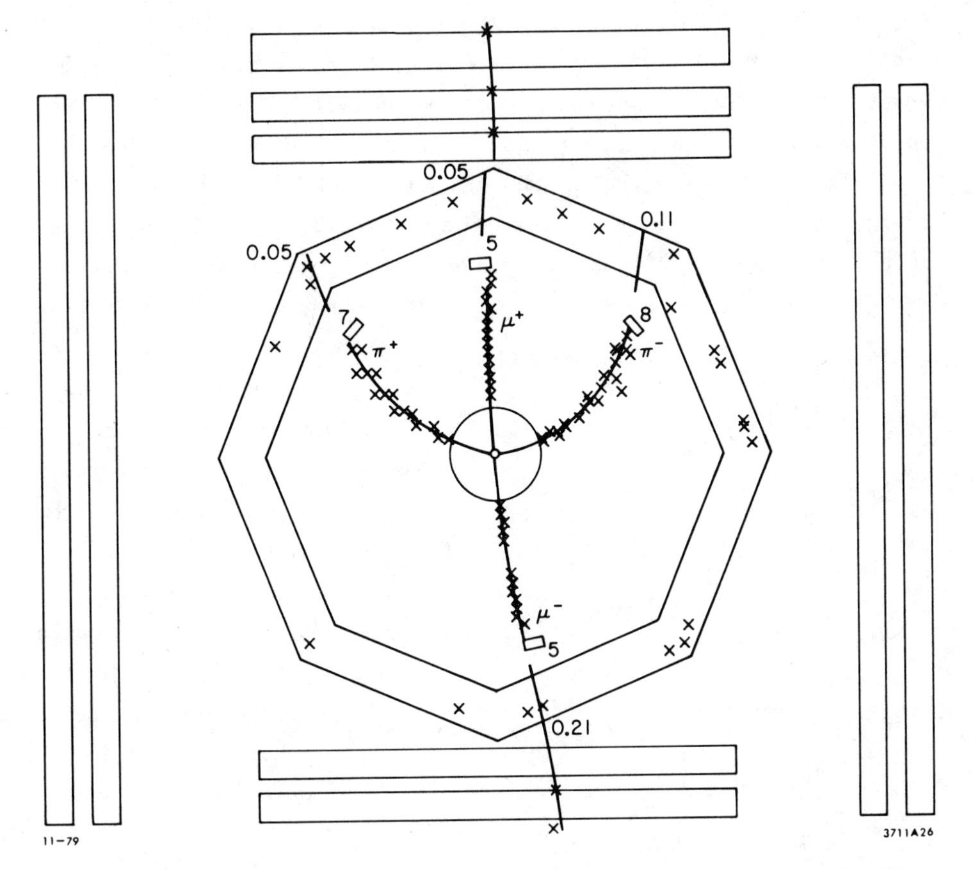

Fig. 4. Computer reconstruction of typical event in the Mark II detector.

As the tracks are all due to the minimum-ionizing particles, no large energy deposits are observed in the LA.

III. RADIATIVE TRANSITIONS FROM THE $\psi(3095)$

Measurements of inclusive photon production at the $\psi(3095)$ by this experiment[11] and the Lead-Glass Wall (LGW) experiment[12] have shown that there is a sizable direct photon component in the momentum spectrum. However, because of the relatively poor photon energy resolutions of our LA shower counter system and the lead-glass counters in the LGW ($\delta E/E \approx 9\%/E^{1/2}$, E in GeV), neither experiment was able to observe any narrow structure in the inclusive photon momentum distribution.

The Crystal Ball detector[13] was designed to provide good energy resolution for electromagnetic showers. The use of NaI(Tℓ) for shower detection presently allows a resolution of $\delta E/E \approx 2.8/E^{1/4}$ (E in GeV) to be obtained.

Figure 5 shows a preliminary measurement of the inclusive γ energy distribution of the ψ from the Crystal Ball.[14] It is plotted as a function of the logarithm of the γ energy (E_γ in MeV) so that the bin width is roughly proportional to the energy resolution at all energies. This distribution is based on a sample of approximately 900,000 events obtained during approximately two weeks of running near the peak of the ψ. Details of the analysis can be found in Ref. 13.

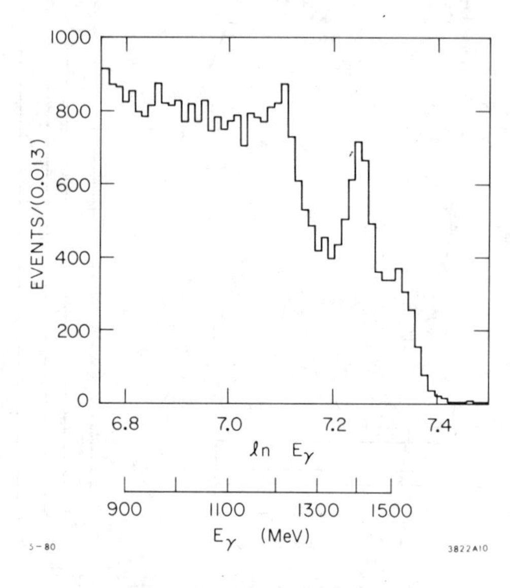

Fig. 5. Inclusive γ distribution at the ψ as a function of the logarithm of the γ energy in MeV (data from the Crystal Ball collaboration).

The structure observed in Fig. 5 is evidence for exclusive processes of the type

$$\psi \to \gamma + X \quad .$$

There is clear evidence for the radiative transitions to the η,[15-17] η'(958),[15-18] and a new state which I will refer to as the E(1420) which has recently been observed by us.[19] [Although I refer to this state as the E(1420), this assignment is still open to question.] An additional transition which has been previously observed is $\psi \to \gamma$ f(1270).[20,21] Because of the relatively small branching fraction for this transition, it is not observed in this inclusive distribution. Each of these four transitions will be discussed in turn in the following sections.

A. $\psi \to \gamma\eta, \gamma\eta'$

As the η and η' are members of the same SU(3) nonet, it makes sense to discuss the radiative transitions to these two states at the same time, along with the transition to the π^o. I will take the extremely naive approach that it is possible to understand these processes in terms of leading-order QCD diagrams. Thus, one can imagine that the radiated photon is produced either from the outgoing quark line (assumed to be u, d, or s) as in Fig. 6(a) or from the initial charmed quark line as in Fig. 6(b). In the first case, the minimal coupling between the charmed quark line and the ordinary quark line requires three gluons. In the second case, two gluons are sufficient.

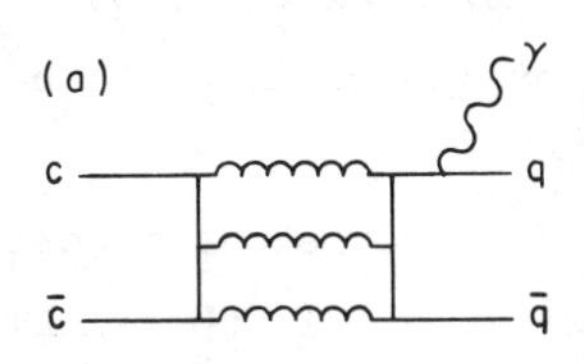

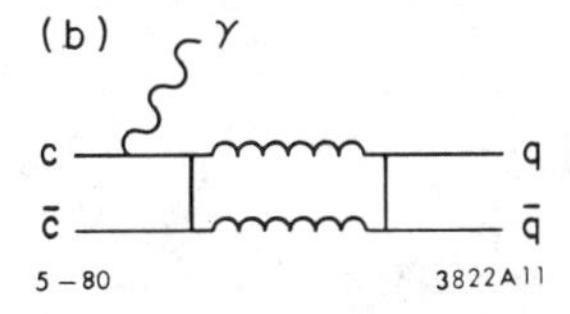

Fig. 6. Leading-order diagrams for radiative transitions from the ψ with a) photon emission from the final-state quark line and b) photon emission from the initial-state charmed quark line.

Let me first consider only the process shown in Fig. 6(a) and assume it is the dominant one. By invoking vector-meson dominance, I can relate the $\gamma\pi^o$ and $\rho^o\pi^o$ decay widths

$$\Gamma(\psi \to \gamma\pi^o) = (\alpha\pi/\gamma_\rho{}^2)\Gamma(\psi \to \rho^o\pi^o) \quad .$$

This leads to a prediction for the $\gamma\pi^o$ branching fraction $B(\psi \to \gamma\pi^o) \approx 2 \times 10^{-5}$ from the measured $\rho^o\pi^o$ branching fraction.[18,22,23] This is consistent with the experimental measurement[16] $B(\psi \to \gamma\pi^o) = (7 \pm 5) \times 10^{-5}$.

The next step is to relate the widths of the $\gamma\eta$ and $\gamma\eta'$ transitions to the width of the $\gamma\pi^o$ transition. The η and η' have the following SU(3) singlet and octet components

$$\eta = \eta_8 \cos\theta + \eta_1 \sin\theta$$

$$\eta' = -\eta_8 \sin\theta + \eta_1 \cos\theta \quad ,$$

where θ is the standard octet-singlet mixing angle. If one assumes SU(3) invariance, only the octet components contribute to the process shown in Fig. 6(a) and one obtains (up to phase space corrections)

$$\Gamma(\psi \to \gamma\pi^o):\Gamma(\psi \to \gamma\eta):\Gamma(\psi \to \gamma\eta') = 3:\cos^2\theta:\sin^2\theta \quad .$$

Using the experimentally determined mixing angle $\theta = -11°$, one calculates

$$\Gamma(\psi \to \gamma\pi^o):\Gamma(\psi \to \gamma\eta):\Gamma(\psi \to \gamma\eta') = 3:0.96:0.04 \quad ,$$

which grossly contradicts the experimental measurements.[15-18] The $\gamma\eta'$ branching fraction has been experimentally determined to be larger than the $\gamma\eta$ branching fraction, and both are at least an order of magnitude larger than the π^o transition. The conclusion is that the process in Fig. 6(b) is the dominant one.

One can proceed with similar calculations for the second process [shown in Fig. 6(b)]. Assuming SU(3) invariance (now only the singlet components contribute) and ignoring phase space corrections, one obtains

$$\Gamma(\psi \to \gamma\pi^o):\Gamma(\psi \to \gamma\eta):\Gamma(\psi \to \gamma\eta') = 0:\sin^2\theta:\cos^2\theta \quad .$$

This is qualitatively in better agreement with the data. However, the predicted ratio $\Gamma(\psi \to \gamma\eta')/\Gamma(\psi \to \gamma\eta)$ is much larger than the experimentally measured ratio.

If one allows for SU(3) symmetry breaking, these results are modified. Fritzsch and Jackson[24] have calculated the relative widths of the $\gamma\eta$ and $\gamma\eta'$ transitions by considering gluon-mediated mixing between the three isoscalar states η, η', and $\eta_c(2980)$. Based on the experimental masses of these states, they find the following admixture of η and η' in the η_c:

$$\eta_c = c\bar{c} + \varepsilon\cdot\eta + \varepsilon'\cdot\eta' \quad ,$$

where $\varepsilon \approx 10^{-2}$ and $\varepsilon' \approx 2.2 \times 10^{-2}$. The decay widths (for M1 transitions) for $\gamma\eta$ and $\gamma\eta'$ are

$$\Gamma(\psi \to \gamma\eta) = \varepsilon^2 \frac{4\alpha}{3m_c^2}\left(\frac{2}{3}\right)^2 k^3\, \Omega^2$$

$$\Gamma(\psi \to \gamma\eta') = (\varepsilon')^2 \frac{4\alpha}{3m_c^2}\left(\frac{2}{3}\right)^2 (k')^3\, \Omega^2 \quad ,$$

where m_c is the charmed quark mass, $k(k')$ is the momentum of the $\eta(\eta')$, and Ω is an overlap integral. If it is assumed that the overlap integral is the same for both the η and η' transitions, one finds for the ratio of the two partial widths

$$\frac{\Gamma(\psi \to \gamma\eta')}{\Gamma(\psi \to \gamma\eta)} = \left(\frac{k'}{k}\right)^3 \left(\frac{\varepsilon'}{\varepsilon}\right)^2 \approx 3.9 \quad .$$

By estimating the overlap integral $\Omega^2 \approx 0.1$, Fritzsch and Jackson also make predictions for the absolute values of the widths, $\Gamma(\psi \to \gamma\eta) \approx 60$ eV and $\Gamma(\psi \to \gamma\eta') \approx 220$ eV.

Branching fractions for the transitions $\psi \to \gamma\eta$ and $\psi \to \gamma\eta'$ have recently been published by the Crystal Ball collaboration.[17] The measurements were based on a sample of decays $\psi \to 3\gamma$. Figure 7 shows the Dalitz plot for this sample of events. Two distinct bands associated with $\gamma\eta$ and $\gamma\eta'$ transitions are observed.[25] The branching fractions for these transitions were determined from a fit to the Dalitz plot. They are $B(\psi \to \gamma\eta') = (6.9 \pm 1.7) \times 10^{-3}$ and $B(\psi \to \gamma\eta) = (1.2 \pm 0.2) \times 10^{-3}$.

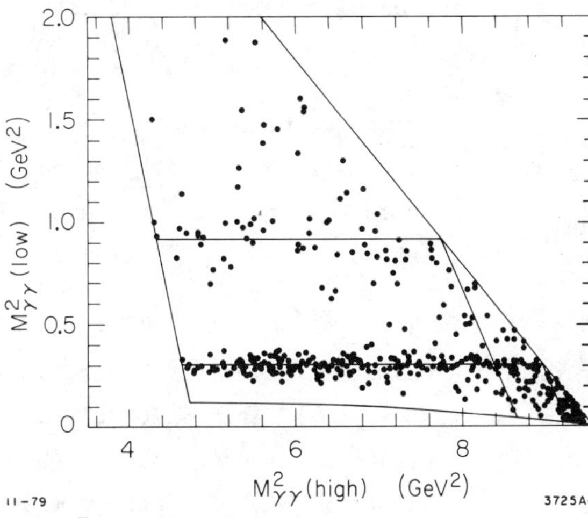

Fig. 7. Dalitz plot for $\psi \to 3\gamma$. Boundary includes effects of both kinematics and $\gamma\gamma$ opening angle cuts (data from the Crystal Ball collaboration).

We have measured the branching fraction for the process[26]

$$\psi \to \gamma\eta' \ , \ \eta' \to \pi^+\pi^-\gamma \quad .$$

The data sample used in this analysis is basically the same as the Crystal Ball data sample, as both experiments were running at SPEAR simultaneously.[27] Events with two oppositely charged tracks identified as pions and two or more photons[20] observed in the LA shower counter modules were fit to the hypothesis

$$\psi \to \pi^+\pi^-\gamma\gamma \quad . \tag{1}$$

Events in which the fitted $\gamma\gamma$ invariant mass was between 0.12 and 0.15 GeV (i.e., consistent with the π^0 mass) were eliminated. The $\pi^+\pi^-\gamma$ invariant mass distribution for the events which remained after the χ^2 and π^0 cuts is shown in Fig. 8. From Monte Carlo calculations of the detection efficiency (which included an assumed $1 + \cos^2\theta$ dependence for the ψ decay, where θ is the angle between the photon and the beam direction), we measured the branching fraction $B(\psi \to \gamma\eta') = (3.4 \pm 0.7) \times 10^{-3}$.

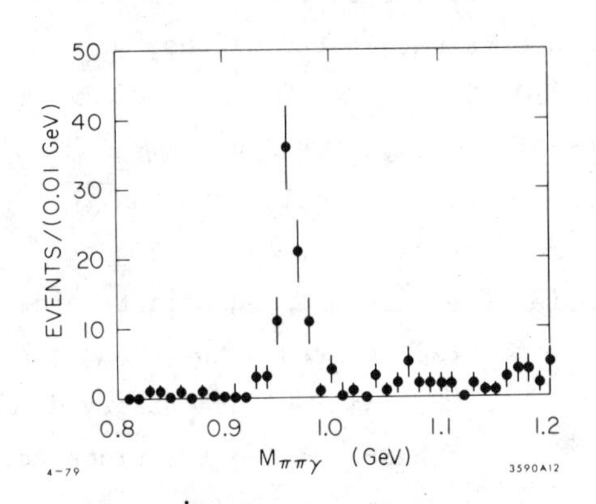

Fig. 8. $\pi^+\pi^-\gamma$ invariant mass distribution for events satisfying (1) with π^0 combinations eliminated.

Due to the bias imposed by the trigger requirement,[29] we are unable to observe the reaction

$$e^+e^- \to \psi \to 3\gamma \quad .$$

In order to measure the $\gamma\eta$ branching fraction, it was necessary to analyze the more complicated process[30]

$$\psi' \to \pi^+\pi^-\psi \ , \ \psi \to 3\gamma \quad . \qquad (2)$$

The $\pi^+\pi^-$ from the ψ' cascade decay provided the trigger. Figure 9 shows the invariant mass of the low-mass $\gamma\gamma$ combinations for the 10 events satisfying fits to (2). Eight are peaked at the η mass. From this, the branching fraction $B(\psi \to \gamma\eta) = (0.9 \pm 0.4) \times 10^{-3}$ is obtained.

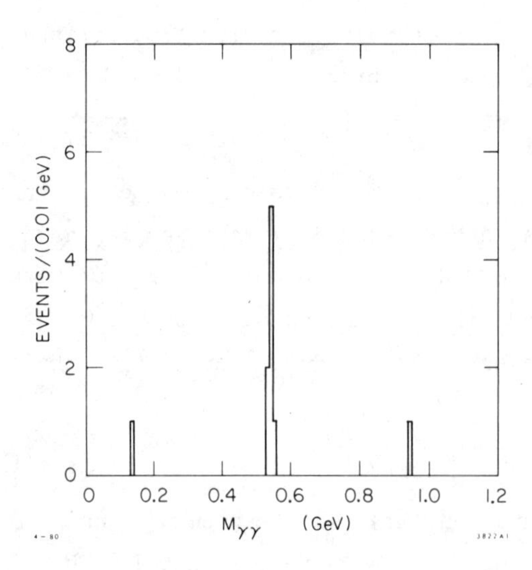

Fig. 9. Low-mass $\gamma\gamma$ invariant mass combinations for events satisfying (2).

In Table I is a compilation of our results and the Crystal Ball results, along with the previous experimental results, for the $\gamma\eta$ and $\gamma\eta'$ branching fractions. Our measurement of the $\gamma\eta'$ branching fraction is somewhat larger than, but still consistent with, the previous measurements. The Crystal Ball finds a branching fraction that

Table I. Branching fractions for radiative transitions from the ψ to the η and η'.

decay	mode	branching fraction	experiment
$\psi \to \gamma\eta'$	$\rho^o\gamma$	$(3.4 \pm 0.7) \times 10^{-3}$	Mark II
	$\gamma\gamma$	$(6.9 \pm 1.7) \times 10^{-3}$	Crystal Ball
	$\gamma\gamma$	$(2.2 \pm 1.7) \times 10^{-3}$	DASP[a)]
	$\rho^o\gamma$	$(2.4 \pm 0.7) \times 10^{-3}$	DESY-Heidelberg[b)]
		3.3×10^{-3}	theory[c)]
$\psi \to \gamma\eta$	$\gamma\gamma$	$(0.9 \pm 0.4) \times 10^{-3}$	Mark II
	$\gamma\gamma$	$(1.2 \pm 0.2) \times 10^{-3}$	Crystal Ball
	$\gamma\gamma$	$(0.8 \pm 0.2) \times 10^{-3}$	DASP[a)]
	$\gamma\gamma$	$(1.3 \pm 0.4) \times 10^{-3}$	DESY-Heidelberg[b)]
		0.9×10^{-3}	theory[c)]

[a)] Ref. 16

[b)] Refs. 15, 18

[c)] Ref. 24

is twice as large as ours. This discrepancy is not totally understood. However, it should be noted that the two measurements are based on different decay modes of the η', and at least part of the discrepancy may come from the uncertainty in the relative branching fractions of the two decay modes. On the other hand, all four determinations of the branching fraction to $\gamma\eta$ are consistent. Also shown in Table I are the theoretical predictions calculated by Fritzsch and Jackson.[24] The excellent agreement between theory and experiment is better than one has a right to expect because of the uncertainties in the calculations.

Table II summarizes the measurements of the ratio $B(\psi \to \gamma\eta')/B(\psi \to \gamma\eta)$. The measured values range from approximately 2 to 6 and the theoretical prediction is 3.9. Thus, I think it is fair to say that we have a reasonable understanding of the M1 transitions from the ψ to the ordinary pseudoscalar meson states.

In order to further explore the properties of the charmonium system, the Crystal Ball and Mark II collaborations have begun similar

Table II. $B(\psi \to \gamma\eta')/B(\psi \to \gamma\eta)$.

ratio	experiment
3.8 ± 1.9	Mark II
5.9 ± 1.5	Crystal Ball
2.8 ± 2.3	DASP[a)]
1.8 ± 0.8	DESY-Heidelberg[b)]
3.9	theory[c)]

[a)] Ref. 16
[b)] Refs. 15, 18
[c)] Ref. 24

studies of radiative transitions from the ψ'. Naively, one would expect these branching fractions to be approximately an order of magnitude smaller than the corresponding branching fractions at the ψ.[31] Presently, no evidence for $\gamma\eta$ or $\gamma\eta'$ production from the ψ' has been observed, with preliminary 90% confidence level upper limits from the Crystal Ball of $B(\psi' \to \gamma\eta') < 8 \times 10^{-4}$ and $B(\psi' \to \gamma\eta) < 10^{-4}$. As these limits are only a factor of eight below the measured ψ branching fractions, there is no reason to worry about the absence of these signals at this time.

B. $\psi \to \gamma f(1270)$

In order to understand the radiative transition to the f(1270), I will once more consider the two processes shown in Fig. 6. The measured branching fraction for the process $\psi \to \omega f$ is approximately 3×10^{-3}.[32] Invoking VMD for the process shown in Fig. 6(a), one is led to expect a rate for the γf transition which is considerably less than the measured value.[20,21] Thus, even in this case, where the final state has $J^P = 2^+$ rather than 0^-, it appears that the process in Fig 6(b) is dominant.

We have measured the branching fraction for the γf transition.[33] Figure 10 shows the $\pi^+\pi^-$ invariant mass distribution (data points with error bars) for events which satisfy a fit to the hypothesis

$$\psi \to \pi^+\pi^-\gamma \tag{3}$$

with $\chi^2 < 15$. Two structures are evident in the mass distribution, one at the ρ mass and the other at the f(1270) mass. Since the decay $\psi \to \rho^o\gamma$ does not conserve charge conjugation parity (C-parity), it is assumed that the events in the ρ^o mass region resulted from $\rho^o\pi^o$ decays in which an asymmetric decay of the π^o led to an acceptable fit to (3). A Monte Carlo was used to determine the $\pi^+\pi^-\pi^o$ feeddown

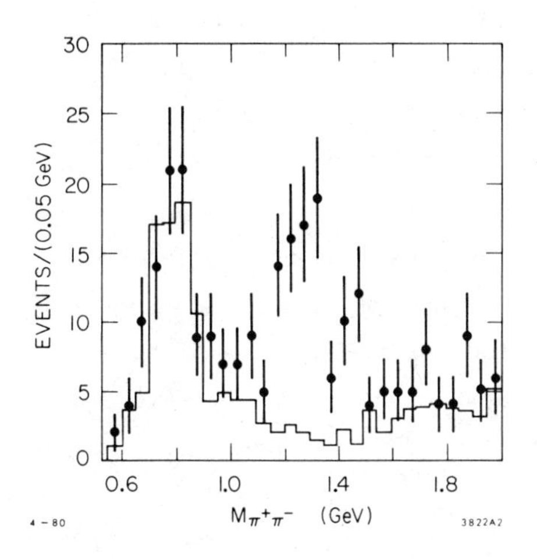

Fig. 10. $\pi^+\pi^-$ invariant mass distribution for events satisfying (3). Histogram shows the expected feeddown from the $\pi^+\pi^-\pi^o$ final state as determined by Monte Carlo.

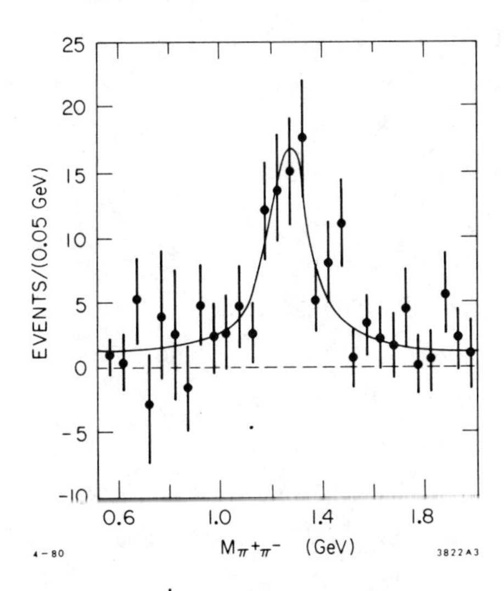

Fig. 11. $\pi^+\pi^-$ invariant mass distribution after subtraction of $\pi^+\pi^-\pi^o$ feeddown. Curve is described in text.

into the $\pi^+\pi^-\gamma$ channel. The resulting distribution (including production of both $\rho^o\pi^o$ and $\rho^{\pm}\pi^{\mp}$) is compared with the data in Fig. 10 and can clearly account for the observed ρ^o peak.

Figure 11 shows the $\pi^+\pi^-$ mass distribution after subtraction of the $\pi^+\pi^-\pi^o$ background. The distribution is dominated by the f. An expression consisting of a Breit-Wigner resonance term plus a flat background term was fitted to this distribution. The curve in Fig. 11 shows the best fit which gave M = 1280 MeV and Γ = 180 MeV for the resonance parameters. The branching fraction for (3) was found to be $B(\psi \to \gamma f) = (1.3 \pm 0.3) \times 10^{-3}$. This branching fraction is consistent with the previously measured values of $B(\psi \to \gamma f) = (2.0 \pm 0.3) \times 10^{-3}$ from PLUTO[20] and $B(\psi \to \gamma f)$ between $(0.9 \pm 0.3) \times 10^{-3}$ and $(1.5 \pm 0.4) \times 10^{-3}$ (depending on the helicity of the f in the final state) from DASP.[21]

As pointed out in the previous section, we seem to have a fairly good understanding of the transitions to the $I_z = 0$ members of the $J^P = 0^-$ nonet. If measurements of the radiative transitions to the f' and A_2^o could be made, we would have an additional check on the theoretical ideas discussed previously. We have preliminary results which show no evidence for transitions to either of these two states. They give 90% confidence level upper limits of $B(\psi \to \gamma f') \times B(f' \to K\bar{K}) < 10^{-3}$ and $B(\psi \to \gamma A_2^o) < 10^{-3}$. Unfortunately, these limits are not yet small enough to provide meaningful constraints on models. As in the case of the $\gamma\pi^o$ transition, one expects to see a very small branching fraction for γA_2^o because of isospin conservation. However, the $\gamma f'$ transition should be observable. Based on a naive calculation assuming SU(3) invariance (similar to the η-η' calculation described earlier,[34] one expects

$$\frac{B(\psi \to \gamma f')}{B(\psi \to \gamma f)} = \frac{1}{2} \quad .$$

Our limit is not yet inconsistent with this prediction.

C. $\psi \to \gamma E(1420)$

As the E(1420) is a fairly obscure resonance, I will briefly review what was known about the E as of the last (1978) Particle Data Group tables[35] before discussing the results on the γE radiative transition. The E is a fairly narrow resonance with width estimates ranging from 40 to 80 MeV. Measurements of the mass lie between 1400 and 1440 MeV. None of the quantum numbers of the E have been firmly established. The isospin is believed to be zero as no charged E has ever been observed; the C-parity is believed to be even; and analyses of the decay Dalitz plot favor an abnormal spin-parity assignment. $J^P = 0^-$ and 1^+ are the preferred values. The principally observed decay mode is $K\bar{K}\pi$, but there is some evidence for an $\eta\pi\pi$ decay mode. Finally, up until 1978, the best signals for the E were observed in $p\bar{p}$ annihilations at rest. I will mention only one of these experiments here. Baillon et al.[36] studied a sample of $p\bar{p}$ annihilations in the CERN 81-cm hydrogen bubble chamber. They did a spin-parity analysis of the E observed in the reaction $p\bar{p} \to E\pi\pi$ and determined $J^P = 0^-$.

We see evidence for the process[19]

$$\psi \to \gamma E \ , \ E \to K_S K^{\pm}\pi^{\mp} \quad . \tag{4}$$

Observation of this transition establishes C = + for the E. Figure 12(a) shows the $K_S K^{\pm}\pi^{\mp}$ invariant mass for events satisfying the 5C fit to (4) with $\chi^2 < 15$.[37] The constraints are the normal ones of energy-momentum conservation with an additional constraint for the K_S mass. A peak is seen near the mass of the E(1420). One is not compelled to interpret this structure as the E(1420), but due to the similar characteristics of this structure and the previously observed E, I will make this tentative assignment.

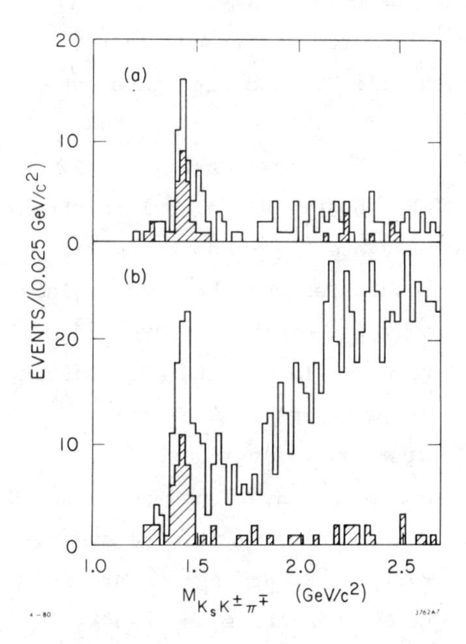

Fig. 12. $K_S K^{\pm}\pi^{\mp}$ invariant mass distributions for events satisfying a) 5C fits and b) 2C fits (i.e., observation of the photon is not required) to (4). Shaded regions have the additional requirement $M_{K\bar{K}} < 1.05$ GeV.

The parameters of the resonance were obtained by fitting the invariant mass distribution to a Breit-Wigner[38] plus a smooth background. We find $M = 1.44^{+0.01}_{-0.015}$ GeV and $\Gamma = 0.05^{+0.03}_{-0.02}$ GeV. These errors include systematic uncertainties due to the functional form used in the fit. The branching fraction product, based on 47 ± 12 observed events, is $B(\psi \to \gamma E) \times B(E \to K_S K^{\pm}\pi^{\mp}) = (1.2 \pm 0.5) \times 10^{-3}$.[39] With the assumptions that the E is an isoscalar and that K_S and K_L production are equal in the decay of the E, one can relate the $K^+K^-\pi^o$, $K^o\bar{K}^o\pi^o$, and $K^oK^{\pm}\pi^{\mp}$ branching fractions and determine the branching fraction product $B(\psi \to \gamma E) \times B(E \to K\bar{K}\pi) = (3.6 \pm 1.4) \times 10^{-3}$.

Previous experiments[35] have found the decay of the E to be associated with a low mass $K\bar{K}$ enhancement which we also observe. If a cut requiring $M_{K\bar{K}} < 1.05$ GeV is imposed on the data, the shaded region in Fig. 12(a) is obtained.

Since the signal is quite clean, it is possible to relax the requirement that the photon be observed. The resulting 2C fit to (4) is shown in Fig. 12(b). Although there is an improvement in statistics,[40] there is also an increase in the background level. However, as shown by the shaded region, the $K\bar{K}$ mass cut again substantially reduces the background.

The Dalitz plot for the sample of events shown in Fig. 12(b) with masses between 1.375 and 1.500 GeV (the signal region) is shown in Fig. 13. The curves show the low-mass and high-mass kinematic boundaries and the dashed lines show the nominal $K^*(890)$ mass values. The points are plotted as functions of the $(K\pi)^o$ invariant mass squared vs. the $(K\pi)^{\pm}$ invariant mass squared. The $K\bar{K}$ axis, if it were shown, would be at an angle approximately bisecting the two $K\pi$ axes. One sees an excess of events in the upper right-hand corner of the Dalitz plot.[41] It is not clear whether these events correspond to a low-mass $K\bar{K}$ enhancement (spread out by the movement of the kinematic boundary as the $K\bar{K}\pi$ mass changes), or to constructive interference where the K^* bands overlap.

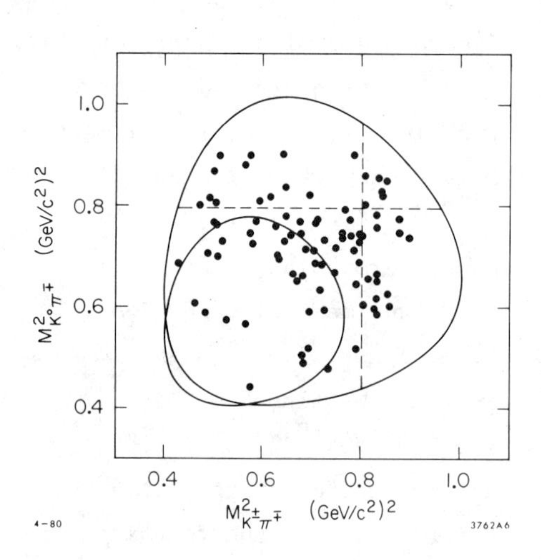

Fig. 13. Dalitz plot for events with $1.375 \leq M_{K\bar{K}\pi} < 1.500$ GeV. Curves show low-mass and high-mass kinematic boundaries. Dashed lines show nominal K^* mass values.

Figure 14(a) shows the $K_S K^{\pm}$ invariant mass distribution for events in the signal region and Fig. 14(b) shows the corresponding distribution for events outside the signal region. There is evidence for a low-mass $K\bar{K}$ enhancement for events in the signal region which is absent for events outside the signal region. One possible interpretation of this enhancement is the $\delta(980)$.

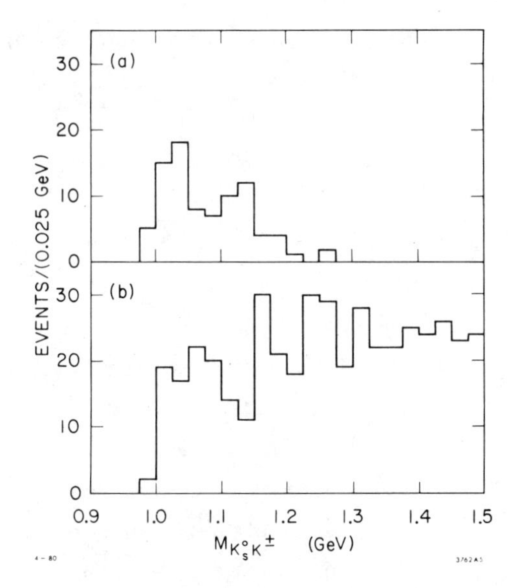

Fig. 14. $K_SK^{\pm}$ invariant mass distributions for events a) in the signal region and b) outside the signal region.

In an attempt to understand the decay mechanism of the E, fits were made to the Dalitz plot which included $K^*\bar{K}$ (the inclusion of both this state and the charge conjugate state are implied by this notation), $\delta\pi$, and phase space contributions. These three contributions were added incoherently, but the $K^*\bar{K}$ contribution included components from both the charged and neutral K^* states, which were assumed to interfere constructively where they cross on the Dalitz plot (as demanded by the even C-parity of the E). The best fit favors $\delta\pi$ as the primary component of the decay with

$$\frac{B(E \to \delta\pi) \times B(\delta \to K\bar{K})}{B(E \to K\bar{K}\pi)} = 0.8 \pm 0.2$$

The quoted error does not include possible systematic errors. One has to be careful in interpreting this result, as the best fit to the Dalitz plot does not completely simulate the $K\bar{K}$ invariant mass distribution. This indicates that the decay mechanism is not completely understood.

An attempt has been made to determine the spin of the E by analysis of the double decay angular distribution for events consistent with

$$\psi \to \gamma E \ , \ E \to \delta\pi \quad .$$

However, the limited statistics do not allow a statistically significant determination of the spin.

Preliminary results from the Crystal Ball also show evidence for the transition $\psi \to \gamma E$.[14] Figure 15 shows the $K^+K^-\pi^o$ invariant mass distribution[42] for events which satisfy the 2C fit to

$$\psi \to \gamma K^+K^-\pi^o \quad , \qquad (5)$$

with $M_{K\overline{K}} < 1.1$ GeV. Although the Crystal Ball detector has excellent energy resolution for photons, the absence of a magnetic field does not allow a momentum measurement for charged particles. This reduces the constraint class for (5) from 4 to 2. Evidence for an E signal is seen in this distribution. As the Crystal Ball efficiency calculations are still in a very preliminary state, estimates of the branching fraction are only good to a factor of two at best. When corrections are made for the K^+K^- mass cut[43] and the unobserved decay modes of the E, they find $B(\psi \to \gamma E) \times B(E \to K\overline{K}\pi) \approx 2 \times 10^{-3}$.

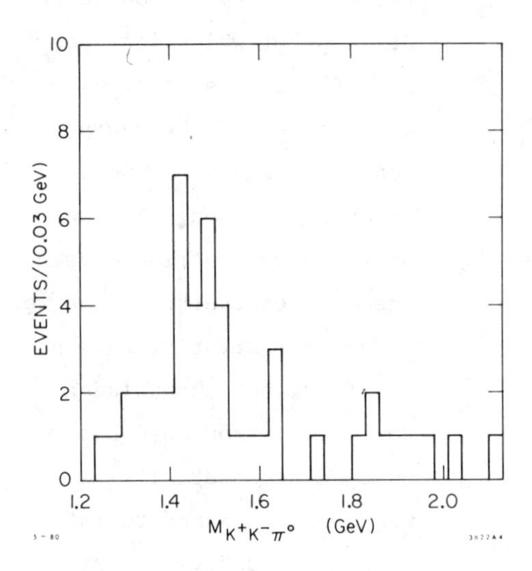

Fig. 15. $K^+K^-\pi^o$ invariant mass distribution for events satisfying (5) with $M_{K\overline{K}} < 1.1$ GeV (data from the Crystal Ball collaboration).

As was mentioned earlier, there is some evidence for the decay of the E into $\eta\pi^+\pi^-$. Figure 16 shows the $\eta\pi^+\pi^-$ invariant mass distribution (from the Crystal Ball) for events satisfying fits to

$$\psi \to \gamma\eta\pi^+\pi^- \quad . \qquad (6)$$

In addition to the η' signal, there is evidence for a peak in the E mass region. A preliminary estimate of the branching fraction product $B(\psi \to \gamma E) \times B(E \to \eta\pi\pi)$ finds it to be smaller than the corresponding number for $K\overline{K}\pi$, but a firm number will have to wait until calculations of the efficiencies are made.

To summarize our results, the E is observed very strongly in radiative transitions from the ψ. The only other transition that has

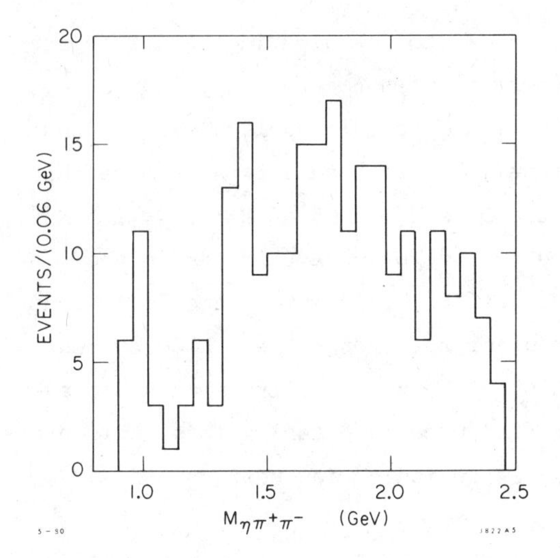

Fig. 16. $\eta\pi^+\pi^-$ invariant mass distribution for events satisfying (6) (data from the Crystal Ball collaboration).

been observed with a comparable branching fraction is the $\gamma\eta'$ transition. Observation of this transition has established the C-parity of the E as even. Unfortunately, a determination of the spin is impossible with the present statistics. Finally, we find the $K\bar{K}\pi$ decay mode of the E to be predominantly $\delta\pi$.

Recent results on the E have come from two hadronic experiments which I would like to discuss briefly, as these results are relevant to the interpretation of the E. The first results are from a high statistics (90 events/μb) bubble chamber experiment in which the reaction

$$\pi^- p \to K_S K^{\pm} \pi^{\mp} n$$

was studied at 3.95 GeV/c.[44] They observe significant E production, and measure values for the E mass and width of M = 1426 ± 6 MeV and Γ = 40 ± 15 MeV. A partial-wave analysis of the data determined the spin-parity of the E to be $J^P = 1^+$, and also determined the branching fraction ratio

$$\frac{B(E \to K^*\bar{K})}{B(E \to K^*\bar{K} + \delta\pi)} = 0.86 \pm 0.12 \quad .$$

However, it should be pointed out that their E signal is over a relatively large background which has a significant $K^*\bar{K}$ component, so that one should regard this result with caution.

In experiment E110 using the Fermilab multiparticle spectrometer, the reaction

$$\pi^- p \to K_S K^{\pm} \pi^{\mp} + X$$

was studied at 50 and 100 GeV/c.[45] Without kinematic cuts, they see no evidence for an E signal. However, if a δ cut is applied, a prominent E signal is observed. If a K^* cut is applied rather than a δ cut, one still sees an E signal, but with considerably more background. They find a value for the mass of the E of M = 1440 ± 6 MeV. The width is not well determined because of uncertainties in the shape of the background. On the surface, this data seems to indicate a preference for the $\delta\pi$ decay mode of the E over the $K^*\bar{K}$ decay mode. However, questions of kinematic overlap in the Dalitz plot and phase space boundaries have not been considered in detail. Thus, this preference should be considered only as an indication until a more sophisticated analysis is done.

Despite all the new information on the E from recent experiments, the situation is not much clearer than it was in 1978. One point of controversy is whether the E decays predominantly into $\delta\pi$ or $K^*\bar{K}$. This experiment (and possibly also the Fermilab experiment of Bromberg et al.[45]) seems to favor the decay $E \to \delta\pi$. On the other hand, Dionisi et al.[44] see little evidence for $\delta\pi$ and find the predominant decay of the E is into $K^*\bar{K}$. As for the spin, Dionisi et al. find $J^P = 1^+$ which agrees with some earlier results, but disagrees with others. However, their determination of the spin goes hand-in-hand with the determination of the predominance of the $K^*\bar{K}$ decay mode. Since this predominance is not firmly established, I think that one should still consider the spin of the E to be an open question until the decay mechanism is understood better.[46]

To understand my reasons for this excessive interest in the quantum numbers of the E, let me refer for the last time to Fig. 6(b). As discussed by numerous people,[47] if gluonium states[48] exist, the process shown in Fig. 6(b), after elimination of the outgoing quark lines, would be an ideal process for production of such states. I would like to suggest the possibility that the E might be such a gluonium state, rather than an ordinary $q\bar{q}$ resonance. Although there is certainly no real evidence for this hypothesis, there are a few peculiarities associated with the γE radiative transition from the ψ which I would like to point out.

First, the branching fraction for $\psi \to \gamma E$ is larger than the corresponding branching fractions for transitions to other ordinary hadrons, with the possible exception of the η'. This is in contrast to hadronic experiments where E production is in general small compared to the production of other resonances. This would lead one to infer a connection between the E and the 2-gluon intermediate state in Fig. 6(b). Whereas the production of gluonium states is expected to be significant in ψ radiative transitions, there is no reason to expect significant production of such states in hadronic reactions.

Second, whereas in most hadronic experiments in which an E is observed to decay into $K\bar{K}\pi$, one observes roughly comparable D(1285) production. Neither this experiment nor the Crystal Ball experiment sees much evidence for D production. We have calculated an upper limit for D production from our data of
$B(\psi \to \gamma D) \times B(D \to K\bar{K}\pi) < 0.7 \times 10^{-3}$ at the 90% confidence level. This might be taken as strong evidence for a difference in the production mechanisms involved in the two different processes, and hence an indication of a large gluonium component in the E. However, if one assumes that the D and E are both members of the standard $J^{PC} = 1^{++}$ nonet, and the E is the primarily singlet state and the D is the primarily octet state,[49] one would expect D production to be suppressed relative to E production because of SU(3) symmetry arguments. Thus, this suppression may not be relevant to the gluonium question at all.

In my opinion, the most important question which should be resolved regarding the E is its spin. If the E can be firmly established as an axial vector state, there is no reason not to make the standard $q\bar{q}$ meson interpretation and put it in the same nonet as the D(1285), A_1, and Q_A. If, on the other hand, the E is finally established as a pseudoscalar, it is difficult to interpret it within the standard quark model. The $J^P = 0^-$ nonet is complete, and one would have to consider the existence of another 0^- nonet, possibly a radial excitation of the ground state, in order to accommodate the E. However, I think it is equally plausible to interpret the E as a gluonium state.

IV. PROPERTIES OF D MESONS

Evidence for production of charmed D mesons by reconstruction of exclusive final states came first from the SLAC-LBL magnetic detector collaboration at SPEAR.[50] I will present some recent results based on a sample of events taken at the $\psi''(3772)$ which corresponds to an integrated luminosity of 2850 nb^{-1}.[51]

In the charmonium model,[52] the ψ'' is presumed to be the $^3D_1$ level of the $c\bar{c}$ system, lying between $D\bar{D}$ and $D\bar{D}^*$ threshold. Because it is less than 100 MeV above the $\psi'(3684)$, but has a total width about two orders of magnitude greater than the ψ' width, the width is generally attributed solely to the strong decay of the resonance into the newly-opened $D\bar{D}$ channel. If the ψ'' has a unique isospin (either 0 or 1), then it couples equally (within phase space factors) to pairs of charged and neutral D mesons. Thus, a measurement of the ψ'' resonant line shape permits a local evaluation of the inclusive D^o and D^+ production cross sections. Our results from a fine scan over the ψ'' give

$$\begin{aligned} \sigma_{D^o} &= 8.0 \pm 1.0 \pm 1.2 \text{ nb} \\ \sigma_{D^+} &= 6.0 \pm 0.7 \pm 1.0 \text{ nb} \end{aligned} \qquad (7)$$

at $E_{c.m.} = 3.771$ GeV.[53] The first error is statistical and the second is systematic. Knowledge of these inclusive cross sections allows a determination of absolute branching fractions for observed D decays. In addition, the proximity of the ψ'' to $D\bar{D}$ threshold results in low momentum, 2-body production, which permits a further reduction of background through kinematic constraints.[54]

A. Exclusive Decays

The branching fractions for various Cabibbo-favored decay modes (i.e., modes containing either one charged or neutral kaon) have been measured.[55] Figure 17 shows the beam energy-constrained invariant mass distributions for the D^o decays $K^-\pi^+$, $K_S\pi^+\pi^-$, and $K^-\pi^+\pi^+\pi^-$. (For clarity, references to D^o or D^+ and their decay modes will be assumed to refer to both the state and its charge conjugate.)

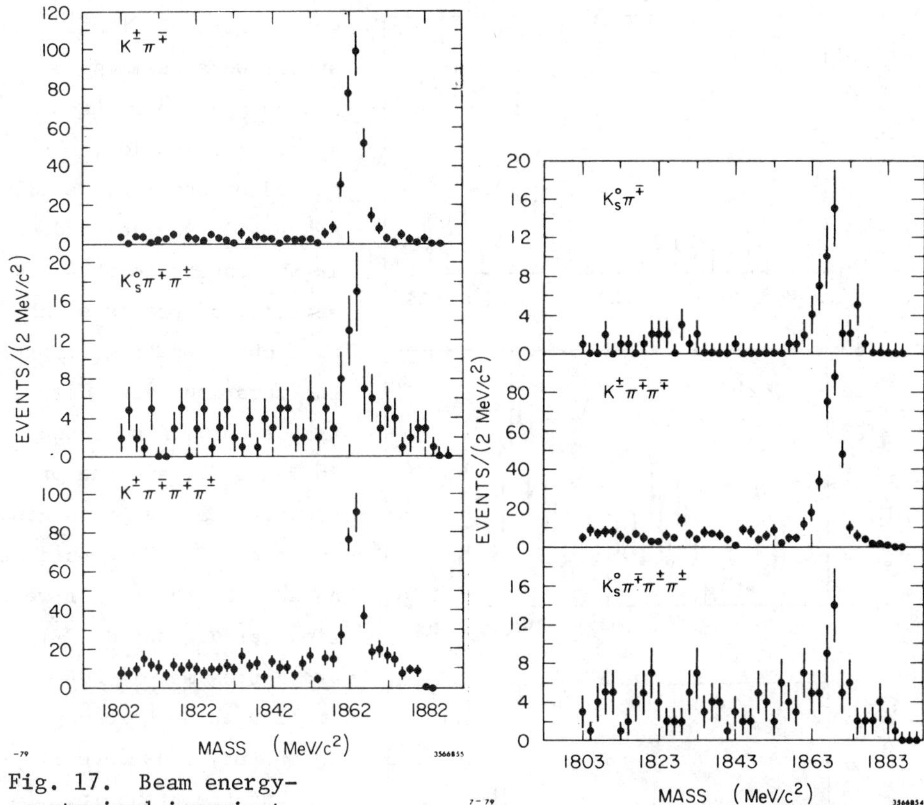

Fig. 17. Beam energy-constrained invariant mass distributions for D^o decays detected in all charged particle modes.

Fig. 18. Beam energy-constrained invariant mass distributions for D^+ decays detected in all charged particle modes.

Figure 18 shows the invariant mass distributions for the D^+ decays $K_S\pi^+$, $K^-\pi^+\pi^+$, and $K_S\pi^+\pi^+\pi^-$. Figure 19 shows the invariant mass distributions for the decays $K_S\pi^o$, $K^-\pi^+\pi^o$, and $K_S\pi^+\pi^o$, each of which contains a single π^o. These distributions provide evidence for the previously unmeasured decays of the D^o into $\overline{K}^o\pi^o$ and D^+ into $K^o\pi^+\pi^o$ and $\overline{K}^o\pi^+\pi^+\pi^-$.

Table III gives the calculated branching fractions for these nine decay modes. The detection efficiency for each decay mode was determined by Monte Carlo calculation. Except for the channels $\overline{K}^o\pi^+\pi^-$ and $K^-\pi^+\pi^o$, where the measured resonant substructure in the decay (to be discussed shortly) was used in the Monte Carlo event

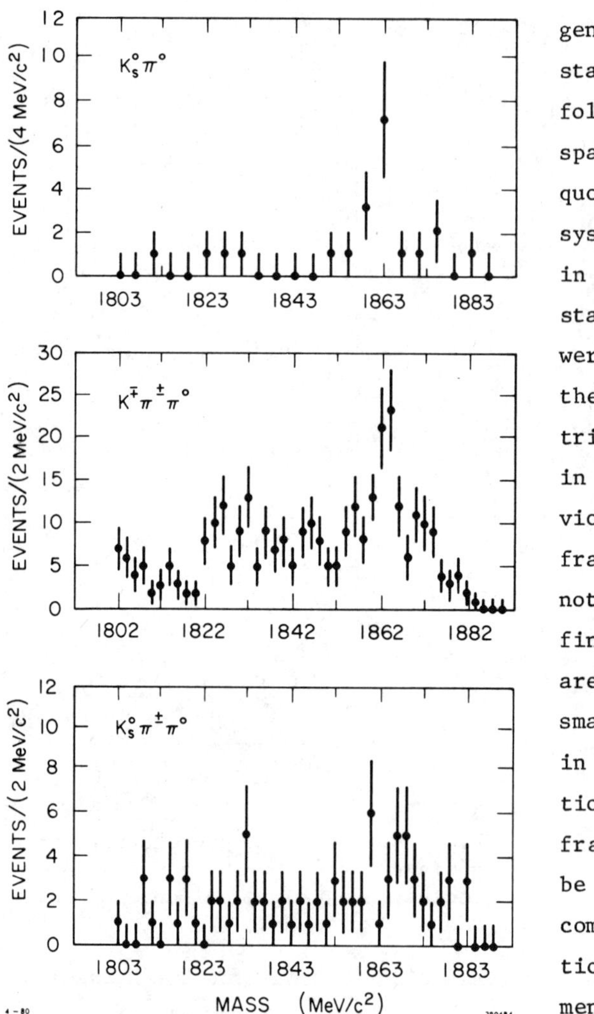

Fig. 19. Beam energy-constrained invariant mass distributions for D^o and D^+ detected in modes with a single π^o.

generation, all final states were assumed to follow a uniform phase space distribution. The quoted errors include all systematic sources added in quadrature to the statistical errors which were obtained from fits to the invariant mass distributions. Also shown in Table III are the previously measured branching fractions.[56] It should be noted that the values we find for σ_{D^o} and σ_{D^+}[53] are approximately 30% smaller than those employed in the previous determination of the branching fractions.[56] This should be taken into account when comparing branching fractions from the two experiments.

We have analyzed the Dalitz plot distributions for the $K^-\pi^+\pi^o$ and $\overline{K}^o\pi^+\pi^-$ decays of the D^o and the $K^-\pi^+\pi^+$ decay of the D^+. The apparently large branching fraction of the D^o into the $K^-\pi^+\pi^o$ relative to the D^+ decay into $K^-\pi^+\pi^+$ has always been theoretically hard to understand.[57] However, our new results on the resonant substructure in these 3-body decays combined with the apparent difference in total widths of the D^+ and D^o (to be discussed later) allows a resolution of the problem.[58]

Table III. Branching fractions for Cabibbo-favored D decays.

decay mode	branching fraction (%) Mark II	LGW[a]
$K^-\pi^+$	3.0 ± 0.6	2.2 ± 0.6
$\overline{K}^o\pi^o$	2.2 ± 1.1	--
$\overline{K}^o\pi^+\pi^-$	3.8 ± 1.2	4.0 ± 1.3
$K^-\pi^+\pi^o$	8.5 ± 3.2	12.0 ± 6.0
$K^-\pi^+\pi^+\pi^-$	8.5 ± 2.1	3.2 ± 1.1
$\overline{K}^o\pi^+$	2.3 ± 0.7	1.5 ± 0.6
$K^-\pi^+\pi^+$	6.3 ± 1.5	3.9 ± 1.0
$\overline{K}^o\pi^+\pi^o$	12.9 ± 8.4	--
$\overline{K}^o\pi^+\pi^+\pi^-$	8.4 ± 3.5	--

[a] Ref. 56

Figure 20(a) shows the Dalitz plot for the decay $D^o \rightarrow \overline{K}^o\pi^+\pi^-$. Figure 20(d) shows the low-mass $\overline{K}^o\pi^\pm$ mass-squared projection. One sees a significant population of the K^* hands, but no strong evidence for ρ production. To determine the amount of resonant substructure in the decay, we performed a maximum-likelihood fit to the data in the Dalitz plot. We used a density function which represented the allowed final-state channels plus background, with corrections for detector acceptance made at each point. In the fit, the ρ and K^* amplitudes were represented as p-wave Breit-Wigner line shapes, with energy dependent widths and with the appropriate decay angular distributions. The amplitudes of all indistinguishable resonant final states that were accessible in the decay were allowed to interfere with arbitrary phase. The non-resonant components (i.e., pure 3-body decays and background events) were added incoherently to the density function used in the fit. The fractions of $\overline{K}^o\pi^+\pi^-$ events in each final state (as determined by the fit) are

$$f(K^{*-}\pi^+) = 0.70 \begin{matrix} +\ 0.13 \\ -\ 0.15 \end{matrix} \begin{matrix} +\ 0.05 \\ -\ 0.06 \end{matrix}$$

$$f(\overline{K}^o\rho^o) = 0.02 \begin{matrix} +\ 0.08 \\ -\ 0.02 \end{matrix} \pm 0.03$$

$$f(\overline{K}^o\pi^+\pi^-) = 0.30 \begin{matrix} +\ 0.16 \\ -\ 0.13 \end{matrix} \pm 0.05$$

The first set of errors in each case is the statistical error derived from the likelihood function in the fitting procedure. The second set of errors is the estimated systematic uncertainty from the

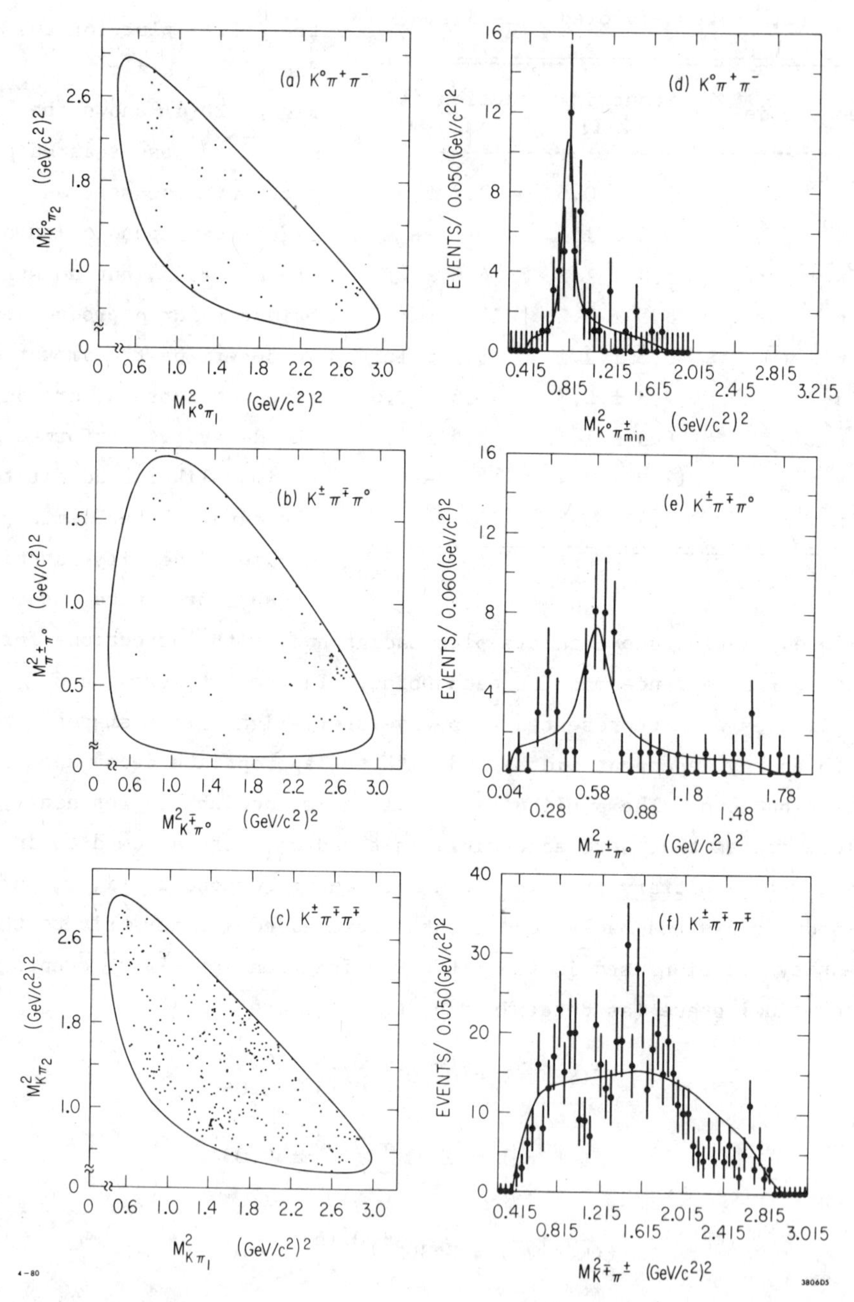

Fig. 20. Dalitz plots for the decays a) $D^o \to \bar{K}^o\pi^+\pi^-$, $D^o \to K^-\pi^+\pi^o$, and c) $D^+ \to K^-\pi^+\pi^+$. Invariant mass-squared projections for d) low-mass $\bar{K}_o\pi^\pm$ from $\bar{K}^o\pi^+\pi^-$, e) $\pi^+\pi^o$ from $K^-\pi^+\pi^o$, and f) $K^-\pi^+$ from $K^-\pi^+\pi^+$. Curves represent fitted solutions discussed in text.

Monte Carlo statistics and acceptance calculation, and our assumptions about backgrounds and resonance line shapes. (The values we quote represent the fractions of channels in the absence of interference, and hence the fractions will not necessarily sum to unity.) The curve in Fig. 20(d) represents the fitted solution.

Figure 20(b) shows the $K^-\pi^+\pi^o$ Dalitz plot, with $\pi^+\pi^o$ mass-squared projection shown in Fig. 20(e). The ρ peak is evident in the $\pi^+\pi^o$ mass projection, but it is only seen on the right side of the Dalitz plot. One would expect the ρ band to extend across the plot from one boundary to the opposite boundary. However, the π^o detection efficiency varies rapidly from the upper right edge of the plot (where it is relatively large) to the lower left corner (where it is very small). This results in good detection efficiency for ρ's only near the right side of the Dalitz plot. The results of a fit to the Dalitz plot (done in an identical manner to that for the previous Dalitz plot) are

$$f(K^-\rho^+) = 0.85 \begin{array}{l} +\ 0.05 \\ -\ 0.11 \end{array} \begin{array}{l} +\ 0.09 \\ -\ 0.10 \end{array}$$

$$f(\overline{K}^{*o}\pi^o) = 0.11 \begin{array}{l} +\ 0.12 \\ -\ 0.06 \end{array} \pm 0.10$$

$$f(K^{*-}\pi^+) = 0.07 \begin{array}{l} +\ 0.06 \\ -\ 0.04 \end{array} \begin{array}{l} +\ 0.05 \\ -\ 0.02 \end{array}$$

$$f(K^-\pi^+\pi^o) = 0.06 \pm 0.04 \pm 0.05$$

An independent estimate for the $K^{*-}\pi^+$ fraction of this decay mode can be made from the measured $K^{*-}\pi^+$ fraction of the $\overline{K}^o\pi^+\pi^-$ decay, the $K^-\pi^+\pi^o$ and $\overline{K}^o\pi^+\pi^-$ branching fractions from Table III, and the charged K^* branching fractions. We expect a value of $f(K^{*-}\pi^+) = 0.15 \pm 0.07$, which is consistent with the fraction obtained from the direct fit to the $K^-\pi^+\pi^o$ Dalitz plot.

The Dalitz plot for the $K^-\pi^+\pi^+$ decay of the D^+ is shown in Fig. 20(c). Our large sample (292 events with an estimated background of approximately 12%) provides the first evidence for a non-uniform population of the Dalitz plot. The detection efficiency is very uniform across the plot, dropping only near the three corners where π

and K momenta fall below approximately 100 MeV/c. Figure 20(f) shows the sum of the two $K^-\pi^+$ invariant mass-squared combinations and a curve indicating the distribution expected for a purely uniform phase space decay. While no evidence for a significant K^* signal exists, we note again the large deviation of the data from uniform phase space. We present only an upper limit on the $\overline{K}^{*o}\pi^+$ channel in this decay by assuming that all of the events observed in the resonance region ($0.685 \leq M_{K\pi} < 0.905$ GeV^2) arise from the decay $\overline{K}^{*o}\pi^+$. This assumption results in a 90% c.l. upper limit of 0.39 for the fraction of this decay mode in $\overline{K}^{*o}\pi^+$. (Additional systematic errors may change this limit by ±0.06.)

B. Inclusive Properties

The uniqueness of the $D\overline{D}$ final state at the ψ'' provides a means of studying the inclusive properties of D decays by use of "tagged" events. In a tagged event, one D ($\overline{D}$) is identified by the observed decay into $\overline{K}^o\pi^+$, $K^-\pi^+$, $K^-\pi^+\pi^+$, or $K^-\pi^+\pi^-\pi^+$. The recoiling system provides a pure $\overline{D}$ (D) sample. This sample of D's is unbiased by the trigger requirement which is satisfied in all cases by the tagged D. Cuts on the beam energy-constrained masses of ±6 MeV around the nominal D mass (±4 MeV in the case of the $K^-\pi^+\pi^-\pi^+$ decay mode) were made to select the sample of tagged events used in the analysis. The resulting sample consists of approximately 300 D^+ and 480 D^o events over a background equal to about 12% of the signal. In this section, I will discuss the inclusive charged particle multiplicity and strangeness associated with D decays. In the next section, I will discuss the individual D^o and D^+ semileptonic decay fractions.

For each tag, the multiplicity observed in the recoiling system is plotted, with no attempt at particle identification. (For this analysis only, no $\overline{K}^o\pi^+$ tags were used.) These observed charged particle multiplicity distributions are shown in Figs. 21(a)-(c). The shaded regions represent the background multiplicity distributions estimated from events with invariant mass between 1.800 and 1.855 GeV, normalized to the expected number of background events contaminating the tagged D sample. The produced charged particle multiplicity distributions (where $K^o \to \pi^+\pi^-$ decays are counted as two tracks)

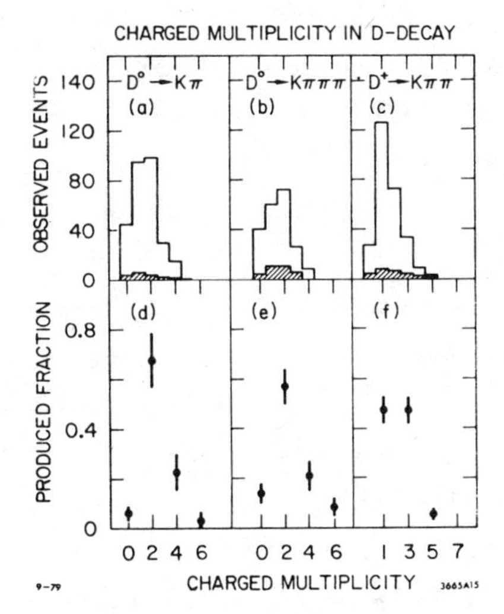

Fig. 21. Observed charged particle multiplicity distributions for a) $D^o \to K^-\pi^+$, b) $D^o \to K^-\pi^+\pi^+\pi^-$, and c) $D^+ \to K^-\pi^+\pi^+$ tagged events. Produced (unfolded) charged particle multiplicity distributions for d) $D^o \to K^-\pi^+$, e) $D^o \to K^-\pi^+\pi^+\pi^-$, and f) $D^+ \to K^-\pi^+\pi^+$ tagged events. Shaded regions are discussed in text.

were obtained by a numerical unfold procedure.[59] Briefly, a solution is sought for $\tilde{N}_p$ in the over-constrained linear system of equations

$$N_q = \sum_p \varepsilon_{qp} \tilde{N}_p + B_q \quad ,$$

where N_q is the number of events detected with q charged prongs, $\tilde{N}_p$ is the number of events produced with p charged prongs, B_q is the number of background events with q charged prongs, and ε_{qp} is the probability (determined by Monte Carlo) that a produced event with p prongs will be detected with q prongs. The produced (unfolded) multiplicity distributions are shown in Figs. 21(d)-(f), where the error bars reflect only the statistics of the unfold procedure. The systematic errors are comparable. The mean charged particle multiplicities are calculated to be $\langle N_{ch} \rangle_{D^o} = 2.46 \pm 0.14$ (based on both samples of D^o tags) and $\langle N_{ch} \rangle_{D^+} = 2.16 \pm 0.16$, where the errors include estimates of the systematic uncertainties. These results are in good agreement with the previously reported value for $\langle N_{ch} \rangle$ of 2.3 ± 0.3 for both the D^o and D^+.[60]

Theoretical estimates of the average D multiplicities have generally been larger than these measured values.[61] However, at least one type of statistical model discussed by Quigg and Rosner[61] (the constant-matrix-element model) predicts average multiplicities which are close to the experimental values. They predict $\langle N_{ch} \rangle_{D^o} = 2.4$ and $\langle N_{ch} \rangle_{D^+} = 2.5$ for hadronic decays of the D.

However, it should be noted that our measurements include contributions from the semileptonic decays, which comprise a large fraction of the D^+ decays (as will be discussed in the next section). If a large fraction of the semileptonic decays are $D^+ \to K^o \ell^+ \nu_\ell$, one would expect a smaller average multiplicity for D^+ decays than for D^o decays, as is observed in the data.

The tagged D samples were chosen to have unique strangeness, so that charged kaons in the recoiling system could be characterized as having either the same or opposite strangeness. The Cabibbo-favored decays should produce one kaon whose strangeness is opposite that of the tag, while Cabibbo-suppressed decays are expected to exhibit either no strange particles, or two of opposite strangeness. $D^o - \overline{D}^o$ mixing and doubly suppressed decays can produce a particle with strangeness equal to the strangeness of the kaon in the tag, but these effects are expected to be very small.[62]

Table IV summarizes the results of an analysis to determine the kaon fractions in the systems recoiling against the tagged D's. These numbers have been corrected for kaon detection efficiency, misidentification of $\pi^\pm$ as $K^\pm$, and kaons from the background events which

Table IV. Kaon multiplicities in D decays.

decay mode	branching fraction (%) D^o	D^+	experiment
$D \to K^- X$	56 ± 11	19 ± 5	Mark II
	36 ± 10[a)]	10 ± 7	LGW[b)]
$D \to K^+ X$[c)]	8 ± 3	6 ± 4	Mark II
	--[a)]	6 ± 6	LGW[b)]
$D \to K^o X$[d)]	29 ± 11	52 ± 18	Mark II
	57 ± 26	39 ± 29	LGW[b)]

[a)] The LGW did not separate the same-strangeness and opposite-strangeness events; both are combined under $D \to K^- X$.

[b)] Ref. 60.

[c)] These decays have kaons with the same strangeness as the kaon in the tag.

[d)] Here the strangeness cannot be determined.

contaminate the tagged D sample. The errors include both statistical errors and estimated systematic errors. Also shown in Table IV is a comparison of our results with those of Ref. 60.

It is expected that if one could extract from the data the sample of events that correspond only to Cabibbo-favored decays, the kaon fraction (including both charged and neutral kaons) should be 100%. This is not possible, but one can make assumptions about the Cabibbo-suppressed decays, and estimate this fraction. If it is assumed that $B(D \to K\bar{K} + n\pi) \approx 0.05$[63] and $B(D \to \pi\text{'s only}) \approx 0.05$, one finds that $86 \pm 17\%$ of Cabibbo-favored D^o decays contain a kaon and $70 \pm 21\%$ of Cabibbo-favored D^+ decays contain a kaon. These numbers were derived independently of the measured wrong-sign kaon fractions, and the errors do not include systematic uncertainties in the assumptions made about the fraction of Cabibbo-suppressed decays.

C. Semileptonic D Decays and the D Lifetime

The pure leptonic decays of D mesons are expected to be strongly suppressed relative to the semileptonic decays,[62] implying that electrons originating from D decays will come predominantly from the semileptonic modes. Previous experiments[64] have measured the average semileptonic branching ratios for D mesons[65] at several center-of-mass energies in e^+e^- annihilations. However, until recently, no measurements of the individual semileptonic branching fractions of the D^o and D^+ have been made. Whereas it is expected that the semileptonic partial widths of the D^o and D^+ are approximately equal,[66] there is no overriding reason to expect that the total D^o and D^+ widths are equal. This inequality in the total widths would lead to a difference in the D^o and D^+ semileptonic branching fractions. Hence, a measurement of the ratio of the two semileptonic branching fractions allows an estimate of the relative lifetimes to be made

$$\frac{\tau_{D^+}}{\tau_{D^o}} = \frac{B(D^+ \to Xe\nu)}{B(D^o \to Xe\nu)} .$$

In order to measure the D^o and D^+ semileptonic branching fractions, the sample of tagged events described in the previous section was searched for electron candidates. All events with electron candidates were hand scanned to remove visible photon conversions. All remaining tracks were separated by charge relative to the strangeness of the tag. We denote these two groups as "right-sign" (having the expected leptonic charge) and "wrong-sign" (having opposite the expected leptonic charge) candidates. Possible back-grounds were carefully considered. Backgrounds arising from hadronic misidentification are charge asymmetric (i.e., the right-sign and wrong-sign backgrounds differ) and strongly correlated with the strangeness of the tagged D. To estimate this background, the measured D-decay momentum distributions were folded with the known hadron misidentification probabilities to estimate the contamination in each group of electrons. Backgrounds from electromagnetic processes are charge symmetric and uncorrelated with the strangeness of the tagged D. This background was corrected for, after corrections for all other backgrounds were made, by subtracting the wrong-sign rate from the right-sign rate to get the net electron rate. Contamination from leptonic kaon decays was estimated to be negligible. Contamination of the tagged D sample was estimated from events below the D mass ($1.800 \le M_{tag} < 1.855$ GeV) and corrections made for this. Finally, some $K^-\pi^+$ events were mislabeled as π^-K^+, which gave an incorrect strangeness assignment. Monte Carlo estimates were used to correct for this contamination.

A summary of the semileptonic rate calculation is shown in Table V. After correction of the number of net semileptonic events for detection efficiency and the number of tags, we find

$$B(D^+ \to Xe\nu) = 16.8 \pm 6.4\%$$

$$B(D^o \to Xe\nu) = 5.5 \pm 3.7\% \quad .$$

While these values are dominated by the statistical errors, they also include our estimates of the systematic errors. Weighting these values by the relative D^o and D^+ production cross sections at the ψ'' (7), we obtain an average single electron branching fraction of $10.0 \pm 3.2\%$. This is consistent with $8.0 \pm 1.5\%$ measured by DELCO[64] and $7.2 \pm 1.8\%$ measured by the LGW[64] at the same center-of-mass energy.

Table V. D semileptonic rate calculation.

$D^o - K^-\pi^+$ and $K^-\pi^+\pi^-\pi^+$ tags	
net tagged events	477
right-sign electrons observed[a)]	36
background[b)]	17.4 ± 1.0
wrong-sign electrons observed[a)]	18
background[b)]	11.8 ± 0.9
additional backgrounds[c)]	-0.1 ± 1.0
net electrons	12.3 ± 7.6
$D^+ - \overline{K}^o\pi^+$ and $K^-\pi^+\pi^+$ tags	
net tagged events	295
right-sign electrons observed[a)]	39
background[b)]	16.3 ± 1.0
wrong-sign electrons observed[a)]	4
background[b)]	4.2 ± 0.5
additional backgrounds[c)]	0.5 ± 0.8
net electrons	23.3 ± 6.7

a) Particles satisfying electron cuts.
b) Expected background from hadron misidentification.
c) Net background from mislabeled $K^-\pi^+$ tags and false tags.

In the ratio, there is a cancellation of some of the systematic errors and we obtain

$$\frac{\tau_{D^+}}{\tau_{D^o}} = 3.1^{+4.6}_{-1.4} \quad . \qquad (8)$$

This ratio was obtained by performing a maximum-likelihood fit to the relative rates. Figure 22 shows the logarithm of the likelihood as a function of the ratio τ_{D^+}/τ_{D^o}. The errors in (8) represent $\pm 1\sigma$ about 3.1, assuming a local Gaussian form for the likelihood function. Statistically, a change of approximately 2σ is required to obtain equal lifetimes, while the upper limit is poorly defined because of the small number of observed semileptonic D^o decays.

The DELCO collaboration has recently made a measurement of the individual semileptonic branching fractions of the D^+ and D^o based on a measurement of the number of ψ'' decays containing either one or two electrons originating from semileptonic decays of D mesons.[67] The values of the D^o and D^+ semileptonic branching fractions (b^o and b^+, respectively) were determined by finding a common solution to the pair of equations

$$N_{1e} = N_o\, \varepsilon_1^o\, 2b^o(1 - b^o) + N_+\, \varepsilon_1^+\, 2b^+(1 - b^+) + \text{smaller terms in } b^2$$

$$N_{2e} = N_o\, \varepsilon_2^o (b^o)^2 + N_+\, \varepsilon_2^+ (b^+)^2 \quad ,$$

where N_{1e} and N_{2e} are the number of observed one-electron and two-electron events and N_o and N_+ are the number of $D^o\bar{D}^o$ and D^+D^- decays in the data sample. The efficiency ε_1^o (ε_1^+) is the probability of detecting a 1e event from a $D^o\bar{D}^o$ (D^+D^-) initial state in which one D decays to an electron. ($\varepsilon_2^{o,+}$ are defined similarly for 2e events.) They determine

$$B(D^+ \to Xe\nu) = 22.0 \begin{smallmatrix} +4.4 \\ -2.2 \end{smallmatrix} \%$$

$$B(D^o \to Xe\nu) < 4.0\% \text{ (95\% c.l.)} \quad .$$

These values imply that the ratio of lifetimes is

$$\frac{\tau_{D^+}}{\tau_{D^o}} > 4.3 \text{ (95\% c.l.)} \quad .$$

These branching fractions are consistent with the values determined by the Mark II.

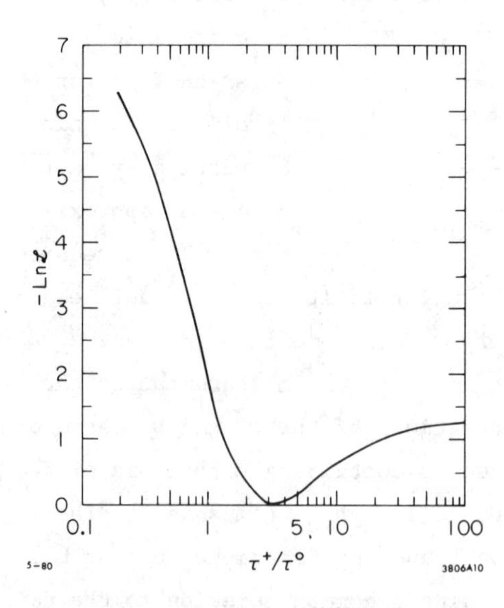

Fig. 22. Logarithm of the likelihood function versus the ratio of the D^+ and D^o total lifetimes.

Based on our branching fraction measurements, the theoretical estimate for the partial width $\Gamma(D \to Ke\nu) = 1.4 \times 10^{11}\ \text{sec}^{-1}$,[68] and the assumption[69]

$$\frac{\Gamma(D \to Ke\nu)}{\Gamma(D \to Xe\nu)} = 0.6 \quad ,$$

we estimate the D^+ and D^o lifetimes

$$\tau_{D^+} \approx 7.2 \times 10^{-13}\ \text{sec}$$

$$\tau_{D^o} \approx 2.4 \times 10^{-13}\ \text{sec} \quad .$$

These estimates compare well with measured lifetimes from the

Fermilab neutrino experiment E531 in which 5 D^+ and 7 D^o candidates have been observed to decay in emulsion.[70] They measure the lifetimes to be $\tau_{D^+} = 10.3 {+10.5 \atop -4.1} \times 10^{-13}$ sec and $\tau_{D^o} = 1.00 {+0.52 \atop -0.31} \times 10^{-13}$ sec.

V. $\tau \to K^*(890)\nu_\tau$

The properties of the heavy lepton τ^- have been carefully studied in a number of experiments.[71] I would like to briefly discuss the decay $\tau^- \to K^{*-}(890)\nu_\tau$ (for simplicity, all references to τ^- imply also the charge conjugate state) which has been observed by the Mark II collaboration. This is the first observation of a Cabibbo-suppressed τ decay.

Figure 23 compares the W-exchange diagrams for the processes $\tau \to K^*\nu_\tau$ and $\tau \to \rho\nu_\tau$. The diagrams are identical except for the coupling of the W to the final-state hadron in each case, where the K^* final state is Cabibbo-suppressed (i.e., the coupling is proportional to $\sin^2\theta_c$) and the ρ final state is Cabibbo-favored (with coupling proportional to $\cos^2\theta_c$). There is no reason not to expect θ_c to be the standard Cabibbo angle.[72]

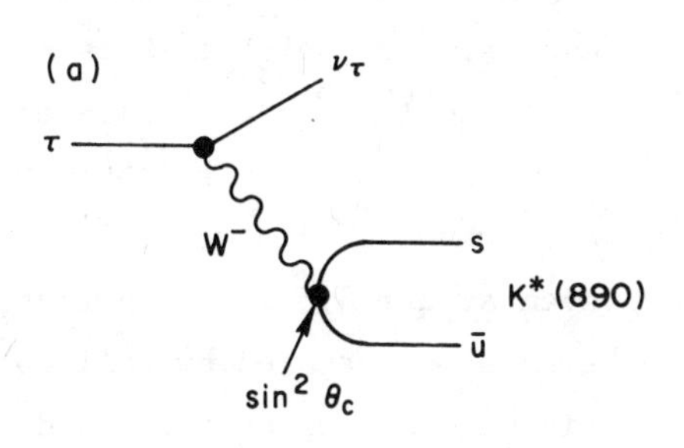

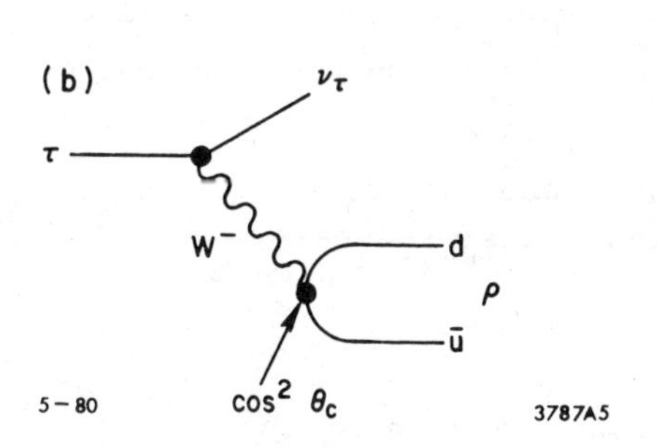

Fig. 23. Diagrams for a) $\tau \to K^*(890)\nu_\tau$ and b) $\tau \to \rho\nu_\tau$.

Our analysis was based on a sample of 40,200 $\tau^+\tau^-$ pairs produced with $E_{c.m.} > 4.2$ GeV. The decay sequence which was analyzed was

$$e^+e^- \to \tau^-\tau^+$$
$$\tau^+ \to \ell^+\nu_\ell\bar{\nu}_\tau$$
$$\tau^- \to K^{*-}\nu_\tau, \quad K^{*-} \to K_S\pi^-, \quad K_S \to \pi^+\pi^-$$

We required four charged particles in the final state, one of which was identified as an e or a μ.

Events in which any of the other three tracks were identified as leptons, kaons, or protons were eliminated. Events with an observed photon with energy greater than 100 MeV were also eliminated.[73] Finally, it was required that two of the oppositely charged pions reconstruct in space to form a secondary vertex at a distance of at least 1 cm from the primary decay point and have a mass consistent with the nominal K^o mass ($0.465 \leq M_{\pi+\pi-} < 0.510$ GeV).

Figure 24 shows the invariant mass spectrum for the $K_S\pi^{\pm}$ events selected above. There are 11 signal events (with $0.825 \leq M_{K\pi} < 0.950$ GeV) with an estimated background of 2.5 events (determined from events in the regions $0.700 \leq M_{K\pi} < 0.825$ GeV and $0.950 \leq M_{K\pi} < 1.075$ GeV). The detection efficiency was calculated by Monte Carlo separately for the events with an e and the events with a μ, yielding $\varepsilon_{eK^*} = 2.1\%$ and $\varepsilon_{\mu K^*} = 1.4\%$. Seven of the eleven events observed in the signal region were electron events. As the statistics were too limited to estimate what fraction of the subtracted background events were electron events, the 8.5 ± 3.6 corrected events were scaled by 7/11 to give the number of corrected electron events and 4/11 to give the number of corrected muon events.

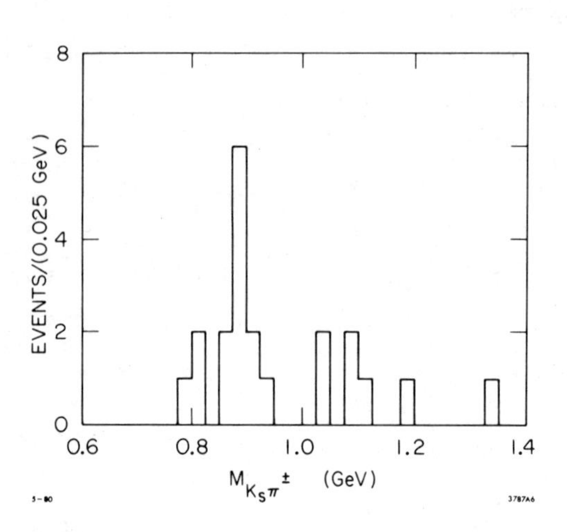

Fig. 24. $K_S\pi^{\pm}$ invariant mass distribution in candidate events for $\tau \to K^*(890)\nu_\tau$.

The branching fractions were calculated separately for the e and μ event samples according to the prescription

$$N_\ell = 2\varepsilon_{\ell K^*} B(\tau \to K^*\nu_\tau) \times B(\tau \to \ell\bar{\nu}_\ell\nu_\tau) \quad .$$

For the leptonic branching fraction, we used a value $B(\tau \to \ell\bar{\nu}\nu_\tau) = 18.5 \pm 1.5\%$ obtained from analysis of $e\mu$ events in the same data sample. Our best estimate for $B(\tau \rightarrow K^{*}\nu_{\tau})$ is a weighted average of the two determinations:

$$B(\tau \rightarrow K^{*}\nu_{\tau}) = 1.6 \pm 0.8\% \quad .$$

This branching fraction can be compared with theoretical estimates based on the measured branching fraction for $\tau \rightarrow \rho\nu_{\tau}$ of $20.5 \pm 4.1\%$ from our experiment.[74] Based on calculations of Tsai[72]

$$\frac{B(\tau \rightarrow K^{*}(890)\nu_{\tau})}{B(\tau \rightarrow \rho\nu_{\tau})} = f(M_{\tau}, M_{\rho}, M_{K^{*}}) \cdot \tan^{2}\theta_{c} \quad ,$$

where f is a phase space factor and is approximately 0.93. The value used for the Cabibbo angle is $\sin^{2}\theta_{c} = 0.07$ rather than the customary value of $\sin^{2}\theta_{c} = 0.05$ in order to take into account the difference between the coupling constants f_{K} and f_{π}.[75] From this, we expect $B(\tau \rightarrow K^{*}\nu_{\tau}) = 1.4 \pm 0.4\%$ which is in good agreement with the experimental measurement.

VI. CONCLUSIONS

I have presented some recent results from three different areas of $e^{+}e^{-}$ annihilation physics. The first part of the talk dealt with radiative transitions from the ψ to ordinary hadrons. The new results on the η, η', and f confirm earlier results, but there is still a possible factor-of-two discrepancy involved with the $\psi \rightarrow \gamma\eta'$ branching fraction. I tried to emphasize that these transitions can be understood in terms of minimal gluon-coupling ideas, with mixing between the different isoscalar states in the SU(4) multiplet playing a fundamental role in quantitatively understanding the results. The results on the $\gamma E(1420)$ transition are particularly interesting as the possibility exists that the E is a gluonium state. However, it is doubtful whether it will ever be possible to show convincing evidence for this interpretation.

Although it was not emphasized during the talk, there has been some effort by the Mark II collaboration to look for other radiative transitions from the ψ. All states with reasonable acceptance in the Mark II detector (i.e., states decaying into combinations of $\pi^{\pm}$, $K^{\pm}$, K_{S}, p, and $\bar{p}$), and even some with poor acceptance (e.g., states with

π^o's or η's in the final state), have been considered. No statistically significant signals aside from those shown today have been observed. Thus, if the E is not a gluonium state, neither the Mark II nor the Crystal Ball sees any evidence for such a state in radiative transitions from the ψ.[76]

A significant amount of information on the properties of D mesons has come from the Mark II, only part of which could be discussed here. In terms of exclusive channels, a few new decay modes have been observed, and a detailed analysis of the resonant substructure of the $K\pi\pi$ decay modes has been made. These results are important in understanding the mechanisms involved in D decays. Of the inclusive properties discussed, the most interesting is the result on the D^o and D^+ semileptonic decays rates. The combined results from the Mark II, DELCO, and the ν emulsion experiments provide fairly conclusive evidence that the D^+ lifetime is considerably longer than the D^o lifetime.

Finally, the measured branching fraction for $\tau \rightarrow K^* \nu_\tau$ is consistent with expectations from standard Cabibbo theory.

REFERENCES

1. Members of the SLAC-LBL Mark II collaboration: G. Abrams, M. Alam, C. Blocker, A. Boyarski, M. Breidenbach, D. Burke, W. Carithers, W. Chinowsky, M. Coles, S. Cooper, W. Dieterle, J. Dillon, J. Dorenbosch, J. Dorfan, M. Eaton, G. Feldman, M. Franklin, G. Gidal, G. Goldhaber, G. Hanson, K. Hayes, T. Himel, D. Hitlin, R. Hollebeek, W. Innes, J. Jaros, P. Jenni, D. Johnson, J. Kadyk, A. Lankford, R. Larsen, V. Lüth, R. Millikan, M. Nelson, C. Pang, J. Patrick, M. Perl, B. Richter, A. Roussarie, D. Scharre, R. Schindler, R. Schwitters, J. Siegrist, J. Strait, H. Taureg, M. Tonutti, G. Trilling, E. Vella, R. Vidal, I. Videau, J. Weiss, and H. Zaccone.

Members of the Crystal Ball collaboration. California Institute of Technology, Physics Department: R. Partridge, C. Peck and F. Porter. Harvard University, Physics Department: D. Andreasyan, W. Kollman, M. Richardson, K. Strauch and K. Wacker. Princeton University, Physics Department: D. Aschman, T. Burnett,

M. Cavalli-Sforza, D. Coyne, M. Joy and H. Sadrozinski. Stanford Linear Accelerator Center: E. D. Bloom, F. Bulos, R. Chestnut, J. Gaiser, G. Godfrey, C. Kiesling, W. Lockman and M. Oreglia. Stanford University, Physics Department and High Energy Physics Laboratory: R. Hofstadter, R. Horisberger, I. Kirkbride, H. Kolanoski, K. Koenigsmann, A. Liberman, J. O'Reilly and J. Tompkins.

2. D. Coyne, invited talk this conference.
3. J.-E. Augustin et al., Phys. Rev. Lett. 34, 233 (1975; J.-E. Augustin et al., Phys. Rev. Lett. 34, 764 (1975).
4. W. Davies-White et al., Nucl. Instrum. Methods 160, 227 (1979).
5. This measurement error is obtained when the tracks are constrained to pass through the known beam position.
6. G. S. Abrams et al., IEEE Trans. Nucl. Sci. 25, 309 (1978).
7. Neutral pions are reconstructed from pairs of γ's detected in the LA. Pairs with invariant mass between 0.075 and 0.200 GeV are considered to be π^o candidates. The π^o signal is extracted after subtraction of the combinatorial background.
8. R. L. Ford and W. R. Nelson, Stanford Linear Acclerator Center Report No. SLAC-210 (1978), unpublished.
9. H. Brafman et al., Stanford Linear Accelerator Center Report No. SLAC-PUB-2033 (1977), unpublished.
10. This process is not related to any physics I will discuss in this talk. I show this particular event only because of the striking similarity between this event and an event observed in the Mark I detector approximately five years earlier [see Fig. 3 in G. S. Abrams et al., Phys. Rev. Lett. 34, 1181 (1975)].
11. G. S. Abrams et al., Phys. Rev. Lett. 44, 114 (1980); D. L. Scharre et al., Stanford Linear Accelerator Center Report No. SLAC-PUB-2513 (1980), to be submitted for publication.
12. M. T. Ronan et al., Phys. Rev. Lett. 44, 367 (1980).
13. Details of the experimental apparatus can be found in E. D. Bloom, in Proceedings of the Fourteenth Rencontre de Moriond, Vol. II, edited by Trân Thanh Vân (R.M.I.E.M. Orsay, 1979), p. 175; and C. W. Peck et al., California Institute of Technology Report

No. CALT-68-753 to be published in the Proceedings of the Annual Meeting of the American Physical Society, Division of Particles and Fields, McGill University, Montreal, Canada, October 25-27, 1979.

14. D. G. Aschman, to be published in the Proceedings of the Fifteenth Rencontre de Moriond, Les Arcs, France, March 15-21, 1980.
15. W. Bartel et al., Phys. Lett. 66B, 489 (1977).
16. W. Braunschweig et al., Phys. Lett. 67B, 243 (1977).
17. R. Partridge et al., Phys. Rev. Lett. 44, 712 (1980).
18. W. Bartel et al., Phys. Lett. 64B, 483 (1976).
19. D. L. Scharre et al., Stanford Linear Accelerator Center Report No. SLAC-PUB-2514 (1980), to be submitted for publication.
20. G. Alexander et al., Phys. Lett. 72B, 493 (1978).
21. R. Brandelik et al., Phys. Lett. 74B, 292 (1978).
22. W. Braunschweig et al., Phys. Lett. 63B, 487 (1976).
23. B. Jean-Marie et al., Phys. Rev. Lett. 36, 291 (1976).
24. H. Fritzsch and J. D. Jackson, Phys. Lett. 66B, 365 (1977).
25. Because of the six-fold symmetry of the final state, the Dalitz plot has been folded. This results in the observed folding of the η and η' bands at the boundary.
26. D. L. Scharre, in Proceedings of the Fourteenth Rencontre de Moriond, Vol. II, edited by Trân Thanh Vân (R.M.I.E.M. Orsay, 1979), p. 219.
27. However, due to problems with the LA shower counter system, there was no photon detection for approximately half of the ψ running time. Thus, any analyses which required photon detection were based on half of the total data sample.
28. Due to noise in the LA electronics, spurious photons were occasionally reconstructed by the tracking program. In order not to lose good events, extra photons were allowed in candidate events. When analyzing these events, separate fits were attempted for each two-photon combination.
29. In general, two or more charged tracks were required to trigger the detector.

30. Approximately one million ψ' events were taken. From this sample, 92,000 ψ decays were identified by missing mass from the $\pi^+\pi^-$ system.
31. This ratio is expected to be roughly the same as the ratio of the leptonic branching fractions of the ψ' and ψ.
32. J. Burmester et al., Phys. Lett. 72B, 135 (1977), F. Vanucci et al., Phys. Rev. D 15, 1814 (1977).
33. C. Zaiser et al., in preparation.
34. It is expected that the mixing due to gluon exchange will not affect this ratio as severely as in the case of the pseudoscalar nonet.
35. Particle Data Group, Phys. Lett. 75B, 1 (1978) and references therein.
36. P. Baillon et al., Nuovo Cimento 50A, 393 (1967).
37. This distribution includes a few additional events corresponding to the process $\psi' \rightarrow \pi^+\pi^-\psi$, $\psi \rightarrow \gamma E$. Details can be found in Ref. 19.
38. The mass resolution of these constrained events is considerably smaller than the natural line width of the resonance and is ignored in the fit.
39. The efficiency used in the determination of this branching fraction was based on a Monte Carlo analysis which assumed all decay distributions were isotropic. If the spin-parity of the E were 0^-, which results in a $1 + \cos^2\theta$ distribution for the angle of the photon with respect to the beam axis, the branching ratio product should be increased by 19%.
40. The increase in sample size arises principally from the fact that the sample of data in which the LA system was not operational could be used.
41. Monte Carlo analysis shows the acceptance to be roughly flat over the entire Dalitz plot. Hence, the observed structure is not the result of variations in the acceptance.
42. This decay mode is expected to be half the $K_S K^{\pm}\pi^{\mp}$ mode by isospin conservation.

43. The correction was based on the $K_SK^{\pm}$ mass distribution for the sample of $E \to K_SK^{\pm}\pi^{\mp}$ events from the Mark II.

44. C. Dionisi et al., CERN Report No. CERN/EP 80-1 (1980), submitted to Nucl. Phys. B.

45. C. Bromberg et al., California Institute of Technology Report No. CALT-68-747 (1980).

46. It has been suggested that the resonance seen in hadronic experiments and known as the E(1420) is not the same state that has been observed by us in radiative transitions from the ψ. However, because of the consistency of the parameters of the states seen in these two processes and the outward similarity of the Dalitz plots, I think it is logical to consider them to be the same state until some evidence to the contrary is produced.

47. See for example J. Donoghue, to be published in the Proceedings of the VI International Conference on Experimental Meson Spectroscopy, Brookhaven National Laboratory, Upton, New York, April 25-26, 1980.

48. A gluonium state is a bound state of two or more gluons.

49. Unfortunately, the masses of the other members of the nonet (the A_1 and the Q_A) are not well enough known to provide a reliable estimate of the mixing angle.

50. G. Goldhaber et al., Phys. Rev. Lett. 37, 255 (1976); I. Peruzzi et al., Phys. Rev. Lett. 37, 569 (1976).

51. R. H. Schindler et al., Stanford Linear Accelerator Center Report No. SLAC-PUB-2507 (1980), to be submitted for publication.

52. See for example, E. Eichten et al., Phys. Rev. Lett. 34, 369 (1975); E. Eichten et al., Phys. Rev. D 21, 203 (1980).

53. R. H. Schindler et al., Phys. Rev. D 21, 2716 (1980).

54. The energy of the D can be constrained to the beam energy, which results in a significant improvement in the mass resolution.

55. We have previously reported measurements of the Cabibbo-suppressed decay modes K^-K^+ and $\pi^+\pi^-$ in G. S. Abrams et al., Phys. Rev. Lett. 43, 477 (1979).

56. I. Peruzzi et al., Phys. Rev. Lett. 39, 1301 (1977); D. L. Scharre et al., Phys. Rev. Lett. 40, 74 (1978).

57. See for example, M. Matsuda et al., Prog. Theor. Phys. 59, 1396 (1978).

58. The theoretical interpretation of our results is discussed in detail by M. Chanowitz, invited talk this conference.

59. R. H. Schindler, Stanford Linear Accelerator Center Report No. SLAC-219, Ph.D. Thesis, Stanford University, 1979 (unpublished).

60. V. Vuillemin et al., Phys. Rev. Lett. 41, 1149 (1978).

61. See for example, C. Quigg and J. L. Rosner, Phys. Rev. D 17, 239 (1978); M. Peshkin and J. L. Rosner, Nucl. Phys. B122, 144 (1977).

62. M. K. Gaillard, B. W. Lee, and J. L. Rosner, Rev. Mod. Phys. 47, 277 (1975).

63. It is assumed that decays with K^+K^- and $K^o\overline{K}^o$ are produced equally.

64. R. Brandelik et al., Phys. Lett. 70B, 387 (1977); J. M. Feller et al., Phys. Rev. Lett. 40, 274 (1978); W. Bacino et al., Phys. Rev. Lett. 43, 1073 (1979); J. M. Feller et al., Phys. Rev. Lett. 40, 1677 (1978).

65. At center-of-mass energies above F^+ or Λ_c^+ threshold, there will be some contamination from the semileptonic decays of these other charmed particles.

66. A. Pais and S. B. Frieman, Phys. Rev. D 15, 2529 (1977).

67. W. Bacino et al., Stanford Linear Accelerator Center Report No. SLAC-PUB-2500 (1980), submitted for publication to Phys. Rev. Lett.

68. D. Fakirov and B. Stech, Nucl. Phys. B133, 315 (1978).

69. This number is consistent with various determinations by different experiments. See Ref. 59 for details.

70. J. Prentice, to be published in the Proceedings of the VI International Conference on Experimental Meson Spectroscopy, Brookhaven National Laboratory, Upton, New York, April 25-26, 1980.

71. See for instance, M. L. Perl, Stanford Linear Accelerator Center Report No. SLAC-PUB-2446 (1979), to be published in Ann. Rev. Nucl. Part. Sci., for a review.

72. Y.-S. Tsai, Stanford Linear Accelerator Center Report No. SLAC-PUB-2450 (1979).

73. An observed photon with energy less than 100 MeV is likely to be a spurious signal caused by noise in the LA electronics.

74. G. S. Abrams et al., Phys. Rev. Lett. 43, 1555 (1979). A similar measurement by the DASP collaboration [R. Brandelik et al., Z. Phys. C 1, 233 (1979)] finds $B(\tau \to \rho\nu_\tau) = 24 \pm 9\%$.

75. This value of $\sin^2\theta_c$ is obtained from the $K \to \mu\nu$ and $\pi \to \mu\nu$ partial decay widths, with no corrections made for the different K and π form factors. Hence, this value of $\sin^2\theta_c$ is expected to be applicable to the K^* and ρ decay modes without correction for form factor differences.

76. This does not mean that none exist. Neither the masses nor the widths are well defined theoretically, and the decay modes expected for such states are often such that their detection by the Mark II would be difficult. For a brief but excellent review of the possibilities, see J. D. Bjorken, Stanford Linear Accelerator Center Report No. SLAC-PUB-2366, to be published in the Proceedings of the 1979 EPS High Energy Physics Conference, Geneva, Switzerland, June 27-July 4, 1979.

e^+e^- PHYSICS NEAR CHARM THRESHOLD VIA THE CRYSTAL BALL*

D. G. Coyne
(Representing the Crystal Ball Collaboration)[1]
Stanford Linear Accelerator Center
Stanford University, Stanford, California 94305

and

Princeton University, Princeton, New Jersey 08540

ABSTRACT

We discuss the use of the Crystal Ball detector for the study of e^+e^- annihilations near charm threshold: at the $\psi(3097)$, $\psi'(3685)$, $\psi''(3770)$ and in the "continuum" $3770 \leq \sqrt{s} \leq 4500$ MeV. This paper will concentrate on the special techniques available to the Crystal Ball which allow measurement of inclusive γ, π^0 and η cross sections. Preliminary results for these processes at particular center of mass energies are presented. An update of the inclusive and exclusive evidence for the $\eta_c(2980)$, for the sake of completeness, is given in the Appendix.

I. INTRODUCTION

The use of the Crystal Ball detector in the study of the "continuum" region just above charm threshold introduces a number of unfamiliar but useful techniques in e^+e^- physics. The major purpose of this report is to outline those techniques and to discuss areas of pertinence where they will be applied. Another purpose is to present our preliminary results in some of these areas, based on first analyses of partial data samples.

The organization of this report is to describe the detector briefly, and to make a short survey of the state of knowledge of the continuum prior to the Crystal Ball measurements. After an outline of the strategy of data-taking we have used, we present results of the Crystal Ball "baseline" measurements, i.e., the proof of the ability to measure inclusive γ, π^0 and η signals at c.m. energies <u>below</u> charm threshold and/or on- ψ resonances. The resolutions and efficiencies for these modes are discussed. We then move <u>above</u> charm threshold and into the "continuum" region (which itself appears to have imbedded ψ-like resonances). The report concludes

* Supported in part by the Department of Energy, contract DE-AC03-76SF00515, and by National Science Foundation Grant PHY79-16461.

ISSN:0094-243X/80/620253-26$1.50

with our preliminary measurements of selected inclusive modes at interesting c.m. energies.

II. THE DETECTOR

The Crystal Ball is a fieldless, segmented spherical shell of NaI(Tℓ) surrounding chambers having charged-particle tracking capabilities. The detector, built and operated by the Crystal Ball Collaboration,[1] is shown diagrammatically in Fig. 1. A detailed des-

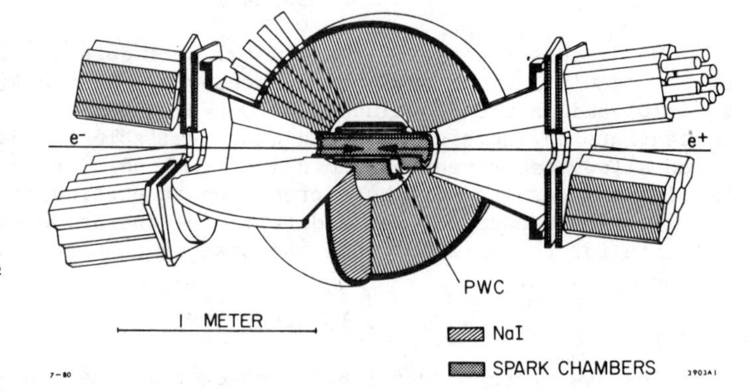

Figure 1. Schematic cutaway view of the Crystal Ball detector.

cription of the apparatus is given elsewhere;[2] for the purposes of this discussion there are several crucial parameters.

A) The energy resolution for photons in the Ball itself (as opposed to the endcaps) is $\sigma_E/E_\gamma \approx 3\%/E_\gamma^{1/4}$ with E_γ in GeV. This holds well for 50 MeV $\leq E_\gamma \leq$ 2 GeV and leads to a Γ(FWHM) of about 10% of E_γ in the regions of interest. This resolution is available over $\sim$85% of 4π radians.

B) The angular resolution for photons is determined by the segmentation of the NaI (720 segments over the complete sphere) and by weighting the crystal segments according to the energy deposition within each participating crystal of a photon shower. This gives $\sigma_{\theta_\gamma} \approx 1$ to 2°, depending on the range of photon energies. (θ_γ is a polar angle relative to the γ true direction.)

C) The angular resolution for charged particles is determined by the number of proportional wire chamber planes and magnetostrictive spark chamber planes through which it passed. We get $\sigma_{\theta_c} \approx .3 - 1^\circ$ depending on the geometry of the track, with smaller values pertinent for the 75% of the solid angle furthest from the endcaps.

The analysis modes most appropriate to this detector are measurement of inclusive γ, inclusive π^0, η and other states decaying into all-γ final states. In addition the Ball can overconstrain any event exclusively for any number of electromagnetically decaying secondaries in the final state, as long as the number of unabsorbed charged particles does not exceed 2. (Zero-constraint "fits" can allow 4 charged particles.) An example of these modes was our detection of a candidate state for η_c in inclusive and exclusive states, as summarized in the Appendix.

We have developed algorithms for the recognition of final-state photons and charged particles, together with cuts designed to take advantage of the portions of the Ball and the event topologies where the best measurements are available. Our general approach is to use highly cut data for the search for new effects, and relatively uncut data to arrive at numerical results in the least-biased manner possible.

The major complication of the analysis of data from this detector is pattern recognition in the presence of strongly interacting particles and shower fluctuations. Residuals from untagged charged particles (at the level of a few percent) also complicate the analysis.

Our basic check of the credibility of the cuts and subsequent pattern recognition is the ability of the detector to reproduce standard QED distributions involving e's and γ's with absolute normalization.[3]

III. STATE OF KNOWLEDGE OF THE CONTINUUM (PRE-CRYSTAL BALL)

While we cannot do justice here to the many papers published concerning the effects within the continuum, we can mention those areas of study most likely to be investigated by the Crystal Ball.

The global behavior of the cross section $e^+e^- \rightarrow$ hadrons has been studied by most of the past detectors at e^+e^- storage rings. A representative sample[4] is shown in Fig. 2. Absolute differences of the order of 20% in R are evident, and point-to-point systematics clearly dominate the statistical errors. Nonetheless, certain features are

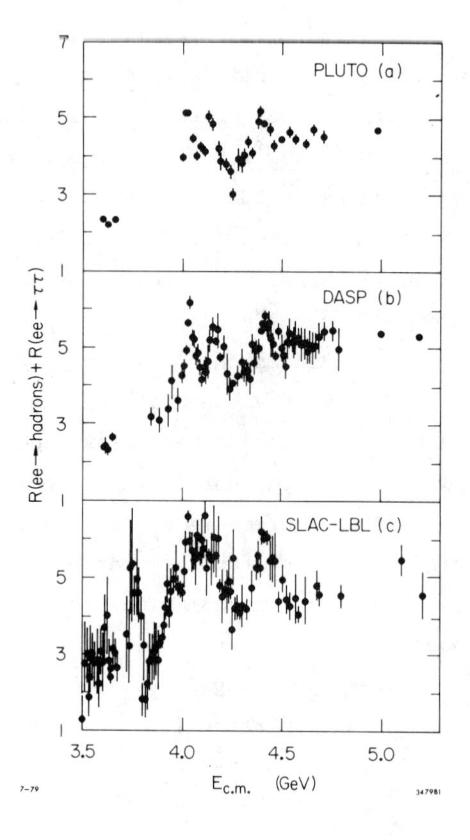

Fig. 2. $R = \sigma(e^+e^- \to \text{hadrons})$ $\sigma(e^+e^- \to \mu^+\mu^-)$ as measured by three detectors in the region $\sim 3.7 < \sqrt{s} < \sim 4.6$ GeV. Radiative corrections but none for τ have been applied (Ref. 4).

invariably present in all samples, such as the ψ"(3770) (when measured), the "D^*" resonance at ~4.03 GeV, and the large bump at ~4.4 GeV. Often but not invariably present are suggestive structures at 3.9 GeV, 4.16 GeV, and details of the rapid fall after 4.2. While the global variations in R are presumed to be related to variations in D production[5] there is no clear picture of the fine structure (if any) of R nor of small processes which may occur in addition to D production.

The existence of thresholds for charmed particles in this region has been a fruitful topic of study. The D's have been studied most intensely, though not exhaustively, at the $D\bar{D}$-factory ψ"(3770) in exclusive final states.[6,7] Of more pertinence to this report are the studies of D and D^* performed at $\sqrt{s} = 4.028$ in quasi-inclusive modes, in which the $D^\pm$ or D^0 is detected and fit from its Kπ or Kππ modes and the spectra of momenta recoiling against the D is used to deduce parameters of the system: D^* masses, $D^* \to D_\gamma$ branching ratio, $D^*\bar{D}^* : D^*\bar{D} + D\bar{D}^*$ production ratios, charged D ratios, etc.[8] Figure 3 shows how the recoil momentum spectrum is affected by the possible variation of these parameters. The method is fairly direct for some of the parameters (such as production ratios) but quite indirect for the D^*-D Q-value and γ branching ratio. The data are slim statistically, but rather good numbers on some of the parameters nonetheless emerge (see the later comparison

Fig. 3. Parameter fitting with D recoil momenta studies (Ref. 8).

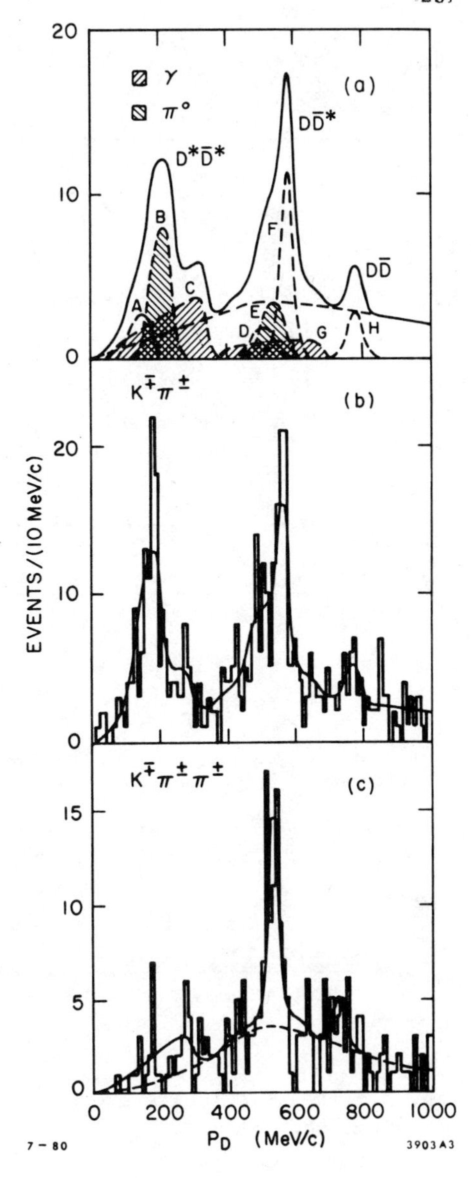

Table III). Figure 4 shows the energy level scheme[9] underlying these spectra.

The situation for strange-charmed mesons, the F/F^*, is much less well-defined. In e^+e^- storage ring investigations, the main evidence for the F comes from two largely independent results of the DASP collaboration.[10,11] The first result, Fig. 5, is that the inclusive cross section for η appears to have one or more thresholds in the vicinity of $\sqrt{s} = 4.16$ and 4.4 GeV. The experiment placed tight limits on the inclusive η cross section at $\sqrt{s} = 4.03$, thus suggesting that neither D's nor non-charm background was an appreciable source of η's; this implied that the eventual appearance of η's was likely evidence for F's, which are expected to have ≈50% of their decays into final states involving an η.[12] Correlations with electrons and low energy γ's (expected from $F^* \rightarrow \gamma F$) strengthened this interpretation. The second result, Fig. 6, was a cluster of events at $\sqrt{s}$ = 4.42 GeV fitting the exclusive hypothesis $e^+e^- \rightarrow F^*F$ or F^*F^* giving unique masses for F and F^* as shown. There are a number of recent results[13] with candidates for F's with a similar mass, and even a lifetime measurement! In this report of Crystal Ball results, we

Fig. 4. The energy level scheme for D^* decays (Ref. 9).

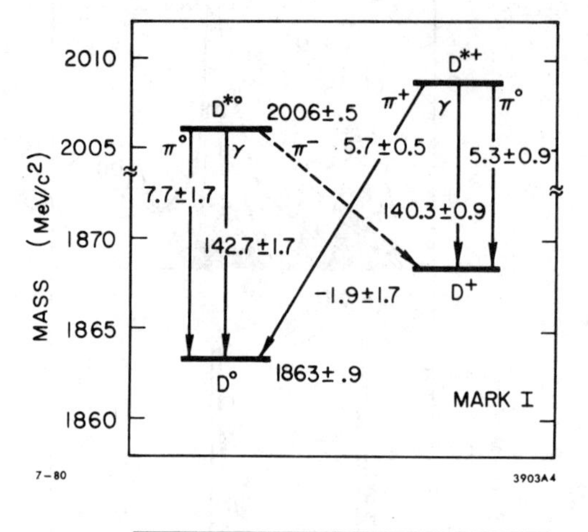

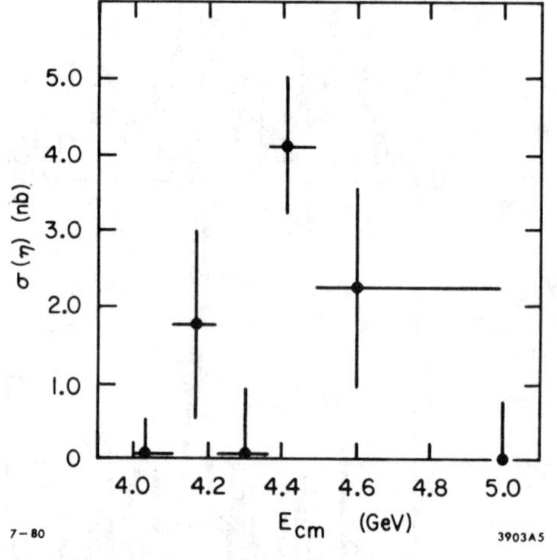

Fig. 5. The η-inclusive cross section (DASP, Ref. 10).

will consider only the η-clusive measurement and its interpretation. Limits on the transition $F^* \rightarrow \gamma F$ and a search for the F via the analogous exclusive modes in the Ball are in preparation and will be discussed in future reports.

Before leaving this subject, we should note that there are several other possible thresholds expected in

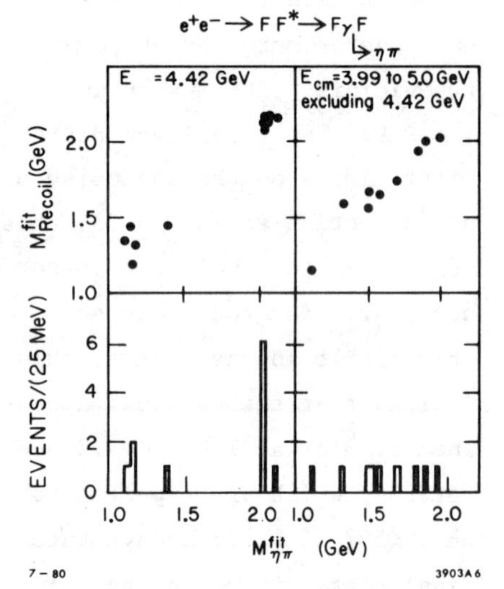

Fig. 6. Exclusive candidates for F,F^* (DASP,Ref. 11).

the range $3.8 < \sqrt{s} < 4.5$ GeV. Figure 7 shows a prediction[14] of the locations of the P-wave excitations of the charmed and strange-charmed mesons, D^{**} and F^{**}. As an example, the threshold for DD^{**} production is ~4.1 GeV and for FF^{**} is ~4.4 GeV. Thus we might expect that if the inclusive modes are examined in detail, many spectral lines characteristic of these γ, π^0, $2\pi^0$, η and K transitions

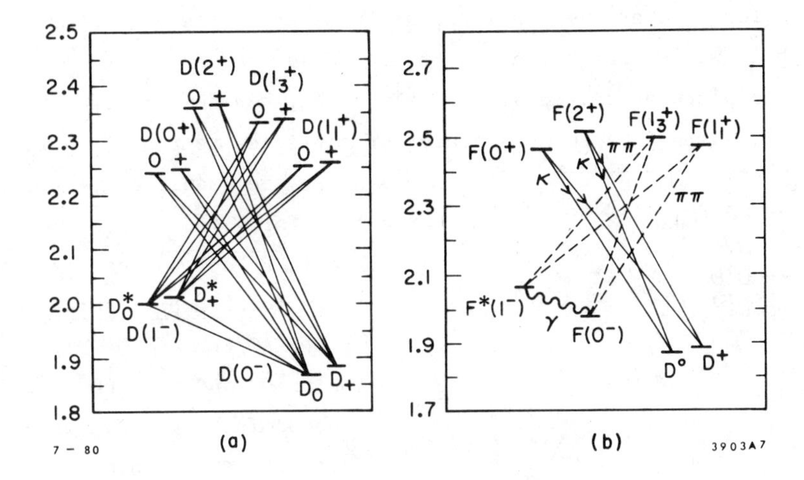

Fig. 7. The energy level scheme predicted for charmed and strange-charmed mesons (Ref. 14).

may be present near the respective thresholds, albeit with small intensities. The detection of such processes and the location of "factories" for new states, if any, are thus outstanding problems in this energy region.

The effect of the possible thresholds on the cross section is in itself an interesting physics question. There exist preliminary theoretical models of the effect of imbedding psi-like resonances in the continuum; the coupled-channel model[15] gives cross sections which are critically dependent on the number and position of thresholds for charmed and strange-charmed mesons. Stated briefly (and imprecisely), the transition of an imbedded psion into two specific mesons is constrained by spin-parity to particular relative angular momenta between the mesons; it is constrained by the masses involved and by $\sqrt{s}$ to particular values of the relative linear momentum between the mesons. This implies that the rate of production of such meson pairs is sensitive to the nodes and antinodes of the spacial wave function of the psion (or to the momentum-space analog). Therefore relatively small changes in $\sqrt{s}$ can make large changes in the $p_{relative}$ of the meson pair near threshold, with concurrent large changes in production rate of meson pairs. The psion channels are coupled in that overlap integrals between two psions interacting through exchanges of the mesons are involved.

This picture seems qualitatively right, but neither theory nor data yet exists in a form adequate for a precise test.[16] Figure 8 shows the quantitative prediction for a model involving D, D^*, F, F^* only. (Masses are taken to be experimental values from SLAC/LBL and DASP). Note that $D^*\bar{D}^*$ and $D^*\bar{D} + D\bar{D}^*$ production dominate the region $3.9 < \sqrt{s} < 4.5$ GeV, with no factories expected for F or F^* and cross sections $\leq .2$ nb. The highest signal/noise for F is expected at the minimum R, $\sqrt{s} \approx 4.25$ GeV. This model can be used as an experimental guide to the rates expected, or inversely, the data can help define what states must be present in the model.

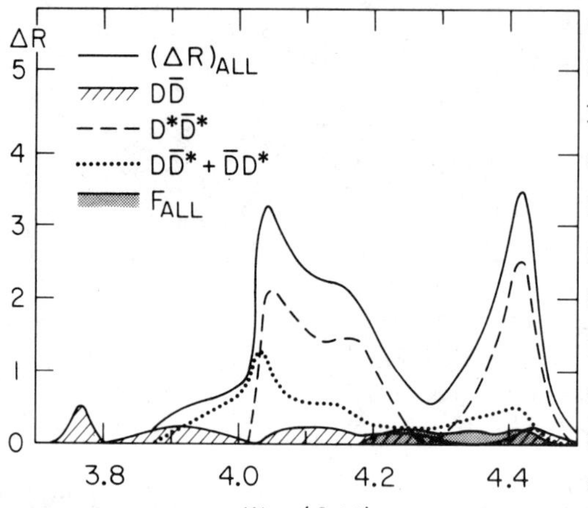

Fig. 8. The partial cross sections expected for production of specific meson pairs (Ref. 15).

Finally, the study of the imbedded states in the continuum for their own sake is of great interest for the testing of models of quark interactions. The quantum numbers, positions and decay properties of the imbedded states should help determine if they are "simple" $q\bar{q}$ resonances (potential model), complex multi-quark states (molecular models) or excitations of the gluon string (vibrational states). An example of a physical quantity testing such details is the angular distribution of the π^0 (relative to the beams) from $D^* \rightarrow \pi^0 + D^0$. It determines the ratio of spin amplitudes in the production of D^*D^*; there is a prediction for this at the $\psi(4028)$ based on resonance dynamics. For details of this theoretical application, see Cahn and Kayser and associated literature.[17]

In summary of this discussion, it is fair to say that the continuum region has had only a preliminary inspection and that many opportunities for confrontation of theory and experiments still exist, dependent on the acquisition of informative data samples.

IV. THE CRYSTAL BALL AS A CONTINUUM DETECTOR: "BASELINE" MEASUREMENTS

A) A Strategy for Continuum Measurements.

We have seen that the collected data in the continuum are sparse; this stems from the fact that there is a relatively wide region, some 1000 MeV, in which new features can appear, but that these new features can persist or fluctuate over c.m. energy ranges as small as 20 MeV. The 4.028 resonance and its attendant structure is such a case. A simple calculation of the integral luminosity required to cover this continuum thoroughly, with sensitivity to phenomena with $\Delta R \approx .2$, yields run plans of about five years duration--a practical impossibility.

The approach used with the Crystal Ball detector is to scan the entire region, using inclusive distributions of γ, π^0, η and R to indicate regions of $\sqrt{s}$ with special interest, and then to rescan for confirmation of the effects. Finally, long fixed-energy runs are made at selected values of $\sqrt{s}$ to provide data sets for exclusive analysis. The calendar periods required for these three steps are about 5, 12 and 6 weeks, respectively. At present, we have finished the initial two scanning stages and one fixed energy (4.028). The data discussed below are mostly from the first stage and 4.028. Table I shows the integral luminosities corresponding to the various scan periods.

Table I: Integral Luminosities Collected by the Crystal Ball in a Scan Mode

Date	Energy Region	$\int \mathscr{L} dt (nb^{-1})$
May 1979	3.88-4.45 GeV	3092
Nov. 1979	3.67,4.03 GeV	300, 840
Mar. 1980	4.04-4.214 GeV	2875
June 1980	3.67,4.22-4.48 GeV	200, 2360
	Total	9667 nb^{-1}

This approach is not new; Fig. 9 shows some inclusive indicators used by Mark I[18] in similar scans. The mean multiplicity and charged energy fractions turned out to be insensitive indicators of changes in the underlying physics--the best indicator was R itself.

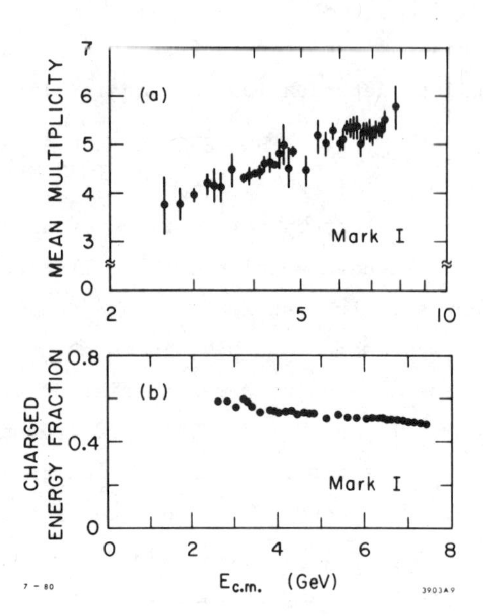

Fig. 9. Inclusive Quantities in the Mark I Scan Mode: a) charged energy fraction, b) mean multiplicity (Ref. 18).

What is new is that the Crystal Ball measures neutral-related quantities such as neutral energy fraction and neutral multiplicity. Fig. 10 shows that the former of these is again insensitive to changes in underlying physics, but that large changes in neutral multiplicity are seen.[19] This implies that the distributions of low-energy γ's, π^0's and η's could be useful indicators, and these were stressed in the scan analysis.

B) Crystal Ball Efficiencies, Resolution and "Baseline" Measurements

As an introduction to the use of these inclusive distributions in a scan, we will discuss the efficiencies and resolution for each particle type, and then look at sample distributions taken at the psions or off-resonance/below charm threshold ($\sqrt{s}$ = 3670). These show the baseline characteristics of the spectra where no D's are present.

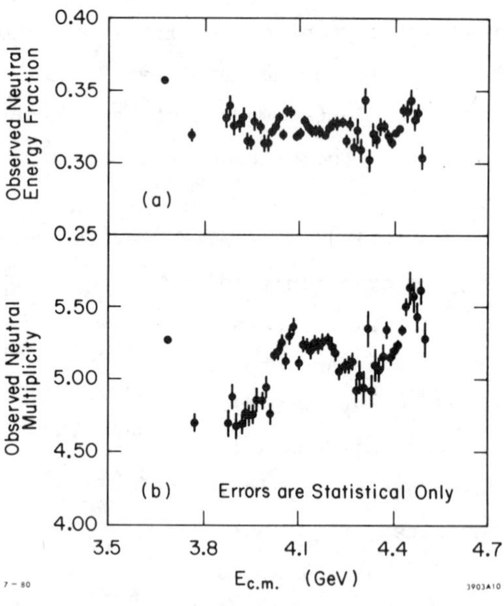

Fig. 10. Inclusive Quantities in the Crystal Ball Scan Mode: a) Neutral energy fraction, b) Neutral Multiplicity (Ref. 19).

Then we use the spectra at $\psi''(3770)$ to see if a D-rich final state distorts the baseline measurement. The ψ'' then often serves as a background curve for the remainder of the scan region.

1) <u>Hadronic cross section</u>: The most basic distribution is R vs $\sqrt{s}$, obtained from the calorimetric properties of the Ball. A simple energy threshold is inadequate to separate hadrons, so cuts on the spatial distribution of energy and track topologies are introduced. The results[19] are additionally τ-subtracted and background subtracted. Radiative corrections depend critically on the structures present and have not been applied to samples in this report. At this juncture we estimate that there may be a 15% absolute systematic error on the average R measured in a local region of $\sqrt{s}$, and a point-to-point systematic of ±5% around this average. Our baseline result for R is

$$R(\sqrt{s} = 3670 \text{ MeV}) = 2.4 \pm .3$$

2) <u>Photons</u>: We have already given the energy and angular resolutions. NaI(Tℓ) has essentially no intrinsic inefficiency for photons of energy > 20 MeV (our software cutoff). Effective inefficiencies are introduced by cuts to avoid overlap of photons with other particles. A precise absolute photon energy calibration can be maintained by interspersing data runs at the ψ', where the observed χ-lines test the overall calibration to $\sim \pm 1$ MeV. As a guide for how monochromatic γ's will appear in inclusive spectra we show Fig. 11, the photon spectrum from ψ' as observed in the counting room after three hours of data acquisition (10.5 nb^{-1}).

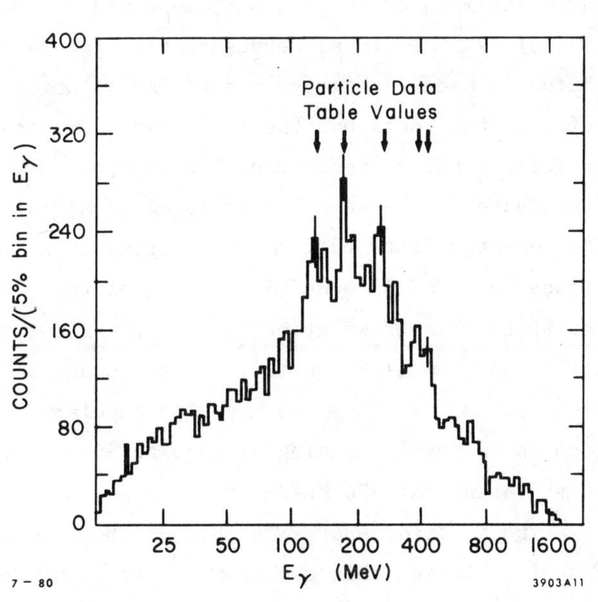

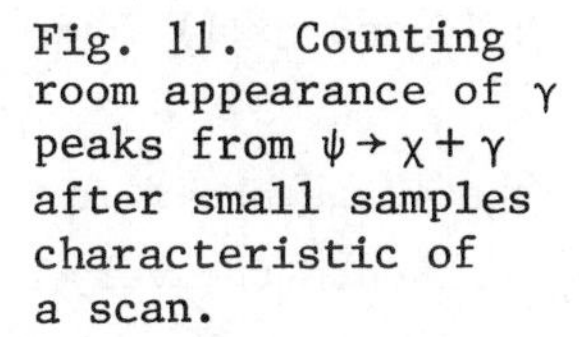
Fig. 11. Counting room appearance of γ peaks from $\psi \to \chi + \gamma$ after small samples characteristic of a scan.

This tiny sample has been so chosen to give the same absolute signal in each peak as would a resonance $F^* \rightarrow \gamma F$ produced with $\Delta R = 1$ and $B_\gamma = 100\%$, where the hypothetical run would last $\sim\frac{1}{2}$ day. (The peak would have two to four times less background under reasonable assumptions).

The low-energy photon spectra actually obtained below charm threshold (3670) and at ψ'' are unremarkable: Fig. 12 shows the "baseline" distribution at ψ''. Note that the choice of a $\ell n\, E\gamma$ scale transforms the spectrum so that the prominent peak at $m_\pi/2 \approx 70$ MeV which we see in a linear plot (arising from $\pi^0 \rightarrow \gamma\gamma$) completely disappears. This means that any π^0-related bump appearing on this type of spectrum near 70 MeV signals an excess of low energy π^0's over the normal population. The value $E_\gamma \approx 200$ MeV is significant in that minimum ionizing charged particles masquerading as photons would produce a peak here.

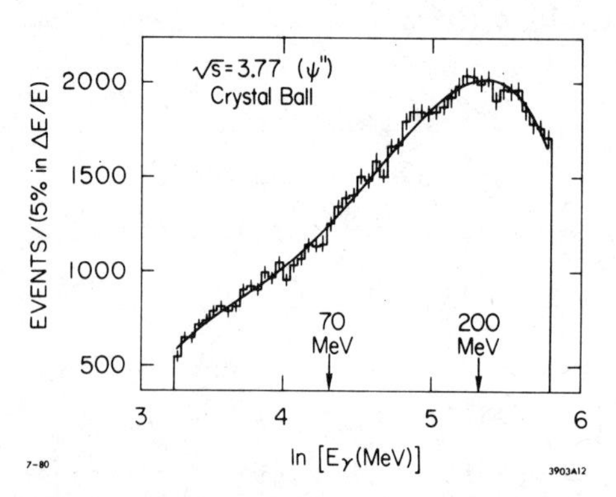

Fig. 12. The inclusive γ energy-spectrum of the ψ'' as obtained by the Crystal Ball.

3) $\underline{\pi^0}$: A global reconstruction for π^0's is used in which each photon is used in at most one π^0 and in which the goodness of fit reflects the total hypothesis for assignments of γ's to π^0's. The efficiency for π^0 reconstruction is $\leq 60\%$, where there is a marked dependence on the severity of topological cuts (such as overlap). The momentum distribution of resulting π^0's is relatively unbiased because the NaI is sensitive to low-energy γ's. From measurements at the ψ, the mass resolution is $\sigma/m_\pi \approx 4\%$, but this depends sharply on the π momentum distribution. In the scan, π^0 selection effectively picks out $m_{\gamma\gamma} = m_{\pi^0} \pm 15\%$. The angular resolution of the π^0 is such as to permit meaningful angular distributions ($\Delta\cos\theta \approx .1$) down to π^0 momenta of ~20 MeV/c.

The π^0 total energy "baseline" spectrum for the ψ'' is shown in Fig. 13. Again, this spectrum is devoid of features; note the very

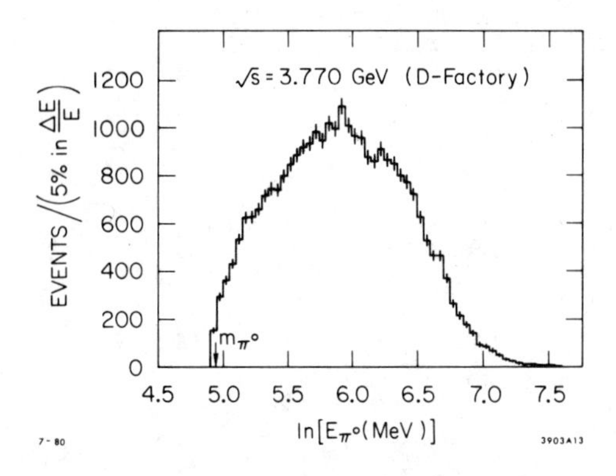

Fig. 13. The inclusive π^0 total-energy spectrum of the ψ'' (Crystal Ball).

smooth rise from m_{π^0}. In these π^0 energy spectra, the π^0 mass is constrained to the usual value after the global assignment of γ's to π^0: this has the effect of adjusting the γ energies and opening angle to more precisely determined values.

4) <u>η^0</u>: The efficiency for detection of the decay $\eta \to \gamma\gamma$ is about 30% after all topological cuts. The resolution for the total inclusive η signal observed in the scan is $\sigma/m_\eta = 4\%$ (see later discussion in Section V.C and Fig. 24). An improvement in the signal:background for η's can be gained from π^0 subtraction or from energy/topology cuts (requiring mainly m_γ > 140 MeV and particle multiplicities between two and six for charged and neutrals separately--the so-called DASP cuts). Such improvements decrease the efficiency and introduce larger errors in the calculation of the efficiency.

It is something of a surprise to see how weakly a reasonable η-signal should appear in an $m_{\gamma\gamma}$ plot. Figure 14 shows a very large data sample taken at the ψ, where the average number of η's/event is about 1/4. The η signal is quite evident, but only by virtue of the good statistics (note the error bars). It behooves us to find cuts clearly defining the η for use in less abundant data sets. Figure 15 shows the upper end of this same sample, with π^0 subtraction, and Fig. 16 shows it with the DASP cuts but no π^0 subtraction. Both techniques work, but the former is better statistically for the Ball and is less biasing.

We now consider the "baseline" η measurements at $\sqrt{s}$ = 3670 (off-ψ') and at the D-factory ψ''(3770). There are clear (and strong) η signals at both points. Figure 17a,b show these results in the π^0-subtracted plots. Fits were made to both the unsubtracted and π^0-subtracted data, with the answers in close agreement after efficiency

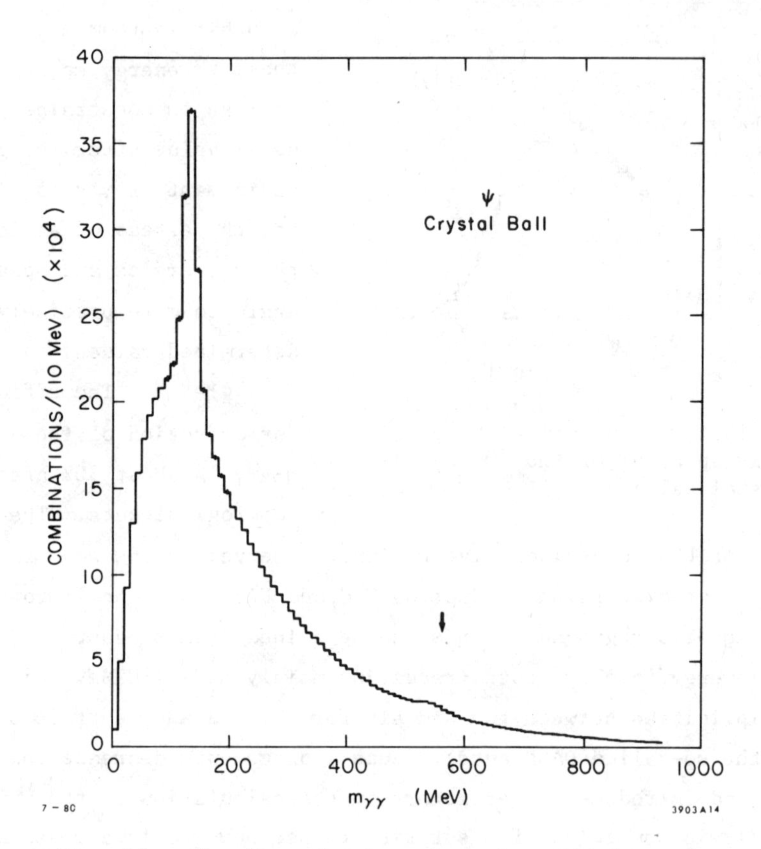

Fig. 14. The distribution of $m_{\gamma\gamma}$ for all photons detected inclusively at the ψ (Crystal Ball).

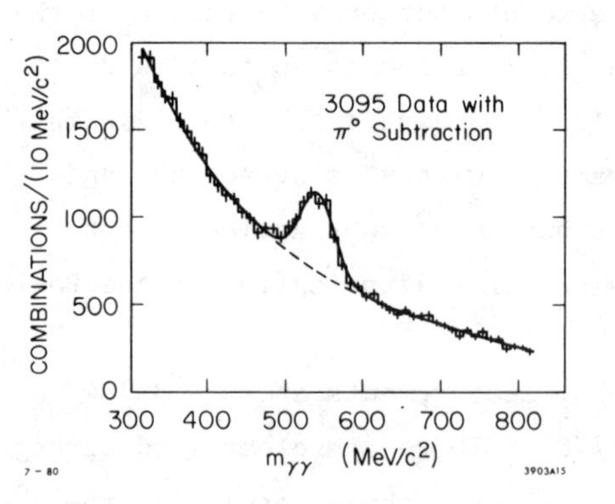

Fig. 15. The π^0-subtracted $m_{\gamma\gamma}$ spectrum of Fig. 14.

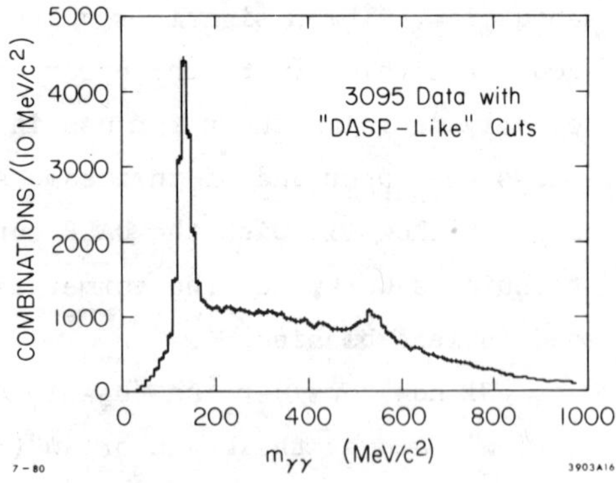

Fig. 16. The DASP-cut $m_{\gamma\gamma}$ spectrum of Fig. 14.

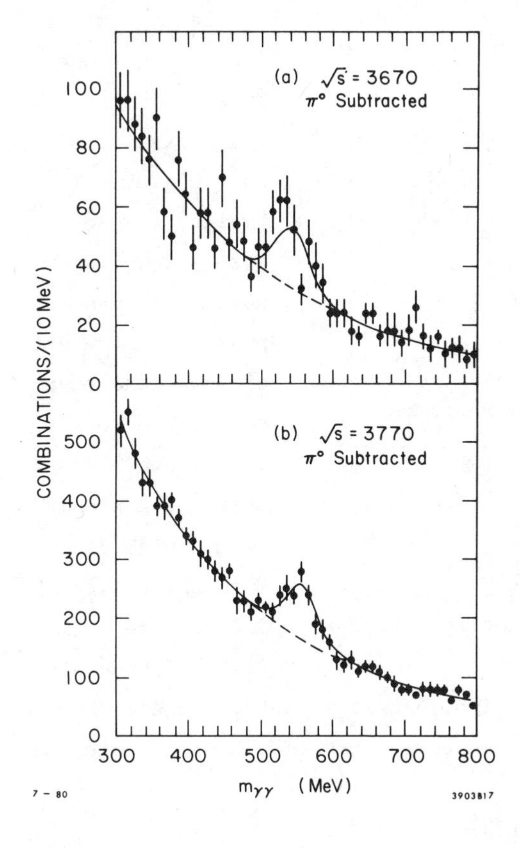

Fig. 17. Eta signals in the Crystal Ball:
a) at $\sqrt{s}$ = 3670 MeV;
b) at $\sqrt{s}$ = 3770 MeV.

corrections (see Section V.C for more detail about the fits). At $\sqrt{s}$ = 3670 we get f_η = .28 ± .06 and at the ψ'', f_η = .16 ± .03 [$R_\eta = f_\eta \cdot R$].

The baseline measurements for γ, π^0 energy and inclusive η have thus been established, with little structure seen for γ and π^0 at and below ψ'' (except for ψ', of course). The strong non-charmed production of η's even compared to ψ'' implies that the old physics can be a major source of η's in the scan region. The ψ'' value shows that D's may also have substantial decay modes involving η's.

V. CRYSTAL BALL RESULTS FOR $\sqrt{s}$ > 3.9 GeV

After this long preparation, we can now discuss the available spectra in the regions most crucial to the physics outlined in Section III.

A) Preliminary Results for R.

Data from the 5-week initial scan (Table I) have been analyzed for R, and the results[19] are displayed in Fig. 18 (note suppressed zero). Even for this small sample the systematics greatly dominate the statistical errors. The usual qualitative picture is confirmed, with clear 3.77, 4.03 and 4.4 peaks. The 4.16 region is consistent with one broad structure if the ±5% point-to-point systematic errors are assumed to be present at their extreme. A less conservative interpretation is that the structure near 4.05 - 4.20 may be more

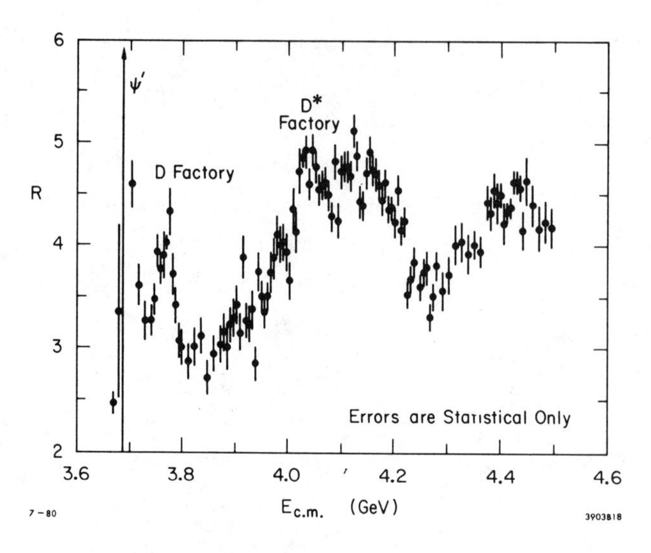

Fig. 18. The behavior of R through the Crystal Ball Scan Region [τ- and background-subtraction have been made but not radiative corrections] (Ref. 19).

complex. Similar fluctuations are present in the sharp rise from 3.9 to 4.03 GeV. The "valley" region 4.2 - 4.3 GeV shows no localized narrow peaks, but the data are not sensitive to the smoothly distributed .2 - .3 units of R predicted in Fig. 8. With the entire data sample (Table I) the statistical precision will be very high, but it is not yet clear that point-to-point systematic error can be reduced enough to exploit the large sample.

B) Results at $\sqrt{s}$ = 4.028 GeV ("D^*-Factory").

The main thrust of the data gathered at this copious source of D^*'s is to refine the parameters related to production and decay of these states. It should be stressed that there are two major <u>independent</u> measures of these quantities: the inclusive, completely π^0-subtracted γ energy spectrum and the pure π^0 energy spectrum. However, because π^0 separation is never complete, the plots will be somewhat correlated. We choose to fit unsubtracted γ and π^0 spectra for the parameters separately, and quote the best results (not averages).

1) <u>The γ energy spectrum</u> for well-measured photons is shown in Fig. 19 (complete data sample). Prominent peaks appear at ~70 and ~135 MeV, with widths corresponding to the Doppler broadening for γ's from π^0's of kinetic energy ~6 MeV (for the lower peak) and to the recoil-shifted, Doppler- and resolution-broadened photon transition

$D^* \to \gamma D$ (upper peak). The effect of (partial) π^0 subtraction is shown in Fig. 20, where the lower peak almost disappears. Figure 21 shows

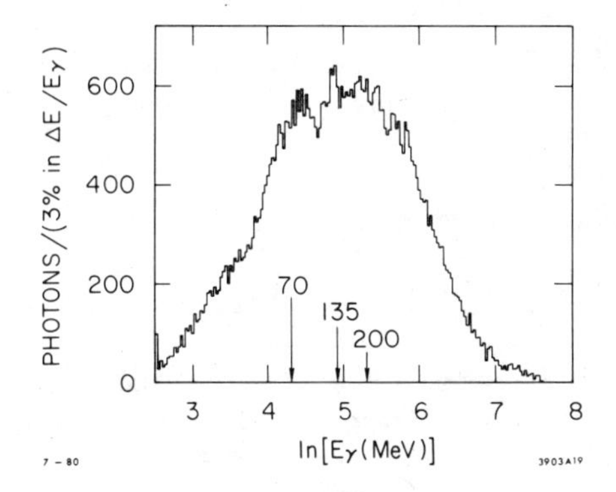

Fig. 19. Inclusive energy spectrum at $\sqrt{s}$ = 4028 MeV from the Crystal Ball.

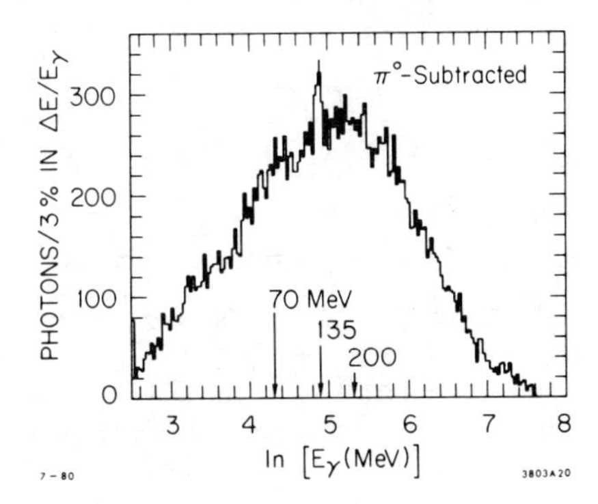

Fig. 20. The π^0-subtracted γ spectrum of Fig. 19.

the details of a multi-parameter fit to this spectrum, where the smooth background is taken from the ψ'' result and the shapes of the bumps from Monte Carlos (checked for effects of resolutions, interactions, inefficiencies and real analysis programs). The variable parameters are the D and D* masses, the peak/background

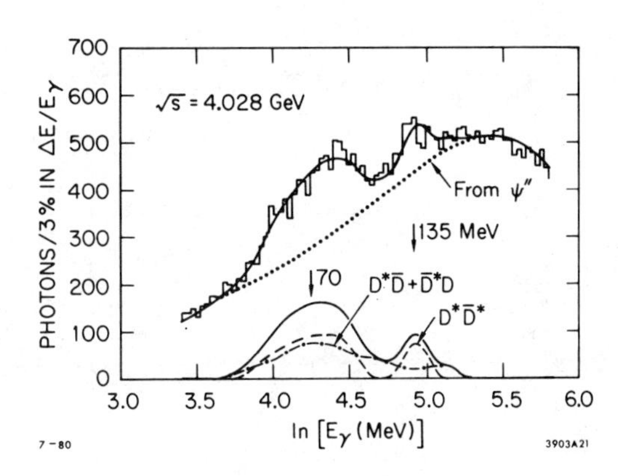

Fig. 21. Components of a fit to the spectrum of Fig. 19.

ratio, the $D^*\bar{D}^*:D^*\bar{D} + D\bar{D}^*$ ratio and the branching ratio of the radiative transition B_γ. Although transitions of neutral D's dominate these numbers, we delete the superscript "o" whenever the fits really measure a sum of neutral and charged D transitions. The results of this particular fit are shown in Table II. Similar fits with variation of cuts, backgrounds and Monte Carlo parameters were done to test the systematic errors. Best values with systematics attached from these and following fits will be shown in Table III.

Table II: Results of a typical fit to the γ-energy spectrum at $\sqrt{s}$ = 4028 MeV

m_{D^o}	= 1863 ± 2 MeV
$m_{D^{*o}}$	= 2005 ± 2
B_γ	= 32 ± 2%
$D^*\bar{D}^*:D^*\bar{D} + D\bar{D}^*$	= 1:1.9 ± .4

2) The π^o energy spectrum can be obtained with or without π^o candidates constrained to the π^o mass. Fig. 22, the former spectrum, distinctly shows that the broad bump seen in the γ spectrum (Fig. 19) is really very monochromatic, almost

Table III: Best Values of the D/D* Production/Decay Parameters[a]

Measurement	C.B. Value	Previous Best Value[b] (Mark I)
m_{D^o}	1864 ± 2 MeV	1863.3 ± 0.9 MeV
$m_{D^{*o}}$	2005 ± 2 MeV	2006 ± 1.5 MeV
$BR(D^* \to \gamma D)$	31 ± 6%	45 ± 15% (for neutral D's)
$Q(D^{*o} \to \gamma D^o)$	142.2 ± 2 MeV	142.7 ± 1.7
$D^*\bar{D}^*:D^*\bar{D} + D\bar{D}^*$	1:1.95 ± .35 ± .7	1:0.95 ± .15 (for neutral D's)

a Errors include systematics.
b See Ref. 8.

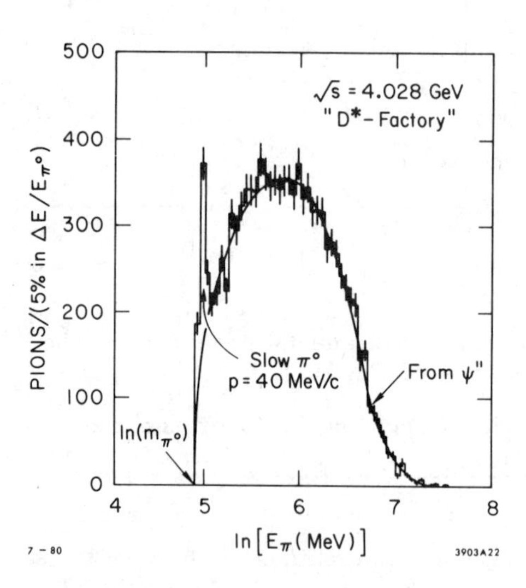

Fig. 22. The π^o total energy spectrum of the Crystal Ball at $\sqrt{s}$ = 4028 MeV with π^o candidates constrained to m_{π^o}.

stationary π^o's. (Note that here the $\ell n\ E_{\pi^o}$ scale is used to spread the peak out to render it distinguishable from $E_{\pi^o}=m_{\pi^o}$!). The dramatic change from the ψ" spectrum assures us that a change from unexcited charm states has occurred. The detailed shape of the spectrum is again very sensitive to the $D^*\bar{D}^*:D^*\bar{D}+D\bar{D}^*$ ratio (because 4028 MeV is ~16 MeV above the first threshold and ~160 MeV above the second). It is also the most sensitive measure of the Q-value $m_{D^{*o}} - m_{D^o}$ because of the spreading of the portion of the spectrum derived from $D^*\bar{D}^*$, though this is correlated to the above ratio. An example of the details of this slow π^0 peak, this time from the sample with the π^0 unconstrained to m_{π^0}, is given in Fig. 23. The fits are good and yield parameters consistent with the other two data sets, though with larger errors.

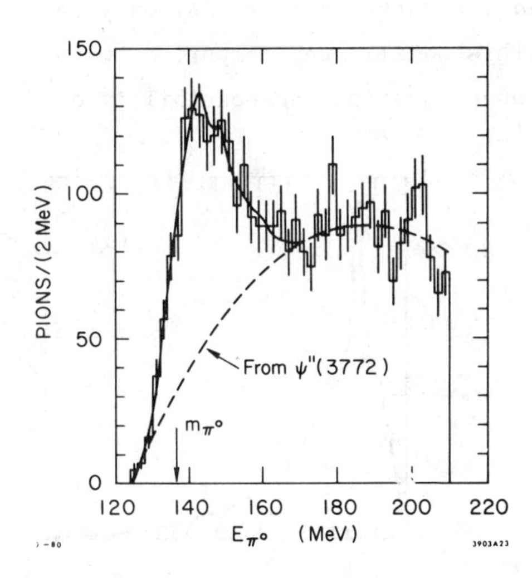

Fig. 23. Details of the slow π^0 energy spectrum, without the π^0 mass constraint of Fig. 22.

Table III gives our best estimates of the parameters of the D^* production and decay as gleaned from the above analysis. The masses have been determined at a level comparable to previous experiments and are consistent. The Q-value is also consistent with the Mark I value and has a similar error; in our data this error is almost entirely systematic. The branching ratio $B_\gamma = 31.0 \pm 6\%$ has been determined much more precisely by this first direct measurement of the transition γ; the error is again almost all systematic. There is a hint of disagreement in the new and old values of the production ratios of D^*'s; although covered by the errors, our independent plots consistently show $D^*\bar{D}^*$ to be a weaker effect than D^*D modes. The Mark I measurement must however be considered a more direct measure at this time. We are investigating the possibility of tagging one slow π^0 and looking at the absolute content of γ and π^0 peaks of the event, a technique which is a direct measure of the proportion of the two modes.

Discussion of the η-inclusive rate at 4.028 is included in the next section.

C) Results of the Inclusive-η Measurement for $3.86 < \sqrt{s} < 4.5$ GeV

A search for F, F^* and F^{**} mesons entails examination of inclusive γ, η and possibly π^0 spectra over the entire scan region. We report here only the inclusive η measurements, and for η's, only an examination in broad steps in $\sqrt{s}$. This facilitates comparison with previous results, but leaves as an open question the possibility of small local enhancements.

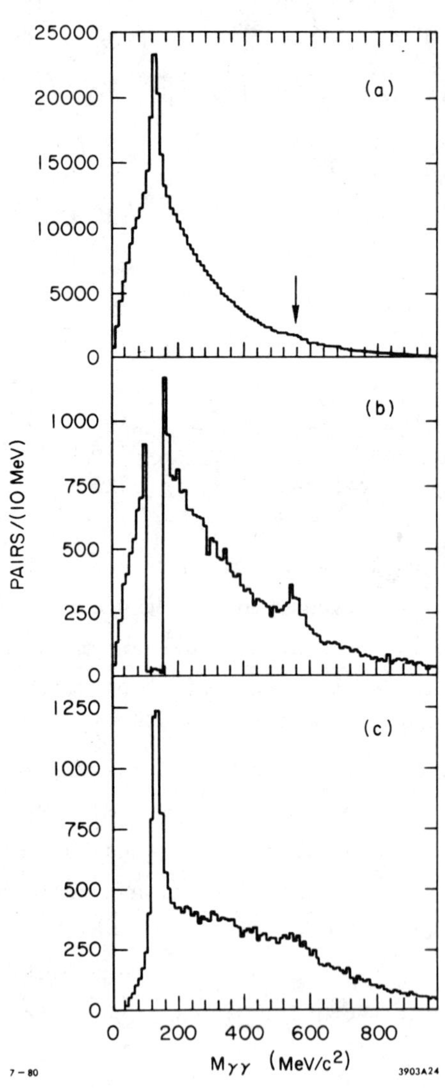

Fig. 24. The $m_{\gamma\gamma}$ mass distribution summed over $3.86 < \sqrt{s} < 4.5$ GeV: a) total sample; b) π^0-subtracted; c) DASP cuts (on Crystal Ball data).

The combined data sample from $3.86 < \sqrt{s} < 4.5$ GeV (1979 data only) shows characteristic η peaks in the $m_{\gamma\gamma}$ distribution as shown in Fig. 24 for the different cuts described earlier. Again, π^0-subtracted plots show the signal best, but both subtracted and unsubtracted plots are used for the fits to estimate systematics. The data are divided into segments closely approximating the choice made by DASP,[10] and fits are performed on each segment. Parameters of the fit are the background shape and amplitude, the η mass, width and amplitude; we find close agreement in the η mass and width in these fits compared to those at ψ'' and elsewhere. The fitted amplitude is translated into an effective number of η's produced by correcting for the detection efficiency (again by Monte Carlo calculation as described earlier) and branching ratio for $\eta \to \gamma\gamma$. Then f_η = (number of effective η)/(number of hadronic events) is the average number of η's/hadronic event. f_η has the advantage of relatively stable behavior in $\sqrt{s}$ although R may oscillate widely at resonances. Of course $\sigma_\eta \equiv R_\eta \sigma_{\mu\mu} = f_\eta R \sigma_{\mu\mu}$ may be used to obtain absolute cross sections.

The preliminary excitation function for f_η is shown in Fig. 25, including values obtained on/off resonances. The numerical values are given in Table IV. The overall systematic error is unknown, but the point-to-point errors are likely dominated by statistics. Previous values, where available, are also shown in both the Figure and Table, where we have used R,[4] σ_η,[10] given by DASP to calculate their f_η. (Limits are 1 - s.d.)

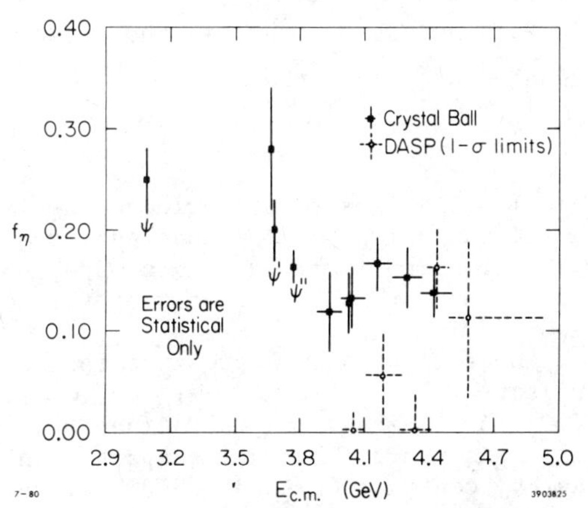

Fig. 25. The excitation of f_η as a function of $\sqrt{s}$: Crystal Ball scan and resonance measurements are solid lines; DASP scan measurements are dashed.

Table IV: Values for f_η as a Function of $\sqrt{s}$: Crystal Ball Scan and Resonance Measurements

c.m. Energy Range in GeV	f_η	
	C. B. Value	Previous Best Value (DASP) [a]
3.670	.28 ± .06	none
3.86 - 4.0	.12 ± .04	none
4.0 - 4.1	.13 ± .03	< .02
4.1 - 4.28	.17 ± .03	.05 ± .05
4.28 - 4.38	.15 ± .03	< .04
4.38 - 4.5	.14 ± .02	.16 ± .04

a See Ref. 10

The trend set by our baseline measurements continues, in that f_η remains large and constant within the errors. We see no evidence that η production shows the thresholds associated with F-production suggested by previous measurements. This rules out neither smaller local η enhancements nor does it rule out the evidence for F, F^* via exclusive kinematic fits in these regions. A subtraction of the noncharmed background is now possible in principle, but awaits amalgamation of the complete sample at $\sqrt{s}$ = 3670 MeV and better control of the systematic errors. The gentle rise of f_η in the R-valley (~ 4.25 GeV) can easily be attributed to a higher f_η (non-charm) than f_η (charm) and the diminution of the charm signal in the valley; we do not need to invoke F signals to explain the data.

ACKNOWLEDGEMENTS

I would like to acknowledge the value of the close cooperation in the management of the continuum scan by my colleague H. F. W. Sadrozinski, but even more, his leadership of the analysis effort which resulted in most of the new physics results in this report.

We also greatly appreciate the special efforts added to this project by D. G. Aschman, M. Joy, G. I. Kirkbride, F. C. Porter, J. C. Tompkins and the Operations Staffs at SPEAR and SLAC.

This report has been supported in part by the Department of Energy, contract DE-AC03-76SF00515, and in part by National Science Foundation Grant PHY79-16461.

APPENDIX

STATUS OF THE η_c CANDIDATE

The major thrust of this report was to introduce the reader to use of the Crystal Ball in the continuum. For completeness, we update another result of topical interest; the η_c candidate (2980 MeV) detected[19b] by the Crystal Ball operating in an inclusive mode. The evidence for the transitions $\psi' \to \gamma\eta_c$ and $\psi \to \gamma\eta_c$ are shown in Fig. 26, where the raw and background-subtracted inclusive γ energy spectra are shown. The new state is determined to be at 2980 ± 16 MeV from both sets of data independently; the joint fit value is 2980 ± 15 MeV. There is no useful η_c width information in the ψ' data because 8% (FWHM) resolution on a ~600 MeV γ yields 50 MeV resolution width. At the ψ, 12% (FWHM) on ~110 MeV γ yields absolute resolutions of ~13 MeV. Fits to this spectrum yield an intrinsic width ~20 MeV, but the error is so large that this is less than two s.d. from an intrinsic width of zero, and even more consistent with a

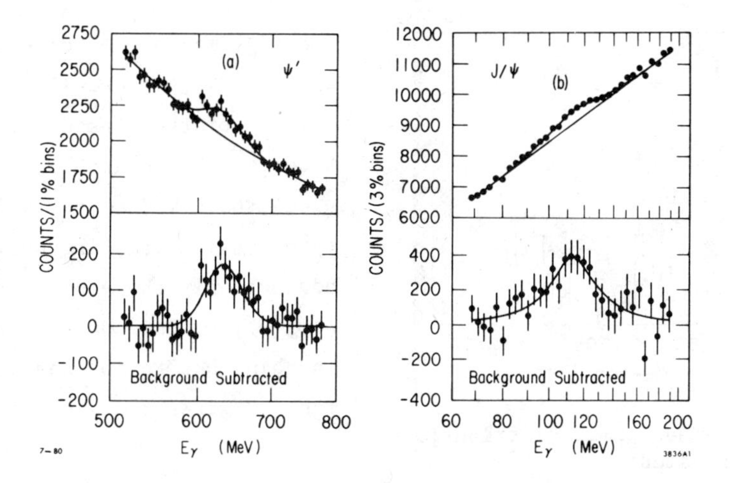

Fig. 26. Inclusive energy spectra from the Crystal Ball: a) $\psi' \to \gamma$ + all; b) $\psi \to \gamma$ + all.

conjectured value of 5 MeV often quoted.[20] Clearly, the η_c width has not yet been measured.

Subsequent to this result, evidence for the $\eta_c(2980)$ in exclusive channels has been found; first by the SLAC/LBL Mark II group[21,22] and shortly after by the Crystal Ball group.[23] The exclusive channels are different: Mark II sums the channels

$$\psi' \to \gamma + \begin{cases} K^{\mp}K^0_S\pi^{\pm} \\ K^+K^-\pi^+\pi^- \\ \pi^+\pi^-\pi^+\pi^- \\ p\bar{p} \\ p\bar{p}\,\pi^+\pi^- \end{cases}$$

(dominated by $\psi' \to (K^{\pm}K^0_S\pi^{\mp})+\gamma$) while the Crystal Ball looks at the pure channel $\psi \to (\eta\pi^+\pi^-)$ $+\gamma$. The fits to these channels are 1C/5C† and 3C, respectively. These new results are shown in Fig. 27 and 28.

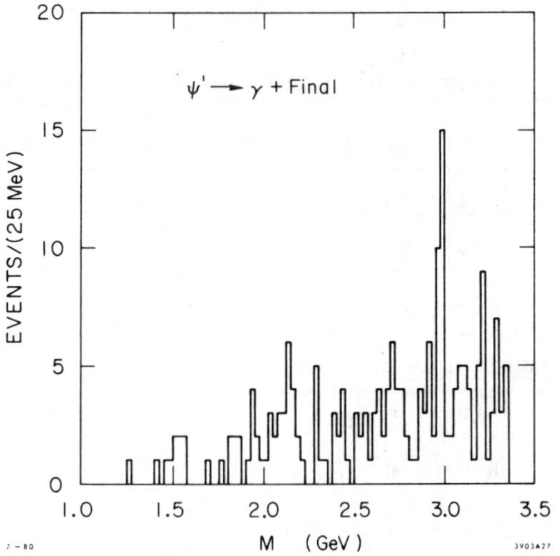

Fig. 27. The m_{final} distribution for the candidates $\psi' \to \eta_c\gamma$, $\eta_c \to$ final.

† The analyses from Ref. 21, 22, and 24 differ.

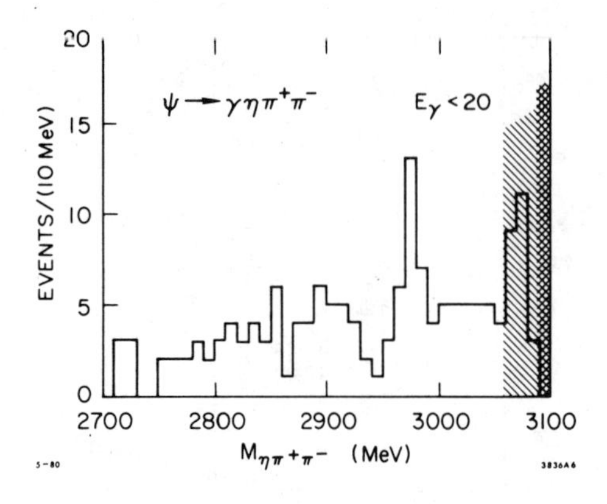

Fig. 28. The $m_{\eta\pi^+\pi^-}$ distribution for the candidates
$\psi \to \eta_c\gamma$
$\hookrightarrow \eta\pi^+\pi^-$

The significance of each peak is >4 s.d. For detailed discussion of these results, see Bloom.[24] The branching fractions, both $\psi/\psi' \to \eta_c\gamma$ and $\eta_c \to$ final state, are estimated, but only the product ratios $\psi/\psi' \to \gamma$ + final state are anything but rough guesses. Table V gives these estimates.

Table V: Estimates and Comparisons of the Product and Single Branching Ratios from $\psi/\psi' \to \eta_c\gamma$ and $\eta_c \to$ Final States

Property	Crystal Ball	Mark II
Measurements:		
Mass	2980 ± 15 MeV	2978 ± 8 MeV
Natural Full Width	< 35 (consistent with 0 to 20)	< 30 MeV
$BR(\psi'\to\gamma\eta_c)BR(\eta_c\to\pi^{\pm}K^{\mp}K_s)$	- - -	$1.5 \pm .6 \times 10^{-4}$
$BR(\psi\to\gamma\eta_c)BR(\eta_c\to\eta\pi^+\pi^-)$	$2.7 \pm 1.5 \times 10^{-4}$	- - -
$BR(\psi\to\gamma\eta_c)BR(\eta_c\to\gamma\gamma)$	$< 4 \times 10^{-5}$ (90% C.L.)	- - -
Rough Indications:		
$BR(\psi'\to\gamma\eta_c)$	~ .5%	- - -
$BR(\psi\to\gamma\eta_c)$	~ 1%	- - -
Implications:		
$BR(\eta_c\to\pi^{\pm}K^{\mp}K_s)$	- - -	~ 3%
$BR(\eta_c\eta\pi^+\pi^-)$	~ 3%	- - -
$BR(\eta_c\to\gamma\gamma)$	~ .4%	- - -

REFERENCES

1. Members of the Crystal Ball collaboration. California Institute of Technology, Physics Department: R. Partridge, C. Peck and F. Porter. Harvard University, Physics Department: A. Antreasyan, Y. F. Gu, W. Kollmann, M. Richardson, K. Strauch and K. Wacker. Princeton University, Physics Department: D. Aschman, T. Burnett (visitor), M. Cavalli-Sforza, D. Coyne, M. Joy and H. Sadrozinski. Stanford Linear Accelerator Center: E. D. Bloom F. Bulos, R. Chestnut, J. Gaiser, G. Godfrey, C. Kiesling, W. Lockman and M. Oreglia. Stanford University, Physics Department and High Energy Physics Laboratory: R. Hofstadter, R. Horisberger, I. Kirkbride, H. Kolanoski, K. Koenigsmann, A. Liberman, J. O'Reilly and J. Tompkins.
2. R. Partridge et al., Phys. Rev. Lett. 44, 712 (1980); Y. Chan et al., IEEE Transactions on Nuclear Science, Vol. NS-25, No. 1. 333 (1978); I. Kirkbride et al., IEEE Transactions on Nuclear Science, Vol. NS-26, No. 1, 1535 (1979).
3. E. D. Bloom, in Proceedings of the XIVth Rencontre de Moriond, Les Arcs, France, ed., j. Tran Than Van (1979).
4. J. L. Siegrist, "Hadron Production by e^+e^- Annihilation at Center-or-Mass Energies between 2.6 and 7.8 GeV," SLAC Report 225, p. 75, Oct. 1979.
5. P. A. Rapidis, "D Meson Production in e^+e^- Annihilation," SLAC Report 220, p. 81, June 1979.
6. R. H. Schindler, "Charmed Meson Production and Decay Properties at the psi(3770)," SLAC Report 219, Ph.D Thesis, May 1979.
7. G. S. Abrams et al., " Observation of Cabibbo Suppressed Decays $D^o \to \pi^+\pi^-$ and $D^o \to K^+K^-$," Phys. Rev. Lett. 43, 481 (1979); also SLAC-PUB-2337 (May 1978).
8. G. Goldhaber, "The Spectroscopy of New Particles," LBL Report 6732, P. 13, Aug. 1977.
9. G. J. Feldman, "Charmed Particle Spectroscopy," SLAC-PUB-2068, Dec. 1977.
10. R. Brandelik et al., Phys. Lett. 70B, 132 (1977).
11. R. Brandelik et al., Phys. Lett. 80B. 412 (1979).
12. C. Quigg and J. Rosner, Fermilab Pub 77/60-TH(1977).
13. R. Ammar et al., Preprint PRINT-80-0338 (Kansas), 1980; to be published in Phys. Lett.; and R. Kumar in Proceedings of the VIth International Conference on Experimental Meson Spectroscopy, (Brookhaven National Laboratory, Upton, N.Y., 1980), ed. S. U. Chung, to be published.
14. A. De Rújula, H. Georgi, S. L. Glashow, Phys. Rev. Lett 37, 785 (1976).
15. E. Eichten, K. Gottfried, T. Kinoshita, K. D. Lane, T. M. Yan, Phys. Rev. D17, 3090 (1978) and CLNS-425 (June 1979).
16. K. Gottfried, in an invited talk at the 1980 Vanderbilt Symposium (in remarks on subjects not to be discussed).
17. R. Cahn and B. Kayser, "Angular Distributions and the Physics of Charmed Meson Production at the 4.028 GeV Resonance," LBL-10692 (1980); Also see A. De Rújula, H. Georgi, S. L. Glashow, Phys. Rev. Lett. 37, 398 (1976).
18. J. L. Siegrist, Op. Cit., p. 39-41.

19. a) J. C. Tompkins, "Recent Results from the Crystal Ball," SLAC Report 224, 578 (1980).
b) E. D. Bloom, Proceedings of the 1979 International Symposium on Lepton and Photon Interactions at High Energies, Aug. 23-29, 1979, Fermilab; SLAC-PUB-2425 (1979).
20. T. Appelquist, R. M. Barnett, K. D. Lane, "Charm and Beyond," Ann. Rev. Nucl. Part. Sci. 28 (1978); also SLAC-PUB-2100 (1978).
21. T. M. Himel, "Decays of the $\psi'(3684)$ to Other Charmonium States," SLAC Report 223, Oct. 1979.
22. G. Trilling, D. Scharre, private communication.
23. D. G. Aschman, "Radiative Decays of ψ and ψ'," SLAC-PUB-2550, (1980) (to be published in Proceedings of the XV Rencotre de Moriond, Les Arcs, France, March 15-21, 1980).
24. E. D. Bloom, "Radiative Transitions to an $\eta_c(2980)$ Candidate State and the Observation of Hadronic Decays of this State," SLAC-PUB-2530 (1980).

FURTHER RESULTS ON THE T MESONS FROM DASP

H. Schröder

Deutsches Elektronen-Synchrotron DESY, Hamburg, Germany

ABSTRACT

The last measurements with the DASP detector at the DESY storage ring DORIS have yielded improved results of the properties of the T meson. The leptonic widths are found to be $\Gamma_{ee}(T) = 1.35 \pm 0.11 \pm 0.22$ keV and $\Gamma_{ee}(T') = 0.61 \pm 0.11 \pm 0.11$ keV. The result on the muonic branching ratio $B_{\mu\mu}$ and the total width Γ_{tot} of the T meson is $B_{\mu\mu}(T) = 2.9 \pm 1.3 \pm 0.5\%$ and $\Gamma_{tot} = 47 {}^{+37}_{-15}$ keV. Evidence for the $T' \rightarrow \pi\pi T$ decay is seen from the inclusive spectra.

At present the T system[1] represents the hydrogene atom of the strong interaction. It is well described as a bound state of a heavy quark b with charge $e_b = 1/3$ and its antiquark[2]. The fact that it contains heavy quarks with masses of about 5 GeV makes the system non-relativistic. Thus the T system can be described in rather simple terms by a two-body Schrödinger equation. Since the T system itself provides information on the potential or the fundamental strong force it is highly desirable to measure its properties accurately. An important quantity which was up to now poorly known is the total width of the T meson. According to QCD the T meson should decay mainly into 3 gluons. This decay is proportional to $\alpha_s{}^3$ leading to a narrow width comparable to that of the J/ψ which is 69 keV. Keeping in mind that up to now the upper limit on $\Gamma_{tot}(T)$ was some few MeV's[3,4] it was quite clear that this situation had to be improved. Thus one topic of my talk will be a better determination of this important quantity, which is given by

$$\Gamma_{tot} = \Gamma_{ee} / B_{\mu\mu} \qquad (1)$$

if one assumes $e - \mu - \tau$ universality. The present large errors on $\Gamma_{tot}(T)$ come mainly from the fact that the leptonic branching ratio $B_{\mu\mu}(T)$ is not well known. Thus measuring this number means essentially measuring $\Gamma_{tot}(T)$ since the leptonic width Γ_{ee} is quite well known[1]. In addition to the determination of $B_{\mu\mu}(T)$ also refined resonance parameters for the T and T' will be discussed. The decay properties of the T and T' will be investigated by means of sphericity

ISSN:0094-243X/80/620279-10$1.50

distributions and inclusive spectra.

The results were obtained by the DASP2 group[5] using the DASP detector at the DESY storage ring DORIS. The DORIS machine was operated in the Υ region in a single-ring-single-bunch mode. With maximum beam currents of typically 18 mA a peak luminosity of 10^{30} cm^{-2} s^{-1} was achieved. The lifetime of the stored beams was about 4 hours and the refill time a few minutes. These conditions allowed us to collect about 30 nb^{-1} day^{-1}. In fall 1979 and winter 1980 luminosity runs were made in the Υ and Υ' region respectively, collecting 825 nb^{-1} resp. 912 nb^{-1} [6].

The DASP detector is well known[7]. After 7 years of successful operation it has now been totally dismounted. I will only sketch the main properties of the detector (fig. 1). It consists of a magnetic double arm spectrometer with excellent particle discrimination abilities but covering only 5 % of 4π. The non magnetic inner detector fig.1a) covers 50% of 4π and detects charged particles as well as photons. TOF-counters on opposite sides of the beam and on the upper iron flux bridge are essential in order to distinguish μ-pairs from the e^+e^- annihilation from cosmic rays.

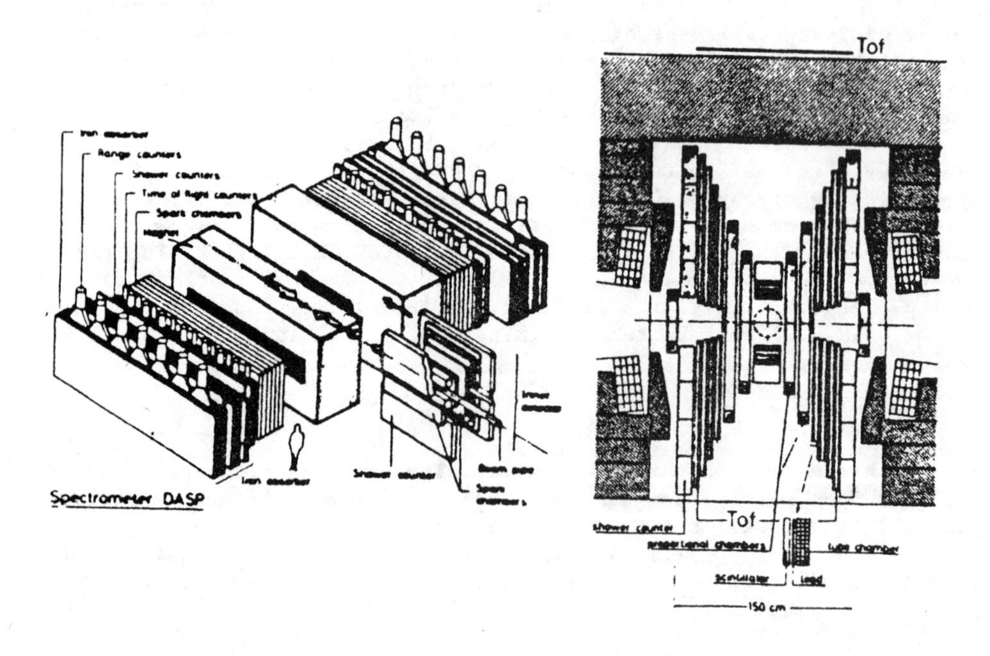

(a)

(b)

Fig. 1

The cross section for the reaction $e^+e^- \rightarrow$ hadrons is displayed in fig. 2. The trigger required 4 or more reconstructed charged tracks or converted photons in the inner detector. The cross section was ex-

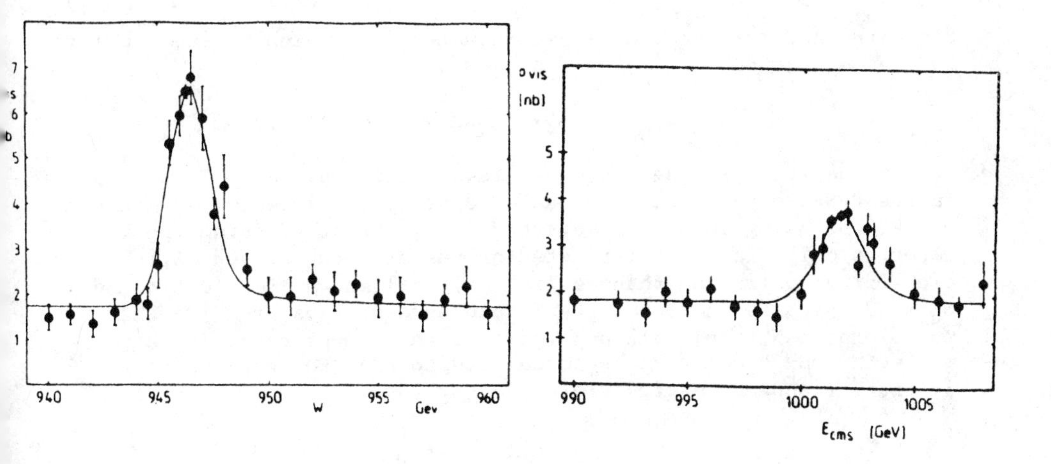

Fig. 2

pressed in terms of a continuum contribution and a Breit-Wigner function for each resonance:

$$\sigma_{had} = \frac{3\pi}{s}\left(\frac{4\alpha^2}{9} R_{had} \cdot \eta_1 + \frac{\Gamma_{ee}\ \Gamma_{had}}{(M-\sqrt{s})^2+\Gamma^2/4} \cdot \eta_2\right) \tag{2}$$

This expression had to be folded with the gaussian resolution of the machine and corrected for radiation in the initial state[8].

The overall efficiencies η_1 for the continuum and η_2 for the Υ meson region were determined by a 2 jet and a 3 jet Monte Carlo[9], respectively, to be η_1 (9.5 GeV) = 42 % and $\eta_2(\Upsilon)$ = 48%. The fit to the data yields the following results (see table I)

TABLE I* Resonance parameters

	Υ	Υ'	
$\Gamma_{ee}(1-3\ B_{\mu\mu})$	1.23 ± 0.09 ± 0.20	0.61 ± 0.11 ± 0.11	keV
M	9463.1 ± 0.7 ± 10	10016.8 ± 1.5 ± 20	MeV
σ_{DORIS}	8.6 ± 0.6	11.0 ± 1.9	MeV
R	4.23 ± 0.21 ± 0.42**		
$M(\Upsilon) - M(\Upsilon')$		553.7 ±1.7 ± 10	MeV

* The first error is the statistical one the second the systematic one

** Radiative corrections are not yet applied

The leptonic branching ratio for the T meson[+] is given by

$$B_{\mu\mu}(T) = \frac{\sigma(e^+e^- \to T \to \mu^+\mu^-)}{\sigma(e^+e^- \to T \to \text{had}) + 3\sigma\,(e^+e^- \to T \to \mu^+\mu^-)} \tag{3}$$

Since the hadronic peak cross section is already known from the fit to the data to be

$$\sigma(e^+e^- \to T \to \text{had}) = 10.1 \pm 0.7 \text{ nb}$$

it remains to determine the peak cross section $\sigma(e^+e^- \to T \to \mu^+\mu^-)$ In the DASP2 experiment this can be done by two independent measurements, one using the outer detector only, the other using the inner detector only. In the outer detector the acceptances and efficiencies for μ- pairs and Bhabhas are very well known. Here one found $N^{on}_{\mu\mu} = 24$ μ-pairs on top of the T-resonance whereas from 135 Bhabhas one would expect $N^{QED} = 14.3 \pm 2$ μ-pairs from the QED-process. The excess $N^{T}_{\mu\mu} = N^{ON}_{\mu\mu} - N^{QED}_{\mu\mu}$ can then be normalized to the QED cross section $\sigma^{QED}_{\mu\mu}$ at the T resonance energy yielding

$$\sigma(e^+e^- \to T \to \mu^+\mu^-) = \sigma^{QED}_{\mu\mu} \cdot \frac{N^{ON}_{\mu\mu} - N^{QED}_{\mu\mu}}{N^{QED}_{\mu\mu}} \tag{4}$$

$$= 0.68 \pm 0.4 \text{ nb}$$

Inserting this number into (3) yields

$$B_{\mu\mu}(T) = 5.6 \pm 3.3 \text{ \%}. \tag{5}$$

In the inner detector μ-pairs are identified as two collinear, minimum ionizing particles. This is justified by the fact that no hadron pair was found in the outer detector. The background consists primarily of cosmic rays which can be separated from the μ-pairs by TOF measurements.

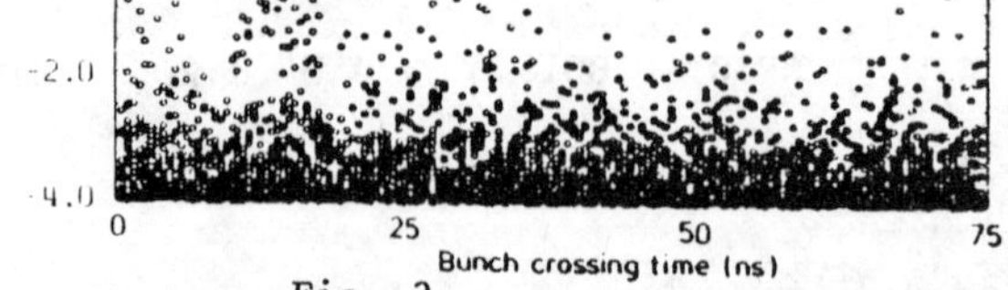

Fig. 3

This is illustrated in fig. 3, where the TOF-difference between two opposite scintillation counters (see fig. 1) is plotted versus the bunch crossing time. μ-pairs from the e^+e^- annihilation should have a zero TOF-difference and should be correlated with the bunch signal (see arrows in fig. 3), whereas cosmic rays are not correlated with the bunch signal and should appear at ΔTOF = -6 ns. In fig. 3 one clearly observes a cluster of μ-pairs in the right region. A projection onto the TOF axis of events correlated within ± 10 ns with the bunch signal is shown in fig. 4a. If one sub-

+ Due to lack of statistics $B_{\mu\mu}$ (T') could not be determined.

tracts the TOF-spectrum of uncorrelated events (fig. 4b) from fig. 4a

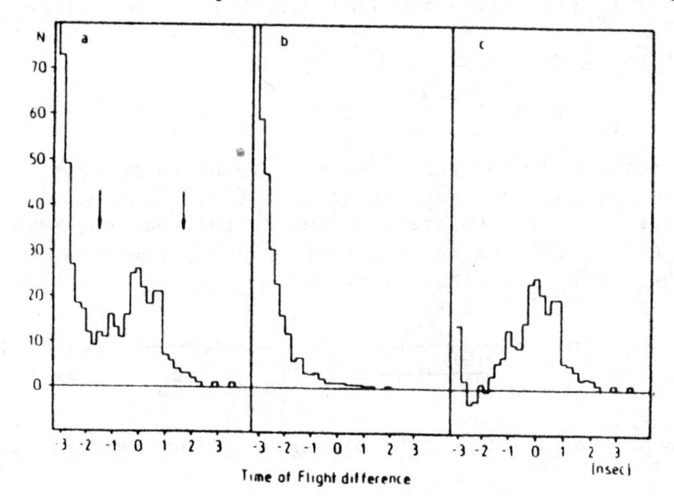

Fig. 4

one gets the spectrum in fig. 4c which exhibits a clear   signal. The angular distribution of these μ pairs shows the expected $(1+\cos^2\theta)$ distribution (fig. 5). From the number of μ pairs on the resonance, N_{on}, and in the continuum, N_{cont}, one gets the peak cross section:

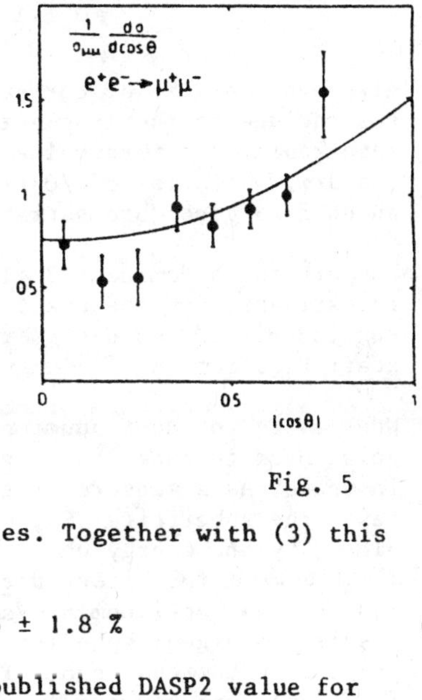

$$\sigma(e^+e^- \rightarrow T \rightarrow \mu^+\mu^-) = \sigma_{\mu\mu}^{QED}\left(\frac{N_{on}}{N_{cont}} \cdot \frac{L_{cont}}{L_{on}} - 1\right)$$

$$= (0.27\pm0.20)\ \text{nb} \qquad (6)$$

Fig. 5

The L's are the corresponding luminosities. Together with (3) this value leads to

$$B_{\mu\mu}(T) = 2.5 \pm 1.8\ \% \qquad (7)$$

Averaging (5) and (7) with the already published DASP2 value for $B_{\mu\mu}(T)$ from 1978 [3] of

$$B_{\mu\mu}(T) = 2.5 \pm 2.1\ \%$$

yields

$$B_{\mu\mu}^{DASP2}(T) = 2.9 \pm 1.3 \pm 0.5\ \%$$

The determination of $B_{\mu\mu}$ allows one to calculate the leptonic width of the T meson :

$$\Gamma_{ee}(T) = 1.35 \pm 0.11 \pm 0.22 \text{ keV}$$

and the total width using relation (1):

$$\Gamma_{tot}(T) = 47 \begin{smallmatrix} +35 \\ -15 \end{smallmatrix} \text{ keV.}$$

This value for the total width of the T meson is found to be very close to that of the J/ψ particle suggesting that their nature is very similar and that their decays are governed by the same mechanism. The dominant decay mode should be in both cases the 3 gluon decay. Using the expression for $B_{\mu\mu}$ in first order QCD

$$B_{\mu\mu} = \frac{e_b^2}{\frac{10(\pi^2-9)}{81\,\pi\,\alpha^2}\,\alpha_s^3 + (R_{had}+3)\,e_b^2} \tag{8}$$

one is able to deduce a value of α_s, the strong coupling constant, for the T meson:

$$\alpha_s(T) = 0.17 \begin{smallmatrix} +0.04 \\ -0.03 \end{smallmatrix}$$

Although the size of corrections to (8) due to higher order processes and due to the fragmentation process is not known, it is quite remarkable that this value fits quite well with the one deduced for the J/ψ, which is $\alpha(J/\psi) = 0.19 \pm 0.02$, and the values of α_s deduced from gluon bremsstrahlung[10].

Support for a dominant 3 gluon decay of the T meson comes from an investigation of the event topologies on the resonance and in the continuum[11-13]. We have performed these investigations with good statistics for the T' meson also.

Whereas in the continuum one expects to see events with two hadron jets going back to back, the 3 gluon decay events should be more extended in space. As a measure of the spatial distribution of an event we have taken the sphericity S, where the momenta of the particles are replaced by the energy which they deposit in the detector. The sphericity distribution for direct decays of the T meson is displayed in fig. 6a and for the continuum at $\sqrt{s} \sim 9.5$ GeV in fig. 6b. Clearly one observes a shift to higher sphericities for the direct decay of the T meson compared with the continuum. The same holds also for the direct T' decay (fig. 7a) when compared with the continuum process at $\sqrt{s} \sim 10$ GeV (fig. 7b). However, the observed shift is not as large as for the T meson (see table II). This may reflect the fact that the T' meson can also decay by photon emission into P-states with even spin. These states should decay mainly into 2 gluons giving rise to 2 jets similar to the continuum process. On the other hand the $T' \to 2\pi T$ decay would make the events' shape more spherical.

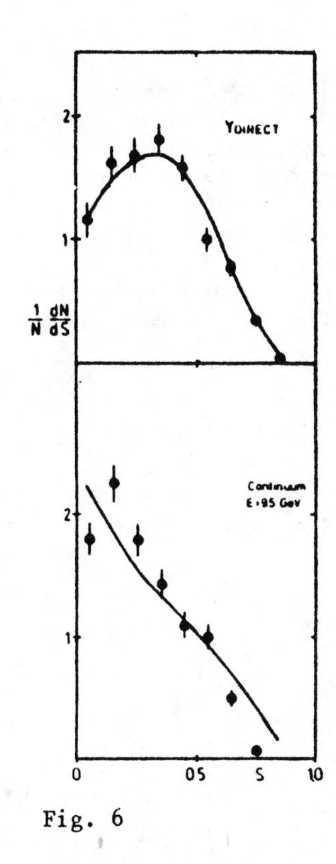

Fig. 6

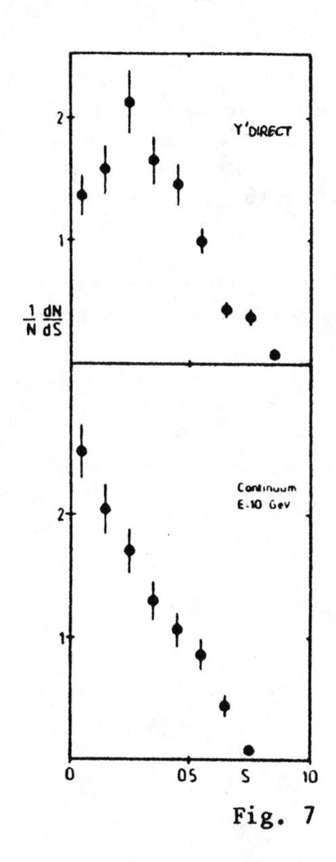

Fig. 7

Table II: Mean sphericity < S>

D a t a		Monte Carlo at $\sqrt{s}$ = 9.5 GeV	
Continuum			
$\sqrt{s}$ = 9.5 GeV	0.29 ± 0.01	2-jet	0.30 ± 0.01
Υ_{direct}	0.36 ± 0.02	3-jet	0.36 ± 0.01
Continuum			
$\sqrt{s}$ = 10 GeV	0.28 ± 0.02	phase space	0.40 ± 0.01
Υ' direct	0.32 ± 0.03		

Since at the moment no quantitative prediction can be given by any theory the data were compared with Monte-Carlo calculations by using three different models:

1. a 2-jet model[14,9] for the description of the continuum,
2. a 3-jet model[9] to describe the expected 3 gluon decay of the meson assuming that the gluons fragment like quarks into hadrons[14], and
3. a multipion phase space model for further comparison.

The results of the Monte-Carlo calculations are shown in Table II. The continuum data compare very well with the 2-jet Monte-Carlo and the same holds true when comparing the Υ_{direct} data with the results from the 3-jet Monte-Carlo. Not only the mean values but also the differential distributions are reproduced (see solid curves in fig. 6). The naive phase space model fails to describe the data in both cases.

Further information on the event topology is provided by the inclusive spectra which are displayed in fig. 8. Parametrizing the cross section as

$$E \frac{d^3\sigma}{dp^3} \propto e^{bE}$$

one gets the following results upon fitting this equation to the data (see Table III) where for the Υ and Υ' meson no continuum contributions are subtracted.

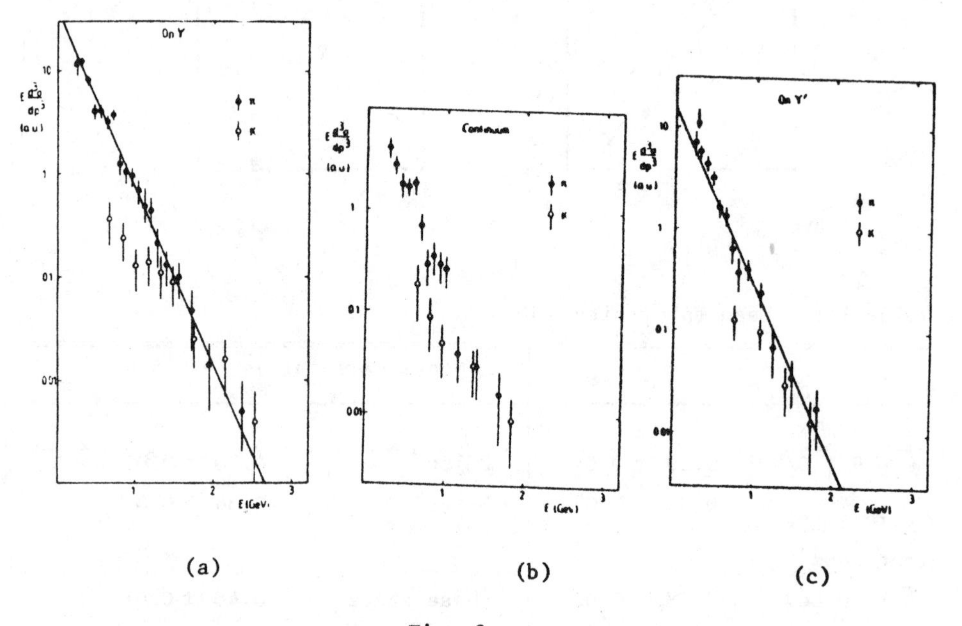

(a) (b) (c)

Fig. 8

One observes that the slope of the spectra does not change when going from the continuuum (fig. 8a) to the peak of the Υ meson (fig. 8b). With a rise of the charge multiplicity of about 20 % in the Υ resonance[11,13,15] one would expect however, that the mean momentum for these events should become smaller and therefore the slope steeper as it is seen in the Monte-Carlo calculations. It might be that this effect is not seen in the data because the errors in the slope parameter are still too large.

Table III: Slope parameter b (GeV^{-1})

D a t a		Monte Carlo	
	b ($\pi^{\pm}$)	b ($\pi^{\pm}$)	
Continuum	- 3.93 ± 0.35	2 jet	-3.78 ± 0.11
Υ_{on}	- 3.94 ± 0.18	3-jet	-3.46 ± 0.10
Υ'_{on}	- 4.94 ± 0.32	phase space	-3.63 ± 0.03
b ($K^{\pm}$)	For all spectra ~ - 2.3		

The inclusive pion distribution from the Υ' meson (fig. 8c) is clearly different from all other spectra. The steeper slope of the spectrum indicates that more pions with small energies are present. This is expected from the $\Upsilon' \to \pi\pi\Upsilon$ decay where the accepted pions have energies between 240 and 400 MeV. By normalizing the fitted curve in fig. 8b to the Υ' spectrum for energies above 450 MeV one observes an excess of pions at low energies which should come from the $\Upsilon'\to\pi\pi\Upsilon$ decay (see arrow in fig. 8c).

ACKNOWLEDGEMENTS

I wish to thank my colleagues from the DASP2-group who helped in preparing this talk.

REFERENCES

1. S. Herb et al., Phys. Rev. Lett 39,252 (1977)
   C. Berger et al., Phys. Lett. 76B,243 (1978)
   C.W. Darden et al., Phys. Lett. 76B,246 (1978) and Phys. Lett. 78B,364 (1978)
   J.K. Bienlein et al., Phys. Lett. 78B,360 (1978)
   D. Andrews et al., Phys. Rev. Lett. 44,1108 (1980)
   T. Böhringer et al., Phys. Rev. Lett. 44,1111 (1980)
2. see for example
   J.D. Jackson, C. Quigg and I.L. Rosner, 'New Particles - Theoretical', in Proc. of the 19th Intern. Conf. on High Energy Phys., Tokyo 1978, p. 391-408.
3. C.W. Darden et al., Phys. Lett. 80B,419 (1979)
4. C. Berger et al., Z. Phys. C1,343 (1979)

5. H. Albrecht, G. Drews, H. Hasemann, E. McCliment, W. Schmidt-Parzefall, H. Schröder, H.D. Schulz, F. Selonke, E. Steinmann, R. Wurth (DESY), W. Hofmann, A. Markees, D. Wegener (Dortmund), K.R. Schubert, J. Stiewe, R. Waldi, S. Weseler (Heidelberg), P. Böckmann, L. Jönsson (Lund), M. Danilov, Yu. Semenov, I. Tichomirov, Yu. Zaitsev (Moscow), R. Childers and C.W. Darden (South Carolina)

6. H. Albrecht et al., DESY 80/30, April 1980, to be published in Phys. Lett.

7. R. Brandelik et al., Z. Phys. C1,233 (1979)

8. J.D. Jackson and D.L. Scharre, Nucl. Instr. Meth. 128,13 (1975)

9. T. Nakada, Heidelberg reports /HEP-HD/79-1/ARGUS and /HEP-HD/79-2/ARGUS, unpublished

10. TASSO collaboration, DESY 80/40, May 1980
M. Chen, 'Results from MARK J', at this conference
R. Heuer , 'Results from JADE', at this conference
G. Knies, 'Results from PLUTO', at this conference

11. W. Schmidt-Parzefall, Proc. of the 19th Intern. Conference on High Energy Phys. Tokyo 1978, p.260

12. PLUTO-collaboration, 'Experimental Search for Υ Decay into 3 Gluons', Intern. Conf. on High Energy Phys., Geneva , 27 June - 4 July, 1979

13. F.H. Heimlich et al., Phys. Lett. 86B,399 (1979)

14. R.D. Field and R.P. Feynman, Nucl. Phys. B136,1 (1978)

15. PLUTO-collaboration, Phys. Lett. 82B,449 (1979)

RESULTS FROM THE LENA DETECTOR ON THE Υ AND Υ' RESONANCES

B. Niczyporuk and T. Zełudziewicz
Inst. of Nuclear Physics, Cracow, Poland

H. Vogel[(a)] and H. Wegener
Physikalisches Inst. der Univ. Erlangen, Nürnberg, Germany

K. W. Chen and R. Hartung
Michigan State Univ., East Lansing, MI

J.Bienlein, R.Graumann, J.Krüger, M.Leissner, M.Schmitz, C.Rippich[(b)]
Deutsches Elektronen-Synchrotron DESY, Hamburg, Germany

E.Filos, F.Heimlich, R.Nernst, A.Schwarz, U.Strohbusch, P.Zschorsch
I.Inst. für Experimental Physik der Univ. Hamburg, Germany

A.Engler, R.W.Kraemer, F.Messing, T.Ridge, B.J.Stacey, S.Youssef
Carnegie-Mellon Univ., Pittsburgh, PA

A. Fridman
DPhPE, CEN, Saclay, France

G.Alexander, A.Av-Shalom, G.Bella, Y.Gnat, J.Grunhaus
Tel-Aviv Univ., Israel

E.Hörber[(c)], W.Langguth[(d)], and M.Scheer
Physikalisches Inst. der Univ. Würzburg, Germany

ABSTRACT

Using the LENA detector at the DORIS e^+e^- storage ring we have measured the hadronic cross section and the μ-pair branching ratio of the Υ (9.46) resonance. We obtain the electronic width Γ_{ee} = (1.23±0.10±0.14) keV and the mass M = (9 461.3±0.7±10.0) MeV, the μ-pair branching ratio $B_{\mu\mu}$ = (3.5±1.5±0.4)% and the total width $\Gamma_{tot} = (35^{+25}_{-10}{}^{+9}_{-7})$ keV. For the Υ' resonance, the mass is found to be (10014.0±1.0±10.0) MeV, and $\Gamma_{ee}\cdot\Gamma_{had}/\Gamma_{tot}$ = (0.51±0.07±0.06) keV.

The discovery of Υ (9.46) in proton nucleus reactions [1] and its subsequent confirmation in e^+e^- annihilations [2,3] opened the study of a new family of particles. The properties of the members of this family provide a testing ground for the many models which have been used to explain the J/ψ family. Of particular interest are the electronic and total widths (Γ_{ee} and Γ_{tot}) for the Υ which can be obtained from the Υ hadronic cross section σ_h(Υ) and the leptonic branching ratio $B_{\mu\mu}$. We present a measurement of $B_{\mu\mu}$ which differs from zero by more than two standard deviations.

The data were taken with the LENA (Lead Glass NaI) detector (Fig.1) at the DORIS e^+e^- storage ring. The LENA detector was

ISSN:0094-243X/80/620289-09$1.50

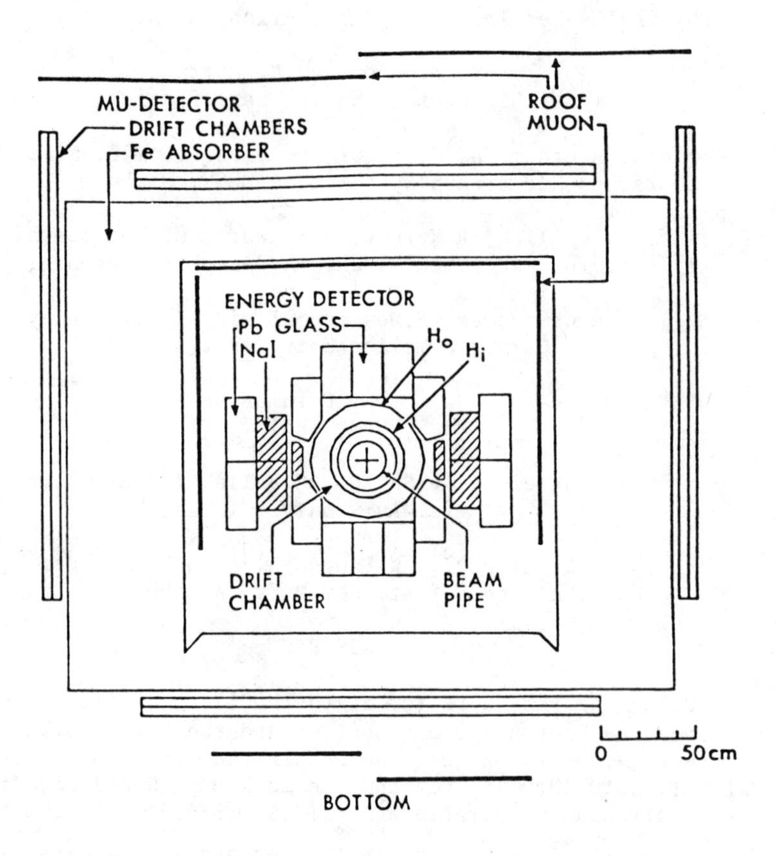

Fig. 1. The LENA detector seen along the beam direction. For details of the detector components see text.

constructed by the DESY-Heidelberg collaboration and has been described elsewhere [4]. Briefly, the inner detector consists of three double layers of cylindrical drift chambers with Z-strip cathode readout and two cylindrical hodoscopes (H_1,H_0) each with 32 elements. The outer hodoscope counters H_0 are used as time of flight (ToF) counters. The inner detector which covers a solid angle of 86% of 4π is surrounded by the energy detector consisting of 178 blocks of lead glass and NaI, providing more than 10 radiation lengths of absorber over 70% of 4π. In order to obtain accurate photon directions, there were active converters placed between the two sidewalls and the inner detector, consisting of 1.8 radiation lengths of active NaI converters followed by drift chambers. For the top and bottom regions, there was 1 radiation length of lead placed between the middle and outer double drift chambers instead. Surrounding the energy detector are additional ToF counters, steel absorber, and muon drift chambers. The ToF counters are labelled "Muon", "Roof", and "Bottom" in the drawing and are at minimum distances of 1.0, 1.9, and 1.5 m, respectively, from the intersection region. The steel absorber is 60 cm thick on the sides and 30 cm thick on the top and bottom.

The experiment was triggered by a coincidence between tracks in the inner detector and pulse height in the energy detector. Since the hodoscopes are highly segmented and the detector is non-magnetic, coincidences between overlapping elements of H_1 and H_0 indicate the presence of tracks leaving the beam pipe radially. Depending on the number of tracks, different energy thresholds were applied to the total energy observed. For ≥ 3 tracks the threshold was roughly 250 MeV. For 2 (1) tracks the threshold was 300 (800) MeV. A further neutral trigger was not used for the data presented here. For the purpose of finding μ-pairs, an additional muon trigger was provided which formed a coincidence between elements of H_0 and the Muon hodoscope and required exactly 2 tracks. For the muon trigger the energy threshold was 200 MeV which corresponds to one minimum ionizing track in the energy detector. The single-ring-single-bunch operation of DORIS permitted the addition of a beam gate to the trigger which largely reduced the cosmic ray background. A fast online computer trigger checked upon the consistency of the hardware trigger with the inner drift chamber information.

The luminosity was measured in two independent ways. A small angle luminosity monitor (SAB) measured Bhabha events at a scattering angle of 130 mrad. Large angle Bhabha (LAB) events ($|\cos\theta| < 0.8$) were identified in the drift chambers and energy detector. The ratio of the measurements is L(SAB)/L(LAB)=(1.03±0.01) before radiative corrections (statistical error only) for the Υ region, and (1.00±0.01) for the Υ' region.

For the measurement of the hadronic cross section the following conditions were required: (i) the event was within ±12 ns of the bunch crossing, (ii) ≥3 tracks were in the inner detector, (iii)≥1.8 GeV was in the energy detector, and (iv) not all the energy was deposited in one half of the detector.

These computer selected events were all visually scanned by physicists to remove any remaining beam gas, Bhabha scattering or cosmic ray events. The final data consist of 7713 hadronic events accumulated in an energy scan over the Υ resonance. The integrated luminosity was 1158 nb^{-1}, half of which was accumulated at the peak of the resonance. The resulting visible cross section is shown in Fig. 2a. Its two components are the continuum cross section and the resonance contribution as described by the folding of a Breit-Wigner resonance shape with a radiative spectrum and an unknown machine energy resolution [5]. This leads to a 4-parameter fit with the following result:

$$\sigma_{had}^{vis}/\sigma_{\mu\mu} = R_{vis}[9.46\ \text{GeV}] = 2.62\pm0.10$$
$$\sigma_{beam} = 8.3\pm0.6\ \text{MeV}$$
$$M_{res} = 9461.3\pm0.7\ \text{MeV}$$
$$\int \sigma_{res}^{vis}(W)\,dW = 220\pm16\ \text{nb MeV}$$

Comparing our visible R value with the R value as measured by the PLUTO collaboration [6] at 9.4 GeV (R=3.7±0.4) and allowing for a contamination by 0.2 units of our hadronic sample by hadronic decays of τ pairs, we arrive at an efficiency $\varepsilon_{cont} = 0.67$ for observing hadron events in the continuum region. Since the Υ is expected to decay primarily via 3 gluons, while the continuum events are primarily 2 quark jets, a different efficiency ε_{res} was applied to the resonance. A Monte Carlo calculation gives $\varepsilon_{res}/\varepsilon_{cont} = 1.17$. Thus we arrive at $\int \sigma_{res}\, dW = 281\pm20$ nb MeV or, equivalently,

$$\bar{\Gamma}_{ee} := \frac{\Gamma_{had}}{\Gamma_{tot}}\,\Gamma_{ee} = 1.10\pm0.08\ \text{keV}$$

We expect a systematic uncertainty of ±10.0 MeV for M_{res} due to the calibration of the DORIS storage ring, and our normalization to the PLUTO measurement causes an uncertainty of ±0.11 keV for $\bar{\Gamma}_{ee}$.

The data from the Υ' region were treated in completely analogous fashion. We collected 4513 hadronic events from a luminosity of 1218 nb^{-1}, one third of which was taken in the peak region. From the visible cross section, shown in Fig. 2b, we obtain for the Υ' resonance

$$R_{vis}[10.01\ \text{GeV}] = 2.68\pm0.10$$
$$\sigma_{beam} = 10.9\pm1.3\ \text{MeV}$$
$$M_{res} = 10014.0\pm1.0\ \text{MeV}$$
$$\int \sigma_{res}^{vis}(W)\,dW = 91\pm12\ \text{nb MeV}$$

The agreement of R_{vis}(Υ') with R_{vis}(Υ) permits us to use the Υ-efficiencies also for the Υ', and we obtain

$$\bar{\Gamma}_{ee}(\Upsilon') = 0.51\pm0.07\ \text{keV}$$

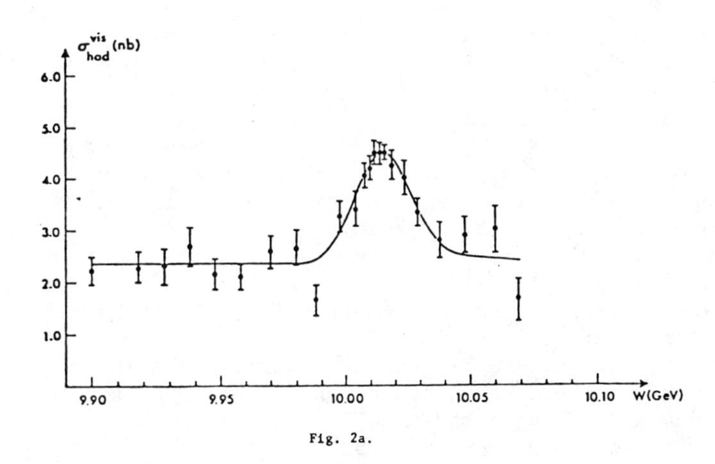

Fig. 2a.

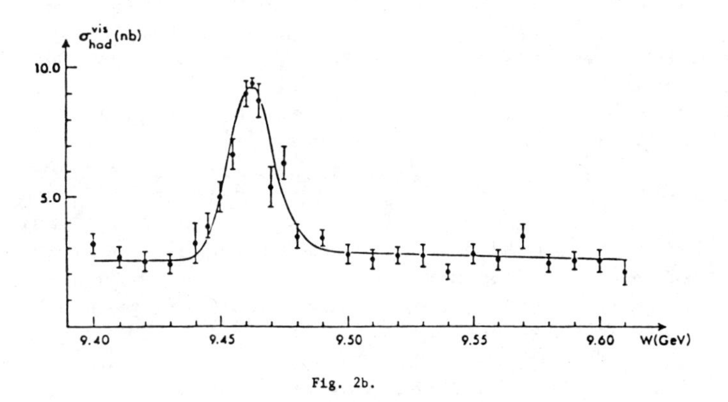

Fig. 2b.

Fig. 2. The visible hadronic cross section $\sigma(e^+e^- \to h)$ in the Y region (a) and in the Y' region (b). The errors shown are only statistical. A systematic error of 10% shifts the curve as a whole. The full line shown is the result of the fit described in the text.

The systematic errors are expected to be equal to those for the Y. However, a normalization-free number is

$$\bar{\Gamma}_{ee}(Y')/\bar{\Gamma}_{ee}(Y) = \Gamma_{ee}(Y')/\Gamma_{ee}(Y) = 0.46\pm0.07$$

assuming the leptonic branching ratios of the Y' and the Y to be the same.

For $B_{\mu\mu}$ we used the data taken with the muon trigger. Presently we show only the $B_{\mu\mu}$ analysis in the Y region. To separate muons from hadron events, we requested two collinear tracks in the inner detector, and Bhabha events were removed by limiting the detected energy to ≤ 1200 MeV. The drift chambers independently determined $\cot\theta$ (where θ is the polar angle of a track with respect to the positron beam direction) and the azimuthal angle ϕ. From cosmic ray muons we found our rms resolutions to be $\delta\phi = 0.015$ and $\delta \cot\theta = 0.070$. We defined for each event a normalized acollinearity $\delta^2 = (\Delta\phi/0.015)^2 + (\Delta \cot\theta/0.070)^2$ where $\Delta\phi$ and $\Delta \cot\theta$ were the measured acollinearities of the two tracks. Two tracks were classified as collinear if $\delta^2 \leq 25$.

Most cosmic ray muons were removed by requiring the event to occur within ±5 ns of the bunch crossing and within a cylindrical region ±60 mm long and having a radius of 6 mm. The final selection was made using the ToF counters. For each event the ToF for the track above the horizon was measured between the counters farthest from and nearest to the intersection region. Figure 3 shows the resulting ToF distributions obtained for the Roof and Muon counters. The signals resulting from cosmic ray muons and muon pairs from e^+e^--annihilation are clearly separated. Cuts were made at the positions indicated. The residual background consists of at most a few events and resulted from drifts in the ToF timings. This was verified independently by running without beam in DORIS and by shifting the bunch crossing cut by ±10 ns away from the true crossing time.

We expect a small amount of hadron background in our muon sample. An estimate was obtained using four-prong hadron events which have tracks pointing to the muon chambers. For these events, we measured the average hadron punch through plus accidentals to be (6.8±0.8)% per track. A direct measure of the residual hadron background was obtained using a subsample of muon events which have both tracks pointing at the muon drift chambers. Within this sample we found (14±2)% of the events have at least one track which does not record a hit in a chamber. Since the muon chamber efficiency, as determined from cosmic ray muons, is consistent with 100%, these must be background events. Tightening the δ^2 cut to 12 reduces this hadron background to 8%. Therefore, to purify the sample, we rejected any event having a track which points at a muon chamber but does not record a hit. For those events in which neither track points at a muon chamber, we cut at δ^2=12. Our final sample consists of 451 events with 359 events having at least one track recorded in a muon chamber. The residual hadron background is (.08) × 92/451 = 1.6% and is not energy dependent. The resulting

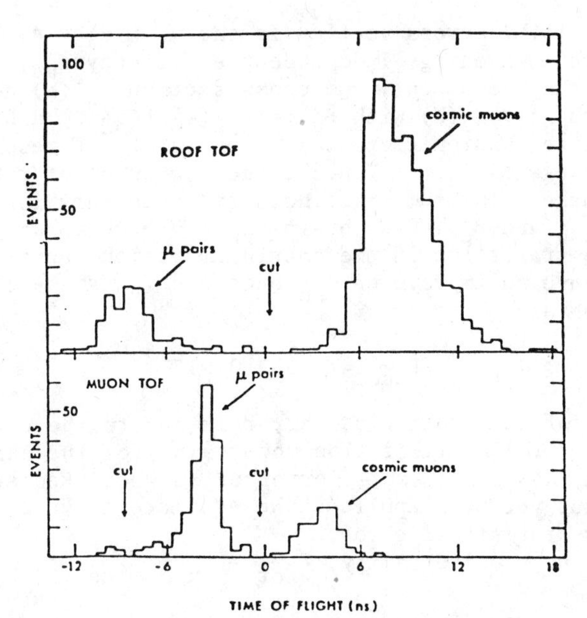

Fig. 3. Time-of-flight distributions for the Roof and Muon sidewall counters. The annihilation μ-pairs are cleanly separated from the cosmic ray muons. The cuts are indicated. The separation of the signals is larger for the Roof counters due to their greater distance from the interaction region.

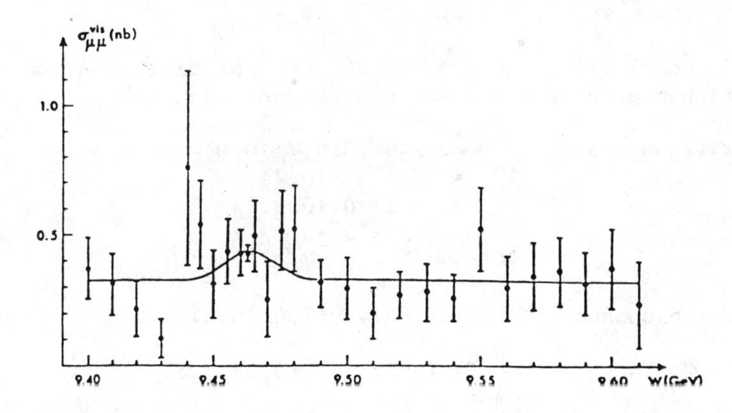

Fig. 4. The visible leptonic cross section $\sigma(e^+e^- \to \mu^+\mu^-)$ in the Υ region. The errors shown are statistical $\sqrt{N}$-type errors. The curve is the result of a maximum likelihood fit.

$e^+e^- \to \mu^+\mu^-$ visible cross section is shown in Fig. 4. Apart from an overall unknown energy independent efficiency $\varepsilon_{\mu\mu}$, it is the addition of the known continuum cross section $\sigma_{\mu\mu}(W)$ and a resonance curve $\bar{B}_{\mu\mu} \cdot \sigma_{res}(W)$ with $\bar{B}_{\mu\mu} := \Gamma_{\mu\mu}(Y)/\Gamma_{had}(Y)$ and $\sigma_{res}(W)$ being the fitted hadronic cross section for the Y resonance. Except for 240 events at the Y peak, the low $\mu\mu$ event rates per energy point make a maximum likelihood fit necessary for the two parameters $\varepsilon_{\mu\mu}$ and $\bar{B}_{\mu\mu}$. We obtain $\varepsilon_{\mu\mu} = 0.33 \pm 0.03$ which is dominated by the statistics in the continuum region containing 102 μ pairs. We hope to improve on $\varepsilon_{\mu\mu}$ once the Y' data are analyzed. For $\bar{B}_{\mu\mu}$ we obtain

$$\Gamma_{\mu\mu}/\Gamma_{had} = 0.039 \pm 0.017$$

A large part of the statistical error is due to the correlation of $\bar{B}_{\mu\mu}$ with $\varepsilon_{\mu\mu}$. The normalization uncertainty of the hadronic data carries over into a systematic error of ± 0.004. Radiative corrections have not yet been applied. We estimate they are small compared to our statistical error.

Using e, μ, τ universality, $\Gamma_{tot} = \Gamma_{had} + 3\Gamma_{ee}$ we obtain for the Y resonance

$$B_{\mu\mu} = \bar{B}_{\mu\mu}/(1+3\bar{B}_{\mu\mu}) = (3.5 \pm 1.5 \pm 0.4)\%$$

$$\Gamma_{ee} = \bar{\Gamma}_{ee}/(1-3\,B_{\mu\mu}) = (1.23 \pm 0.10 \pm 0.14)\ \text{keV}$$

and

$$\Gamma_{tot} = \Gamma_{ee}/B_{\mu\mu} = (35\ {}^{+25}_{-10}\ {}^{+9}_{-7})\ \text{keV}.$$

In conclusion, the results of the LENA experiment on the Y and Y' resonances at present are summarized below:

Y-resonance

$$M = (9461.3 \pm 0.7 \pm 10.0)\,\text{MeV}$$
$$B_{\mu\mu} = (3.5 \pm 1.5 \pm 0.4)\%$$
$$\Gamma_{ee} = (1.23 \pm 0.10 \pm 0.14)\ \text{keV}$$
$$\Gamma_{tot} = (35\ {}^{+25}_{-10}\ {}^{+9}_{-7})\ \text{keV}$$

Y'-resonance

$$M = (10014.0 \pm 1.0 \pm 10.0)\ \text{MeV}$$
$$\bar{\Gamma}_{ee} = \Gamma_{ee} \cdot \frac{\Gamma_{had}}{\Gamma_{tot}} = (0.51 \pm 0.07 \pm 0.06)\ \text{keV}\ (*)$$
$$\Gamma_{ee} = (0.57 \pm 0.08 \pm 0.07)\ \text{keV}\ (*)$$

and further

$$M(Y') - M(Y) = (552.7 \pm 1.2)\ \text{MeV}$$
$$\Gamma_{ee}(Y')/\Gamma_{ee}(Y) = 0.46 \pm 0.07\ (*)$$

((*) assuming $B_{\mu\mu}(Y') = B_{\mu\mu}(Y)$)

We acknowledge the efforts of the DESY directorate, Dr. K. Wille, the DORIS machine group, and the DESY services. We thank the DESY-Heidelberg group which built the detector. The experiment was partially supported by the U. S. Department of Energy and the National Science Foundation.

REFERENCES

(a) Present address: Max Planck Institut, Munich, Germany
(b) Present address: Physics Dept., Carnegie-Mellon University, Pittsburgh, PA 15213
(c) Present address: Messerschmidt Bölkow Blohm, Munich, Germany
(d) Present address: DESY, Hamburg, W. Germany

[1] S. W. Herb et al., Phys. Rev. Letters 39, 252 (1977); W. R. Innes et al., Phys. Rev. Letters 39, 1240 (1977).
[2] PLUTO collaboration, Ch. Berger et al., Phys. Lett. 76B, 243 (1978); Dasp II collaboration, C. W. Darden et al., Phys. Lett. 76B, 246 (1978); ibid 78B, 364 (1978).
[3] DESY-Hamburg-Heidelberg-MPI München collaboration, J.K. Bienlein et al., Phys. Lett. 78B, 360 (1978).
[4] W. Bartel et al., Phys. Lett. 66B, 483 (1976); ibid 77B, 331 (1978).
[5] PLUTO collaboration, Ch. Gerke, Dissertation Universität Hamburg (1979) and DESY Internal Report PLUTO - 80/03.
[6] J. D. Jackson and D. L. Scharre, Nuclear Instr. Methods 128, 13 (1975).

COLUMBIA-STONY BROOK EXPERIMENT AT CESR

R. D. Schamberger Jr.
SUNY at Stony Brook, Stony Brook, N.Y. 11794

ABSTRACT

Using the CUSB layered NaI detector at CESR, we have observed three narrow resonances ($\Upsilon, \Upsilon', \Upsilon''$) in the $\sigma(e^+e^- \to$ hadrons) over the c. of m. energy interval of 9.4 to 10.3 GeV. Their masses are $M(\Upsilon) = 9.4345 \pm .0004$ GeV, $M(\Upsilon') = 9.993 \pm .0010$ GeV, and $M(\Upsilon'') = 10.3232 \pm .0007$ GeV (all in CESR energy scale) and their relative leptonic widths are $\Gamma_{ee}(\Upsilon')/\Gamma_{ee}(\Upsilon) = 0.39 \pm .06$, $\Gamma_{ee}(\Upsilon'')/\Gamma_{ee}(\Upsilon) = 0.32 \pm .04$, where errors are statistical only. We also found a broad resonance whose mass $M(\Upsilon''') = 10.547 \pm .002$ and leptonic width $\Gamma_{ee}(\Upsilon''')/\Gamma_{ee}(\Upsilon) = 0.25 \pm .07$. These values allow us to identify these enhancements as the four lowest triplet S levels of the bound ($\bar{b}b$) quark-antiquark system. The broad natural width ($\Gamma \approx 13$MeV) of the 4^3S_1 state indicates that it lies above the threshold for $B\bar{B}$ production.

INTRODUCTION

Towards the end of November of 1979, CESR initiated its first physics running period, which lasted about five weeks, yielding about 1000 nb^{-1} of integrated luminosity. For two-thirds of that period (corresponding to 600nb^{-1}) data were collected around the c. of m. energies of 9.45 GeV, 10 GeV and 10.3 GeV, i.e. around the mass positions of the first three upsilons. The remainder of that period was devoted to an energy scan around the c. of m. energy of 10.6 GeV, where the fourth Υ was expected to exist. In fact, the existence of this last resonance was not proved till the next running period, which occurred during February-March of 1980, where data were collected in the energy region from 10.46 GeV to 10.6 GeV for 1100nb^{-1} of integrated luminosity. In the following, I will report on the philosophy and performance of the CUSB detector during this early phase of its operations as well as on the data collected on the four Υ's using the CUSB detector.[1,2]

APPARATUS

The CUSB layered NaI-Pb glass $\hat{C}$ detector is located in the 18'x21', "North Area" of CESR. It was designed primarily to study the upsilon family spectroscopy and inclusive photon processes at CESR, i.e., is able to measure with accuracy the soft photons (of 50-200 MeV) arising from the radiative transitions from various S,P states

ISSN:0094-243X/80/620298-13$1.50

to P,S states as well as hard photons arising from some as yet undiscovered upsilon decay modes. However, as the present report will show, this detector also functions extremely well as a hadron detector in total $\sigma(e^+e^- \rightarrow$ hadrons) measurements. In its final design the CUSB detector consists of a system of drift chambers for tracking, which is followed by layered NaI-strip chamber arrays which localize photon conversion vertices and measure over 70% of its energy, and is finally backed by Pb glass Ĉ counters to contain the remainder of the electromagnetic showers. The NaI is divided into 32 azimuthal sectors and 2 polar sectors, thus providing complete azimuthal coverage in the region $45^\circ < \theta < 135^\circ$. Radially, the inner four NaI layers are 1 r.l. thick, interspersed with low mass cathode readout proportional chambers, followed by one four r.l. NaI layer and surrounded by 260 8 rl thick Pb glass blocks. This system will have good spatial resolution (1 mrad in φ, 5 mrad in θ for track position, 0.5 cm for shower conversion point position) and good energy resolution ($\Delta E/E \sim 7\%$ at 0.1 GeV and $\sim 2.5\%$ at 10 GeV). Figure 1 illustrates the end and cut away side view of the NaI components of the CUSB detector. For the

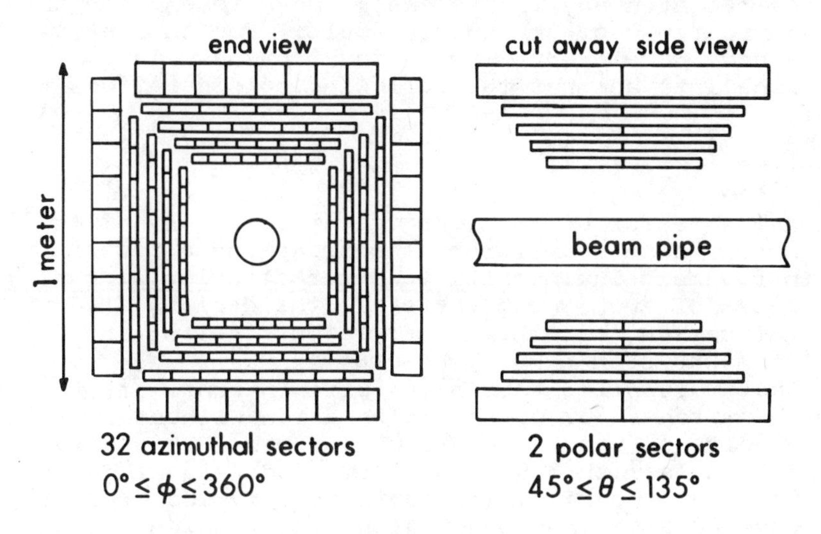

Fig. 1. CUSB detector (NaI).

Nov-Dec run only one of the two polar sections was used and it was centered optimally over the interaction point. In the Feb-March run, both polar sector were used, and the detector was centered along the interaction point.

For the Nov-Dec run, four small strip chambers placed around the beam pipe, localized the beam-beam interaction point. For the Feb-March run, drift chambers were used instead. Charged particles originating from beam interaction point deposit energy in each of the NaI layers, thus yielding five independent dE/dx measurements. Also, charged particles whose path is confined to within one of the 64 NaI sectors subtending a solid angle of 1% of 4π, point to the beam-beam interaction point. Therefore, even though drift and strip chambers were in place during the present runs, the track information obtained from them were used only to confirm the cleanliness of the hadron selection based solely on NaI information.

An enormous amount of electronics design and construction work was needed to make the spectrometer functional. A general idea of the complexity is given by the following numbers:

Analog signals: 3000
Photomultiplier signals: 800
Digital signals: 500

In addition, the signals from the P.M.'s viewing the NaI crystals require extreme care in their handling because of the desired accuracy in the energy measurement and the large dynamic range of the energy scale. For this experiment we have developed a large amount of new electronics taking into account the following considerations foremost:

1)Economy
2)Versatility
3)Reliability
4) At least a factor of ten improvement over conventional systems on noise from internal sources and a factor of 100 improvement in immunity to external noise pickup.

Examples of such work, where all the design, artwork, construction, assembly supervision, testing and installation has been done by us are:

1) Photomultiplier tube bases with internal amplifier, with switches for offset calibration without removing high voltage from the photomultipliers. This allows us to use P. M.'s with 0.5 mA maximum anode signal current to improve linearity and stability. Typical results for this system are: anode equivalent input noise 0.01 pCoul, calibration signal from .66 MeV gamma rays (Cs^{137}) 0.1 pCoul: full scale anode signal (1 GeV) 150 pCoul.

2) Universal differential or single ended input and output line driver-receiver. This is a new circuit used on all our analog signals (4000) with gain adjustable from 0.5 to 20. Use of this circuit permits transmission of precision signals through "twisted-flat" ribbon cable in sets of 16 signals over 32 wires in each ribbon cable. The enormous advantage here is compactness, ease, economy

and simplicity of cable termination.

3) Computer controlled power supplies for P. M.'s. We completely designed these power supplies to meet the stringent stability and settability requirements arising from optimal use of NaI. 800 such channels are in use at present. Each channel is settable locally or by computer control to an accuracy of 0.1 V. Stability is ~50 mv, ripple 5 mv. The supplies are short circuit protected and current limited. The cost per channel including all auxiliary components (crates, low voltage power supplies etc.) is \$50 or about 28% of an equivalent system using helipots.

4) Trigger system. This consists in its simplest form of a total energy trigger. An analog signal corresponding to the total energy deposited in the detector is generated by two levels of linear adders. A typical energy threshold in our runs is 700 MeV, yielding a trigger rate of 0.1 Hz which is only about fifty times the event rate in the continuum and is more than 95% efficient. Note that this analog signal is the sum of 320 input signals. This means that any noise picked up from local sources (coherent in all channels) is amplified by a factor 320 while random electronics noise is amplified by a factor $\sqrt{320}=18$. The noise in this signal is equivalent to 25 MeV. We hope to produce a much more efficient trigger from energy deposition. To this end during this last running period we have been using partial energy sums and requiring appropiate energy distribution over the detector to create flags counted on scalers to compare their efficiency with offline analysis results. (see next section).

5) NaI absolute energy scale calibration. This is a very difficult task because of the large energy range to be covered: 0.66 MeV form Cs^{137} to 5-8 GeV from CESR. For instance, the Crystal Ball experiment uses two channels per P.M. At present we use the following scheme: the differential signals from the line drivers are received by differential input ADC's and are tapped off for calibration, in sets of 16, with differential input nonloading pick-off amplifiers. Cs^{137} sources are interspersed through the thin NaI planes and Co^{60} sources are placed outside the 4 inch thick outer layer crystals. The pick off amplifiers contain the first level summing amplifier for trigger and provide a way for constant monitoring and calibrating the energy scale of each crystal even while running. The calibration is done for 16 crystals at a time by using a set of 16 amplifiers, discriminators, delay lines and ADC's which can be connected to the sixteen outputs of any set of 16 pick-off channels as desired. The position of the Cs^{137} line (0.66 MeV) for the inner crystals and of the sum of the Co^{60} lines (2.5

MeV) for the outer crystals is found using a small stand alone PDP-11/20 which processes the ADC outputs after they have been histogrammed into 128 kBytes of semiconductor memory. The sensitivity of the calibration channels is ~ 70x the sensitivity of the channels for regular running.

6) Photomultiplier tube stability was monitored with light from a spark in argon. The PM gains were stable to within 5% throughout the running period.

DATA COLLECTION

All signals from the PM's were integrated every beam crossing (every 2.56 μs) while a trigger decision was made. If no trigger was present, all integrators were reset to be ready for the next crossing. Only a total energy trigger was used for the present runs: requiring ~420 MeV to be deposited in the outer three layers of the one polar sector of the NaI array for the Nov-Dec run and ~ 700MeV to be depositied in both sectors for the Feb-March run.

Once a trigger was produced, all signals were digitized and recorded on tape. This trigger gave an event rate of 0.1 Hz for a luminosity of 1 $\mu b^{-1}s^{-1}$. Apprximately 15% of the triggers are hadronic events at the Υ and about 10% of the triggers are large angle Bhabha scattering events in our detector. The integrated luminosity for each run was measured by detecting and counting small angle (40 to 80 mrad) collinear Bhabha scatters with lead-scintillator sandwich shower detectors. Hadronic and large angle Bhabha scatter yields in the detector are also monitored on-line by counting on scalers events satisfying appropiate criteria. The Bhabha scattering criterion requires that the total energy deposited in NaI (E_{NaI}) be greater than 5.4 GeV and that two collinear octants each have > 100 MeV in the outer four layers. These online "Bhabhas" comprise approximately 80% of those found later in off-line analysis and serve as an online consistency check of the small angle Bhabha luminosity monitor. The long term stability of the luminosity monitor is confirmed by the yield of the large angle Bhabha events recorded on tape which are also used for absolute normalization. A typical fill of CESR lasts 3 to 5 hours, yielding an integrated luminosity of up to 15nb^{-1}. During such a run, the CUSB R-Meter, a scaler counting events which satisfy the hadron criteria (1.4 GeV$<E_{NaI}<$4.2 GeV in the outer three layers, that two or more octants in each half have >100 MeV deposited in the outer four layers, and that two such octants be collinear, see fig. 2) shows clear Υ, Υ'' signals. The visible cross-section, multiplied by the factor $(M/M_{\Upsilon})^2 \propto (1/s)$, as a function of mass (i. e. $2E_{beam}$) as obtained from the on-line R-meter scaler for the Feb-

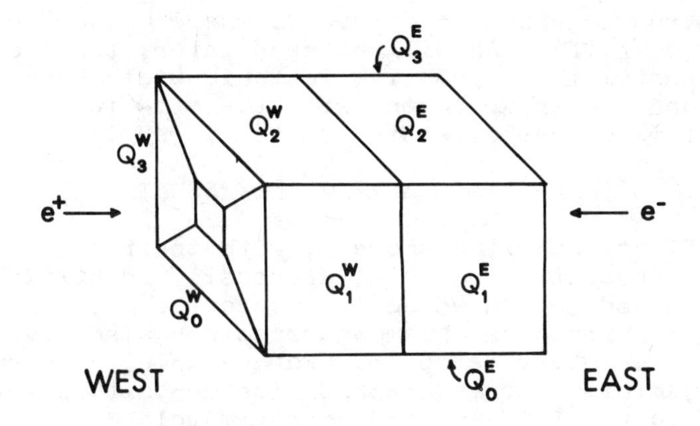

R Meter requires:

1) Discriminate: $E_Q > 100$ MeV

2) Logic: (2 out of 4 in Q^E)$\cdot$(2 out of 4 in Q^W)$\cdot$
(at least 1 collinear E–W pair)$\cdot$
($1.4\ \text{GeV} \leq E_{total} \leq 4.2\ \text{GeV}$)

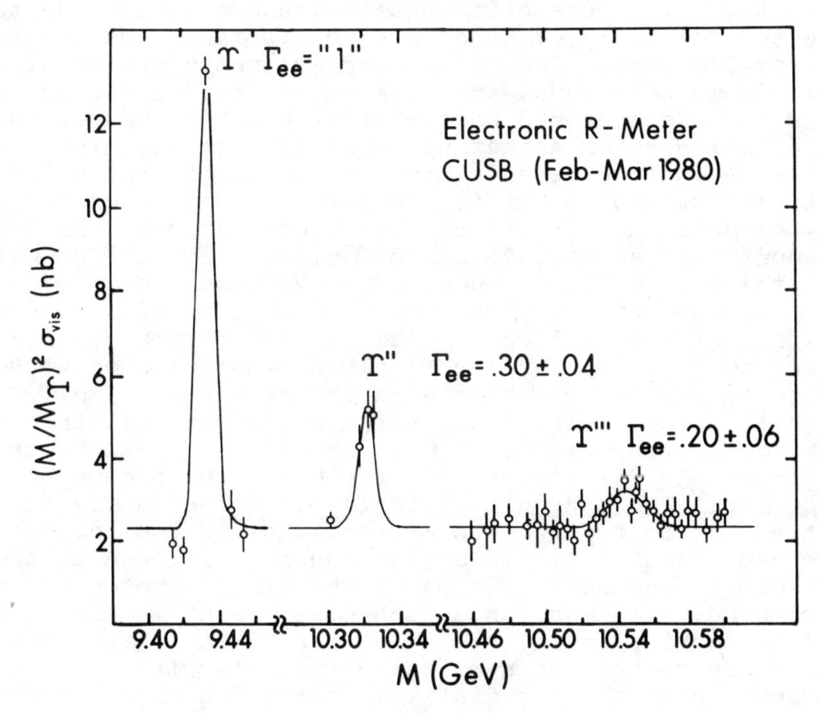

Fig. 2. CUSB On-Line R-Meter Logic and Results

March run is shown in figure 2, where Γ_{ee} are given relative to $\Gamma_{ee}(\Upsilon)$. As will be seen later, the accuracy in mass position and relative leptonic width determination was good and in agreement with the off-line analyses. The live time of our detection system is ~99%.

DATA ANALYSIS AND RESULTS

Electromagnetic showers, with their characteristic energy deposition pattern, are easily identifiable in our layered and segmented active converter array. Charged particles from beam-beam interactions also give a unique signature in our detector. For example, at normal incidence, minimum ionizing particles deposit 15 MeV in the first four NaI layers and approximately 68 MeV in the last layer of a single sector ("track"). A hadronic event usually includes at least one track, some showers, and a balanced deposition of energy in the two polar sectors. In order to develop hadronic event selection computer algorithms, one or more physicists have examined over 95% of the hadronic event candidates. The various algorithms developed for the Feb-March data have efficiencies for continuum events of from 65% to 75% and with contamination from non beam-beam background ranging from 1 to 5% respectively. These efficiencies include losses due to detector solid angle, hence are considerably smaller for the Nov-Dec data. The background estimates are obtained from single beam runs, and from reconstructed vertex positions for those events having drift chamber information. The efficiency for detecting upsilon events is higher than that for continuum events because the former has higher multiplicity and sphericity than the latter. For the Nov-Dec data the net efficiency of our detector was 28% for continuum events, 37% at the Υ peak; for the Feb-March data the corresponding numbers are 73% and 82%.

The main aim of CESR's Nov-Dec run was to locate the position of the upsilon for calibrating the machine energy scale, to quickly scan the Υ' already established at Doris and to search for higher states in the upsilon family. Our half detector performed efficiently and well in this task, collecting 214 Υ, 53 Υ', and 133 Υ'' events above the continuum and 272 events from the continuum around the three Υ's. The hadronic yield is presented in figure 3, plotted in arbitrary units proportional to the ratio of detected events to small angle Bhabha yield. In this way, the energy dependence ($\propto 1/s$) of the single photon process is removed. The horizontal scale is $M(e^+e^-)$, twice the nominal machine energy. Mass values of the three resonances are determined by fitting the data with a constant continuum plus three radiatively corrected Gaussians[3] whose width represent the machine energy spread with however, only a single beam width parameter used for all

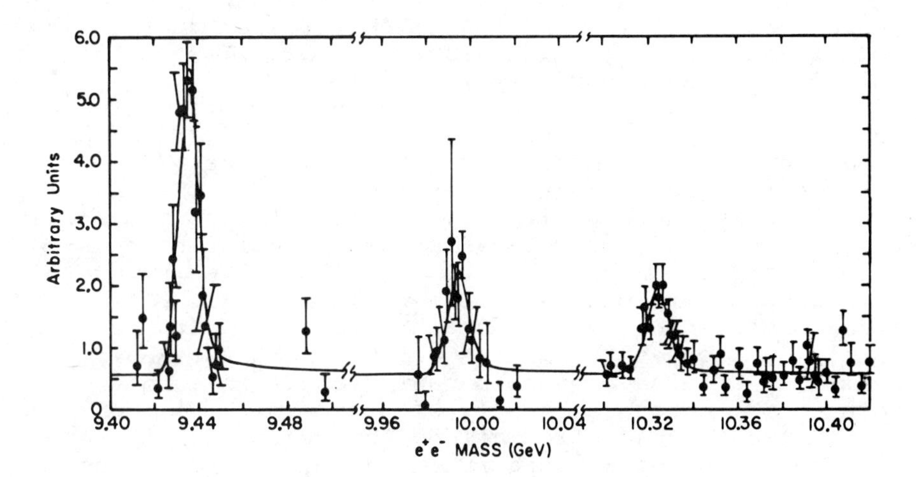

Fig. 3. Hadron yield/Bhabha event vs e^+e^- mass

three resonances, scaling with energy as s, according to machine theory. The machine total energy spread at 9.5 GeV is determined to be 4 ± 0.3 MeV rms from the Nov-Dec data, and 3.7±0.3 MeV from the Feb-March data. These values are consistent with each other and with the value expected from CESR design parameters.

From the fit to the Nov-Dec data we found m(Υ)=9.4345 ±0.004 GeV, m(Υ')=9.993±0.0010 GeV and m(Υ'')= 10.3232± 0.0007 GeV. The mass scale at CESR appears at present to be displaced by 2.7 parts in a thousand with respect to Doris[4]. The determination of which scale is correct will have to await future CESR energy calibrations. The quantities which are relevant to the phenomenonlogy of the bound states (which had been discovered at Fermilab[5] as a narrow enhancement in the dimuon spectrum near 10 GeV) of the b quark are the level splittings and ratios of leptonic widths, for which we obtain M(Υ')-M(Υ)=559±1(±3) MeV and M(Υ'')-M(Υ)=889±1(±5)MeV, where errors in parenthesis represent the aforementioned systematic uncertainties in CESR energy calibration, $\Gamma_{ee}(\Upsilon')/\Gamma_{ee}(\Upsilon)$=.39±.06 and $\Gamma_{ee}(\Upsilon'')/\Gamma_{ee}(\Upsilon)$=0.32± .04. The Υ'-Υ mass level spacing had been measured at Doris and agrees well with our results. Our ratio for the Υ' to Υ leptonic width is different from and is more accurate than Doris's published results. New results from Doris reported at this conference are now in agreement with ours. The Υ'' has been isolated and seen at e^+e^- machine for the first time at CESR, by CUSB and CLEO simultaneously. The errors on the ratios are statistical only; however, they are larger than our estimates of the systematic errors in these ratios due to possible scanning inefficiencies for the three resonances. Our results are in good agreement with many quarkonium model predictions,[6] reinforcing the interpretation that these first three Υ's, (Υ, Υ', Υ'') are the three lowest triplet S states of the $b\bar{b}$ bound system.

In such quarkonium models, the number of quasistable radial excited states increase with the mass of the quark. In particular, for the b quark (m~5 GeV), the 4^3S_1 state should exist with an excitation energy of ~1.15 GeV. The 4^3S_1 state is expected to be very close to the threshold for $B\bar{B}$ production, where B is a pseudo scalar bound system of b and $\bar{u}$ quarks. If the 4^3S_1 state lies below the $B\bar{B}$ threshold, its natural width would be well below 1 MeV, giving observed width dominated by machine width, as for Υ, Υ', Υ'', whereas if it lies above the $B\bar{B}$ threshold, the opening up of decay channels would result in a natural width which increases rapidly with $M(4^3S_1)-2M(B)$.[7] As has been stated in the introduction, the search for the 4^3S_1 state began in the Nov-Dec run; indeed, a hint of the structure was seen by our group by Christmas, but the definitive results were not seen until the Feb-March run. Five thousand $1^3S_1(\Upsilon)$ events and 450 $3^3S_1(\Upsilon'')$ events were also collected during the Feb-March running period. Three thousand events were collected in the region 10.46 to 10.6 GeV. The presence of an enhancement around 10.54 is quite evident, even on the on-line R meter results as was shown in figure 2. In figure 4 is the visible cross-section in

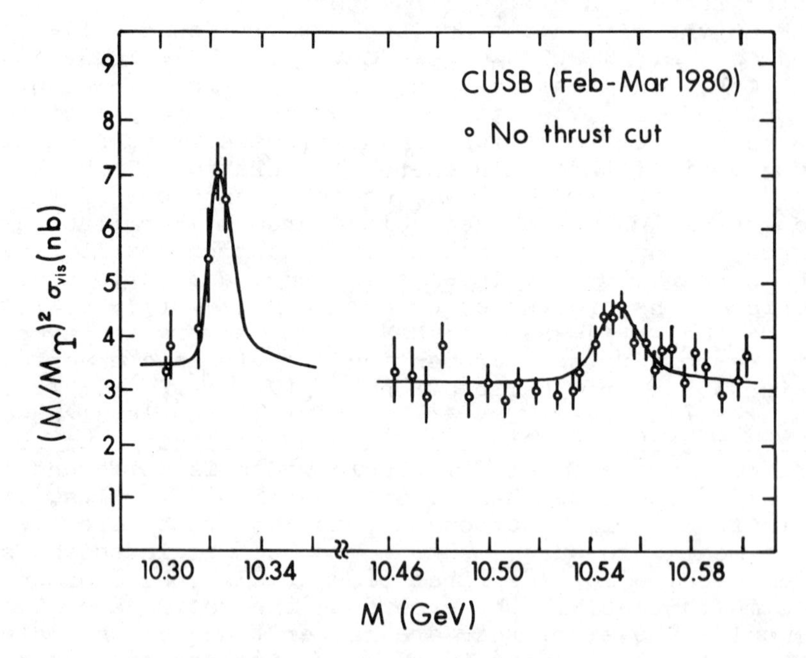

Fig. 4. The observed cross-section for $e^+e^- \rightarrow$ hadrons multiplied by $k=(m/M_\Upsilon)^2$. M is the e^+e^- invariant mass.

our detector for $e^+e^- \Rightarrow$ hadrons, multiplied by the scaling variable $k=(M/M_{\Upsilon})^2$, in the region of the Υ'' and in the energy range between 10.46 and 10.6 GeV. Both the presence of a sizeable enhancement over the continuum and its broadness relative to its neighbor (Υ'') are apparent.
To enhance discrimination between resonance and continuum events we make use of a property first discovered at Doris about the Υ, that the spatial distributions of the continuum and decay events are different. We chose to use a simplified thrust variable, T', define as the maximum of $\Sigma|E_{NaI} \cdot \hat{n}|/\Sigma E_{NaI}$ over all possible n perpendicular to the beam axis. The distributions of T' for Υ events (solid line), continuum events (dashed line), and events in the enhancement peak (data points) are shown in Figure 5.

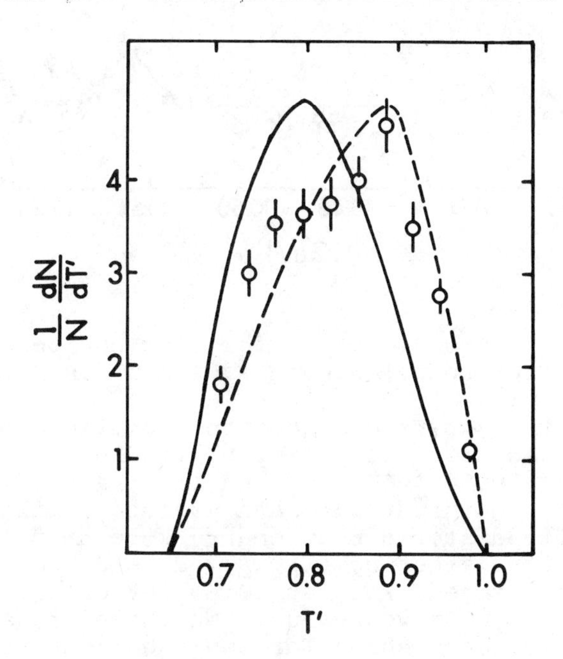

Fig. 5. Pseudo thrust (see text) distribution for Υ events (solid line), continuum events (dashed line), events in the enhancement peak (data points).

The difference between the solid and dashed line figures is quite obvious, confirming the conjecture that continuum events have a two jetlike structure, while resonance decays have a more spherical distribution. It is also clear that events in the enhancement region have contributions from both types of events. A cut at $T' < 0.85$ has been made for all the data, and the result is shown in Figure 6.

We determine the parameters of the Υ''' by fitting the

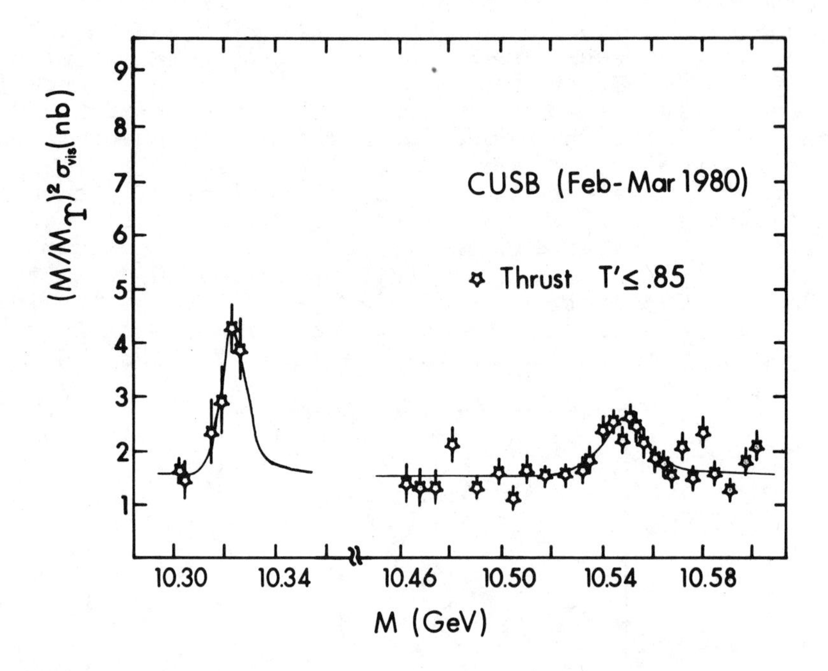

Fig. 6. Observed cross-section for e^+e^- hadrons multiplied by $k+(M/M\ )^2$ after removing events with T' .85.

data sample with T'<.85 to a constant continuum plus a Gaussian with radiative corrections, finding a central position of $M(\Upsilon''')=10.547\pm.002$ MeV and an apparent width of 19 ± 4 MeV. The non-thrust cut data yielded similar results. Scaling our fitted stored beam energy spread at the Υ of 3.7 ± 0.3 MeV by the expected $(E_{beam})^2$ dependence yields an anticipated beam spread FWHM at 10.54 GeV of $10.8\pm.9$ MeV, about half of the observed value. We therefore assume a Breit-Wigner resonance shape for the enhancement, fold in the machine energy spread and radiative corrections, and fit the Υ''' data with the resulting curve. This results in a natural width of 12.6 ± 6.0 MeV. If we constrain the natural width to be much smaller than the machine energy spread, χ^2 increase by 8.3, from 40.3 for 30 degrees of freedom to 48.6 for 29 degrees of freedom. Thus, our value for the natural width is inconsistent with the expected width of less than 1 MeV for a resonance below $B\bar{B}$ threshold. In table I we list the properties of the Υ'''. The mass values are again given in the nominal CESR energy scale. The mass difference is $M(\Upsilon''')-M(\Upsilon)=1114\pm2$MeV, with again a systematic uncertainty of 5MeV. The ratio of leptonic widths calculated from the fitted areas is

Table I. Summary of Υ''' properties

σ_{beam}(at Υ mass)	3.7	± 0.3	MeV
$M(\Upsilon)$	9.433	± 0.001	GeV
$M(\Upsilon''')$	10.547	± 0.002	GeV
$M(\Upsilon''') - M(\Upsilon)$	1.114	± 0.002	GeV
$\Gamma(\Upsilon''')$	12.6	± 6.0	MeV
$\Gamma_{ee}(\Upsilon''')/\Gamma_{ee}(\Upsilon)$	0.25	± 0.07	

Note: Masses shown use local CESR energy scale which has a systematic uncertainty of 0.3%

$\Gamma_{ee}(\Upsilon''')/\Gamma_{ee}(\Upsilon)=.25\pm.07$. Both the mass difference and the ratio of leptonic width are in excellent agreement with many phenomenological calculations of the 4^3S_1 state of $b\bar{b}$[6]. Therefore we conclude that the enhancement we see at M=10.547 is most likely that state. The natural width of ~13 MeV implies that the 4S is above $B\bar{B}$ threshold and that the mass of the B is less than 5.274 GeV. CLEO at the South Area has very similar results and conclusions as we do with regards to all four members of the Upsilon family seen so far at CESR.[9] With the anticipated enhanced production rate for B mesons around the Υ''' region, both groups plan an intense program of B meson study in the next running periods.

ACKNOWLEDGEMENT

The author (R. D. S.) and all the CUSB collaborators (T. Böhringer, F. Constantini, J. Dobbins, P. Franzini, K. Han, S. W. Herb, D. M. Kaplan, L. M. Lederman, G. Mageras, D. Peterson, E. Rice, and J. K. Yoh of Columbia University; G. Finocchiaro, G. Giannini, J. Lee-Franzini, M. Sivertz, L. J. Spencer and P. M. Tuts of SUNY at Stony Brook) are supported in part by the National Science Foundation. I also thank Paula Franzini for typing this manuscript.

REFERENCES

1. T. Böhringer et al., Phys. Rev. Lett. 44, 1111(1980).
2. G. Finocchiaro et al., Submitted to Phys. Rev. Lett.
3. J. D. Jackson and D. L. Scharre, Nucl. Instrum. Methods 128. 13 (1975).
4. Ch. Berger et al., Phys. Lett. 76B, 243 (1978); C. W. Darden et al., Phys. Lett. 76B, 246 (1978); Phys. Lett. 78B, 364(1978); J. K. Bienlein et al, Phys. Lett. 78B, 360(1978).
5. S. W. Herb et al., Phys. Rev. Lett. 39, 252(1977); W. R. Innes et al., Phys. Rev. Lett. 39, 1240, 1640(E) (1977); K. Ueno et al., Phys. Rev. Lett. 42, 486(1979).
6. C. Quigg and J. L. Rosner, Phys. Lett. 71B, 153(1977); E. Eichten et al., Phys Rev. D17, 3090(1978), D21, 203 (1980); G. Bhanot and S. Rudaz, Phys. Lett. 78B, 119 (1978).
7. C. Quigg and J. L. Rosner, Phys. Rpts. 56, #4 (1979).
8. Ch. Berger et al, Phys. Lett. 78B, 176(1978); 82B, 449(1979); F. H. Heimlith et al., Phys Lett. 86B, 399(1979).
9. D. Andrews et al., Phys. Rev. Lett. 44, 1108(1980), and submitted to Phys. Rev. Lett.

RESULTS FROM CLEO

Presented by David L. Kreinick
Cornell University, Ithaca, N.Y. 14853

ABSTRACT

The performance of the Cornell Electron Storage Ring and the solenoidal magnetic detector CLEO are described. Results of the first six months of data taking are presented. Three narrow resonances and a fourth, broader enhancement have been observed in the energy region 9.3 to 10.7 GeV.

INTRODUCTION

On April Fools Day 1979 the first electron beam went around the Cornell Electron Storage Ring CESR. By midsummer there were collisions. Data taking began in earnest in November. CLEO, a solenoidal magnetic spectrometer detector, was built by a collaboration of physicists from Cornell, Harvard, Rochester, Rutgers, Syracuse and Vanderbilt.

This report describes the results obtained in CLEO's first six months of running. It includes:
1) A brief summary of the performance of CESR,
2) A description of the detector CLEO,
3) Results obtained through March 1980 and
4) A comparison of the data with the quarkonium model.

CESR

The current performance of the storage ring CESR is summarized in Table I:

TABLE I: CESR Characteristics

C.M. Energies explored:	9.4 to 11.0 GeV
Currents stored:	8 x 8 mA typically
Luminosity:	1.5 x 10^{30}cm^{-2}sec^{-1} (beginning of fill)
Lifetime:	about 6 hours
Time on one fill:	5 hours typically
Fill time:	about 1 hour

An indication of how well the machine runs is given in Fig. 1, which displays the CLEO gated luminosity as a function of calendar time. The luminosity has been increasing steadily since October. An indication of the reliability of CESR is the clear visibility of scheduled machine study time in the luminosity records for the last three weeks. The machine energy is not limited to the region so far explored. It should go below 8 GeV and, when the full RF complement is installed, up to 16.0 GeV.

ISSN:0094-243X/80/620311-22$1.50

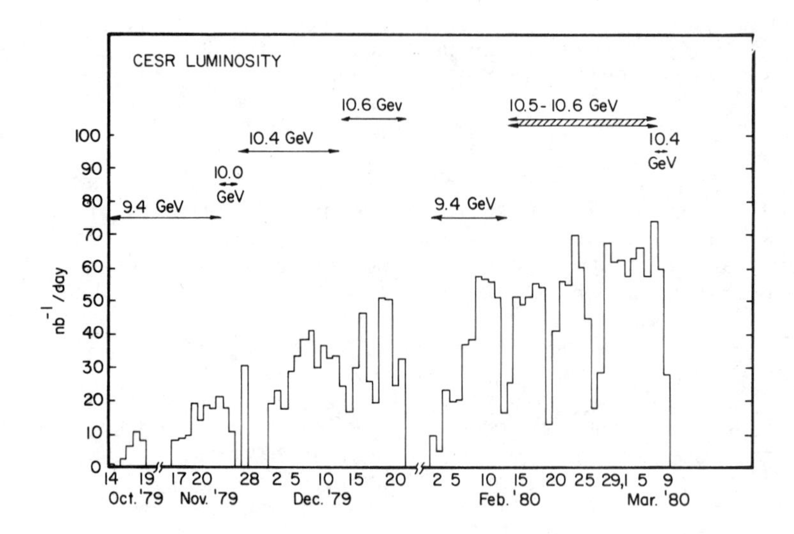

Fig. 1: CESR Luminosity vs. Calendar Date

The absolute determination of the CESR energy will prove important for physics results to be presented later. The energy is obtained by measuring with NMR the magnetic field in a bend magnet "identical" to the ring magnets and powered in series with them. Corrections are made for bending magnet trim currents. At present a large error in the absolute energy measurement is introduced by the "high bend" magnets near the CLEO interaction region. These special magnets are necessary to make a large enough separation between CESR and the synchrotron so that CLEO will fit in. NMR apparatus recently installed in the high bend magnets will permit an improved energy measurement. At present there remains a 0.3% uncertainty in absolute energy calibration-about 30 MeV for our upsilon measurements.

The energy resetability of the machine is not beset with such problems. CESR returns to a resonance peak within 0.3 MeV even after a month long shutdown.

The energy spread of the CESR beams determines the apparent width and height of a narrow resonance. A delta-function resonance will appear to be 4.1 MeV wide at 10 GeV. This width scales proportional to the square of the energy.

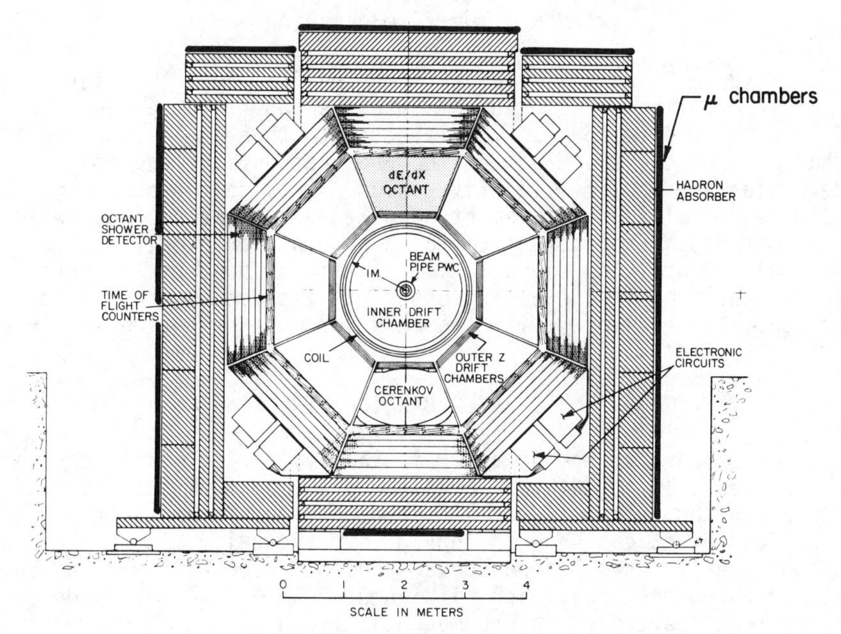

Fig. 2. The CLEO detector as viewed along the beam axis.

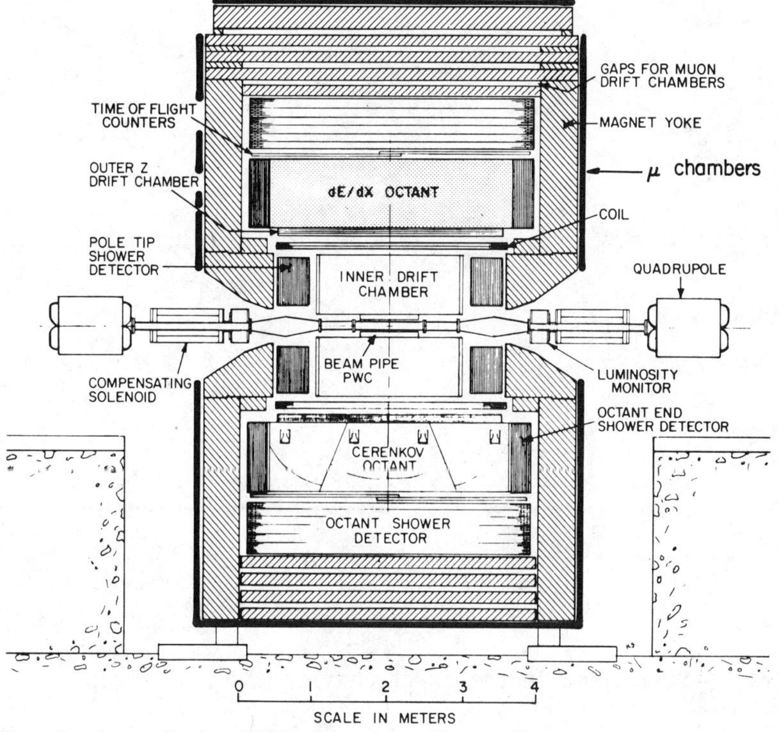

Fig. 3. A view of the CLEO detector perpendicular to the beam axis.

THE CLEO DETECTOR

Figure 2 shows a slice through the CLEO detector[1] perpendicular to the beam line. Positrons go into the paper at the center of the figure. The detector can be divided into three main functions: charged particle tracking (inside and just outside the coil), photon detection and particle identification. Outside the coil, the detector is divided into eight separately removable pieces which we call octants. Each octant contains a thin planar drift chamber, a particle identifying device, time of flight counters and an octant shower detector in increasing order of distance from the interaction region. Large planar drift chambers for muon detection surround the entire apparatus.

CHARGED PARTICLE TRACKING

The heart of the detector is the cylindrical drift chamber. It is used: 1) to obtain a precise measurement of track momentum and 2) in the event trigger. A sample of how it works on a typical "upsilon" event can be seen in Fig. 4. (Actually Fig. 4 shows a Bhabha event with a photon conversion in the beam pipe. The soft electron and positron, each with about 50 MeV/c momentum, demonstrate trapping of particles in the magnetic field.)

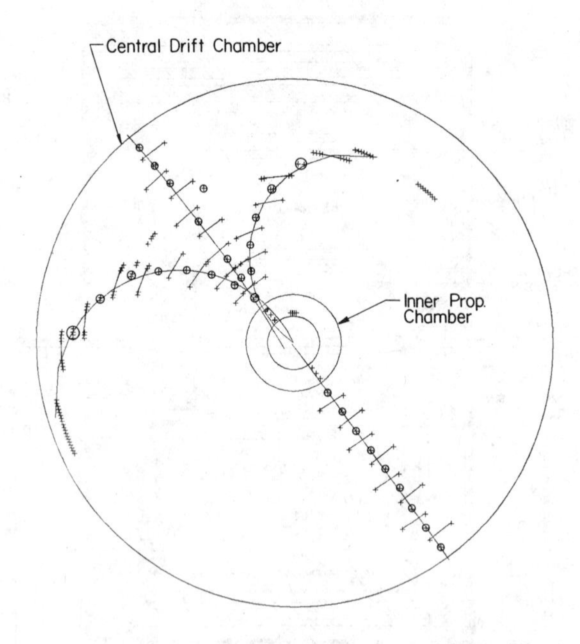

Fig. 4: An event in the drift chamber and inner prop. chamber

The central drift chamber has nine cylinders with wires parallel to the beam direction. These hits are shown as circled X's in Fig. 4. These nine cylinders alternate with eight cylinders in ±3 degree stereo. Hits on these wires show in Fig. 4 as lines because they are viewed obliquely. The cells are approximately 11 mm on a side.

At present a position resolution of 250 microns has been achieved from the central drift chamber. This appears to be limited by software; we have not yet determined drift constants, pedestals etc. well enough. A resolution of 150 to 200 microns is expected when the constants are properly determined. This leads to a momentum resolution of .05 p (p in GeV) at the present magnetic field of 0.37 T. A solid angle of 73% of 4π is covered with full momentum resolution; the innermost cylinder senses particles in 97% of the full solid angle. (A superconducting solenoidal coil capable of producing 1.3 T is now in house at Cornell. The cryostat is being leak tested. We could install the new coil later this year.)

Between the central drift chamber and the beam pipe is the inner proportional chamber. Its function is 1) to measure Z, the position along the beam direction, 2) as a part of the event trigger and 3) to help simplify software in resolving the left-right ambiguity and in identifying beam-wall events.

As can be seen in Fig. 4, the inner proportional chamber has three cylinders of anode wires running parallel to the beam. In the dimension out of the paper, cathode hoops measure the Z position. The inner proportional chamber covers about 91% of the total solid angle.

A genuine upsilon event is depicted in Fig. 5 to demonstrate that the drift chamber and inner proportional chamber are capable of handling many tracks. The tracking program has found 13 tracks, not quite all that the eye can find.

Also visible in Fig. 5 just beyond the solenoidal coil are the outer drift chambers. There are eight chambers, one mounted in each of the octant holders. The function of these devices is 1) to provide additional position information (especially in Z) and 2) to monitor what happens to particles in the 3-inch aluminum coil.

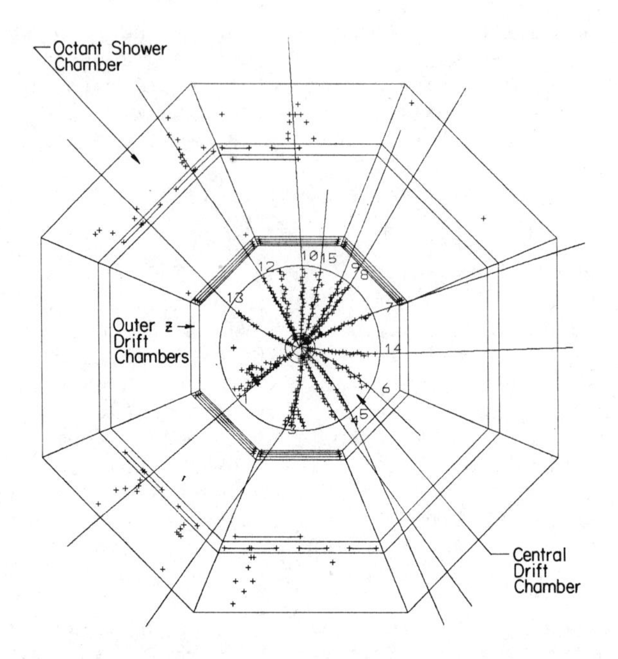

Fig. 5: A high multiplicity hadronic event

PHOTON DETECTION

In both Figures 2 and 5, you can see the octant shower detectors at the back of the octants. Two other shower devices can be seen in the view of CLEO in Fig. 3, a vertical slice through the detector containing the beam line. The octant end shower chambers are visible at the ends of the particle detection devices and the endcap shower chambers are on the magnet pole pieces. All the shower chambers are lead/proportional tube sandwiches. Together they cover 78% of the total solid angle with some sort of photon identification.

The octant chambers are the largest of them, covering with full resolution a total of 47% of 4π. They are used: 1) to detect photons, 2) in the energy trigger and 3) to help discriminate against triggers which don't come from beam-beam collisions. Photon angular resolution is 7 mr; two photons can be separated if they are 170 mr or more apart. The energy resolution is $.18/\sqrt{E}$, E in GeV. Background cutting takes advantage of the fact that throughgoing minimum ionizing particles leave the equivalent of .250 GeV in the octant shower chambers.

Typical performance of the octant shower chambers can be seen in Fig. 6. At the middle left, you can see the two photons from a π°. At about 4 o'clock a minimum ionizing particle leaves a typical narrow, straight track. This upsilon event also contains charged pion decay in the central drift chamber at approximately 12 o'clock.

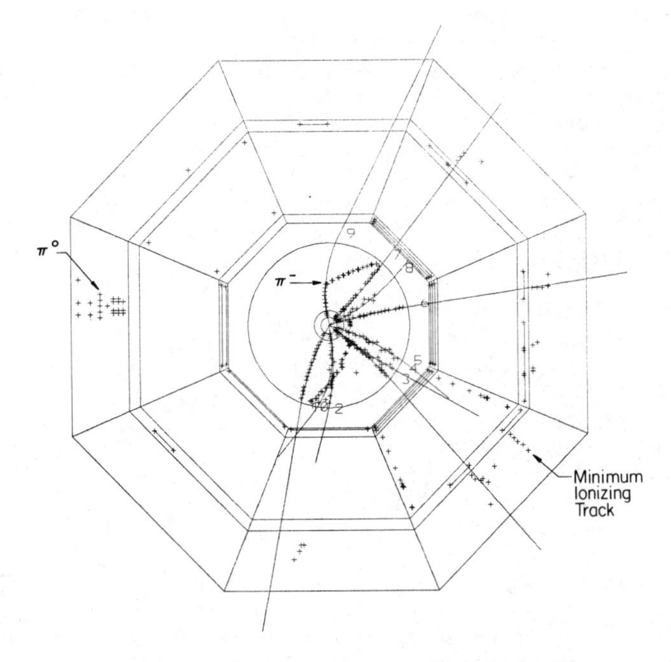

Fig. 6: Hadronic event with a π°

The endcap shower chambers cover the vital polar angle range 13 to 29 degrees. They are used for: 1) photon detection, 2) charged particle detection at polar angles too small for the central drift chamber and 3) obtaining an independent luminosity measurement. They subtend 10% of the solid angle with full resolution. The energy resolution for photons is $.25/\sqrt{E}$.

At the ends of the particle identifiers in each octant are shower chambers to avoid dead spots in photon identification. A total of 11% of the total solid angle is covered with an energy resolution for photons of $.25/\sqrt{E}$.

Also visible in Fig. 3 is the luminosity monitor, which measures Bhabha scatters at 39 to 70 mrad in the lab. It consists of four scintillator and lead-scintillator telescopes which cover nearly the full phi range. A two-telescope coincidence is required. The absolute calibration of the measurement is estimated to be ±8%.

PARTICLE IDENTIFICATION

CLEO uses time of flight, Cerenkov counters, dE/dx chambers and hadron filtering for particle identification.

Time of flight counters are located just in front of the octant shower chambers in all eight octants, 2.1 meters from the interaction point. In each octant 7% of the total solid angle is covered. The counters are constructed of one inch thick scintillator with a phototube at one end. In cosmic ray tests with two overlapping counters a resolution of 270 psec per counter was observed, but so far poor calibration procedures have limited the practical time resolution to about 2 nsec. When the final resolution is achieved, we will be able to separate e/π to 0.5 GeV/c, π/k to 1.0 GeV/c and k/p to 1.7 GeV/c.

Cerenkov counters are installed in six of the octants. Four are one atmosphere Freon 12 counters which are used to separate e/π up to 3.0 GeV/c. These cover 7% of the solid angle per octant. The other two Cerenkov counters operate at 5 atmospheres pressure. Because of the heavy pressure vessel, they cover only 5% of the total solid angle per octant. They separate e/π up to 1.2 GeV/c, π/k from 1.3 to 4.0 GeV/c and k/p above 4.0 GeV/c. The measured efficiency for Bhabha events is 96%.

The final two octants contain dE/dx chambers, the only technologically innovative device in CLEO. D. Hitlin described them in his talk. They sample the energy deposition rate of a particle of measured momentum on 117 wires so that the Landau fluctuations can be accounted for. The tricky part of constructing these devices is achieving sufficiently good mechanical uniformity. Each chamber is filled with 3 atmospheres of Argon-Methane and covers 7% of the solid angle. So far a dE/dx resolution of 6% has been achieved for tracks from hadronic events. The chambers should get better than 5% resolution when wire gain corrections and other calibrations are done properly. Because particle identification is confused when the energy deposition curves for two different particles cross near minimum ionizing, particle identification succeeds at low and high momenta, but fails in some intermediate region. Electrons are identified above 0.2 GeV, π/k separation works below 0.7 and above 1.5 GeV/c and k/p are separated below 1.2 and above 4.0 GeV/c.

Muon identification is accomplished in the standard way, using the magnet yoke and additional iron as hadron absorbers. The muon drift chambers are behind 0.6 to 1.5 meters of iron depending on the part of the detector traversed and the angle of incidence. This corresponds to a minimum momentum limit between 1.0 and 2.0 GeV/c. Muon chambers cover 85% of the full solid angle.

TRIGGER, EVENT SELECTION AND DETECTION EFFICIENCY

Fig. 7 is a block diagram of the trigger logic. The basic hadronic trigger is the track segment processor in coincidence with time of flight counters in two separate octants.

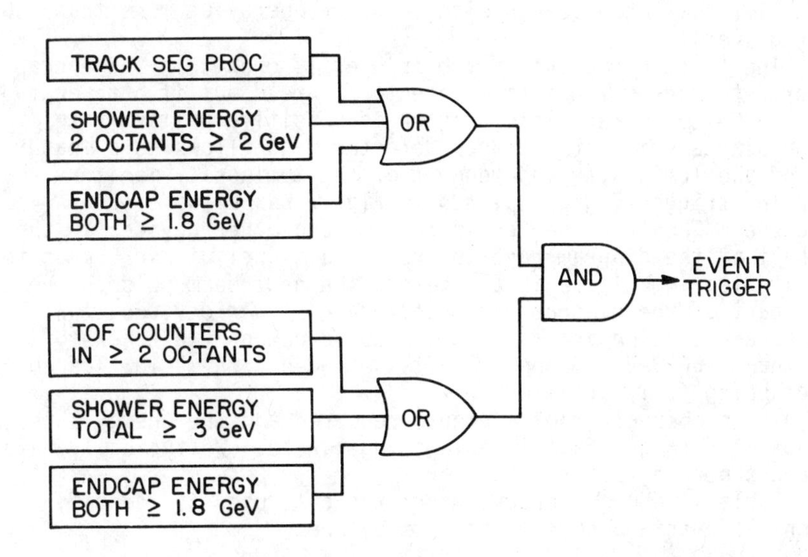

Fig. 7: Simplified diagram of the trigger logic

The track segment processor (TSP) seeks track candidates simultaneously in four sets of 3 cylinders: 1) the 3 cylinders of the inner proportional chamber, 2) drift chamber rings 1, 3 and 5, 3) drift chamber rings 7, 9 and 11 and 4) drift chamber rings 13, 15 and 17. These are all cylinders of charged particle detection with wires parallel to the beam. A "track segment" is found if 2 of the 3 cylinders are hit in a 15 degree phi slice. The charged particle trigger demands that three such segments are found in the inner prop chamber, three in the first three drift chamber cylinders and two each in the remaining two sets of drift chamber cylinders (a so-called 3 3 2 2 trigger). No attempt is made to demand that segments in different detectors line up with one another.

This TSP trigger by itself would admit huge backgrounds from beam wall interactions. Copiously produced soft electrons would get trapped in the magnetic field and make many passes through the drift chamber. It is therefore necessary to demand a signal from some device beyond the coil. We use the time-of-flight counters, which also permit some discrimination against cosmic rays. This introduces a bias against low energy particles in the trigger, since pions of less than 0.2 GeV/c, kaons of less than 0.4 GeV/c and protons of less than 0.6 GeV/c don't make it through the coil.

The 3 3 2 2 trigger rejects such two-track events as Bhabha scatters and muon pair production. An energy trigger wins back the Bhabhas at least, and also gives some efficiency for all-neutral events and hadronic events with a low charged multiplicity.

Basically, two octants each with a shower energy of more than 2 GeV are required.

The third component of the trigger is energy in the endcap shower detectors. The entire detector is read out if both endcaps show 1.8 GeV or more. This trigger is sensitive primarily to Bhabha scatters into the endcap detectors and allows us a measurement of the luminosity independent of the luminosity monitor.

The trigger logic depicted in Fig. 7 has a two tiered structure. The track segment processor and other devices in the top half of the diagram have to provide a signal within 2 μsec to inhibit clearing all registers before the next beam pulse. The lower part of the diagram can accommodate slower devices, for example a precision processor which could demand two tracks of more than .250 GeV/c momentum. At the moment, the "slow" capability is not being fully utilized.

The number of track segments required and the energy thresholds are adjusted to give a tolerable trigger rate. We typically take one event per second.

Table II shows a breakdown of the cuts used to identify hadronic events and how effective they are.

TABLE II: Hadron Event Selection

Total Triggers	1000
Fast preliminary cuts:	
TSP Trigger	-100
2 TOF Counters (tighter time cut)	-205
0.1 GeV in Octant Shower Chambers	-399
Inner PWC beam wall cut	-187
	109 events remain
Final cuts requiring full tracking:	
Three or more tracks from a common vertex	
0.25 GeV in Octant Shower Chambers	
Neutral + Charged Energy more than 15% of C.M. Energy	
Distance of vertex from beam axis less than 25 mm	
Distance along beam axis within 80 mm of crossing point	
	5 events remain

The first three cut criteria introduce losses only via hardware failures and aperture limitations. (A single throughgoing minimum ionizing particle registers in the octant shower chambers as leaving 0.250 GeV). The fourth, which involves track recognition in the inner proportional chamber, kills all but about 6 percent of beam wall events at the cost of about 3 percent of good hadronic events. The remaining cuts utilize precision track processing in the drift chambers and octant shower chambers to further purify the event sample.

Fig. 8 demonstrates the procedure used to cut and subtract beam wall and beam gas interactions using the distance Z from the nominal crossing point along the beam axis. We demand that an event have a vertex with Z within 80 mm of zero. A statistical subtraction is made using the regions between 120 and 300 mm to estimate the residual beam wall and beam gas event rate. The figure shows data run at the third upsilon peak; even in the continuum regions the subtraction is typically less than 9%.

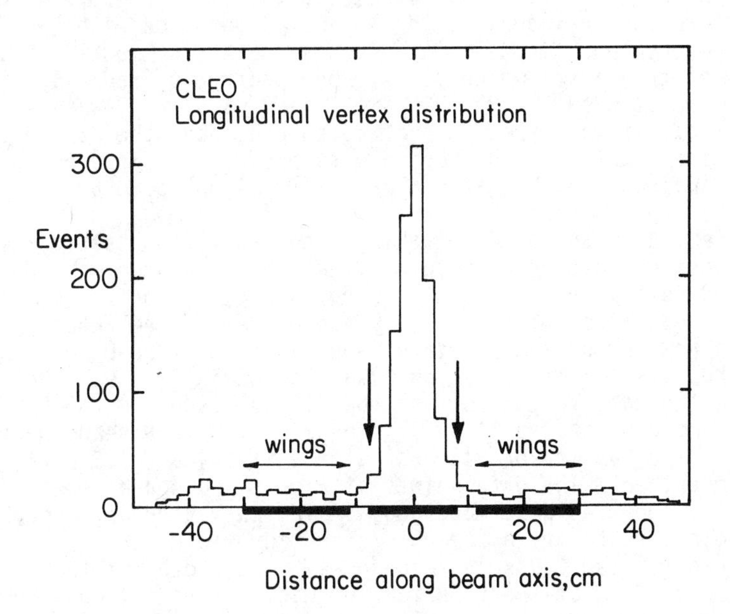

Fig. 8: Distribution along the beam direction of reconstructed vertices for events satisfying all hadron criteria except vertex position. Acceptance cuts and fiducial region for subtraction are indicated.

We use a Monte Carlo to estimate detection efficiency for converting our observed event rates to absolute cross sections. The event generator has several generating hypotheses including a limited transverse momentum two-jet model and a phase space model. The program generates simulated raw data in the eleven different detectors of CLEO. A trigger simulator identifies those events which satisfy the CLEO hardware trigger, and accepted events are put through the real data analysis programs. We estimate an overall efficiency of about 65% for two-jet events and 70% for the more nearly isotropic resonance-like events.

THE UPSILON SPECTRUM

In the summer of 1979, just before we began taking data, it was already clear that upsilon spectroscopy was interesting. Upsilons had been observed at FNAL in p + nucleus --> μ + μ + anything and at DORIS in e^+e^- interactions. The FNAL experiment saw evidence for two upsilon states[2] and possible signs of a third[3]. The DORIS experiments[4-6] confirmed the existence of the first two states and proved them to be narrow.

Meanwhile theorists[7-9] had predicted the existence of two new families of mesons analogous to the ψ family, but made with b and t quarks. The upsilons fit well into such a quarkonium model. Given the masses of the first two upsilons, the spectrum of radially excited S-wave states (1S, 2S, 3S, 4S,...), P-wave and D-wave states was predicted. It was expected that somewhere above the 3S state it would become energetically be possible to produce b quarks bound with light antiquarks, so that higher mass upsilon would have strong decays to B mesons.

Our initial plan for data taking, then, was 1) to observe the first two upsilon resonances to prove that the apparatus worked properly and establish an absolute energy scale, 2) confirm or kill the Υ(10.3) resonance, which was only a shoulder in the FNAL data, 3) seek the new postulated 4S state, and 4) try to find B mesons.

The results for steps 1) and 2) are shown in Fig. 9. These curves show the data for 66 nb^{-1} at the Υ(9.4) resonance, 55 nb^{-1} at Υ(10.0) and 320 nb^{-1} at Υ(10.3). These data have already been published[10]. We have taken 213 nb^{-1} at the Υ(9.4) and 185 nb^{-1} at Υ(10.3) since then. The data points of Fig. 9 are corrected for detector efficiency as described above. The curves drawn through the data points represent a common 1/s continuum plus three radiatively-corrected Gaussian resonances. The widths of the resonances are fixed at the CESR machine width. The fit quality is good; chisquare per degree of freedom is 0.94.

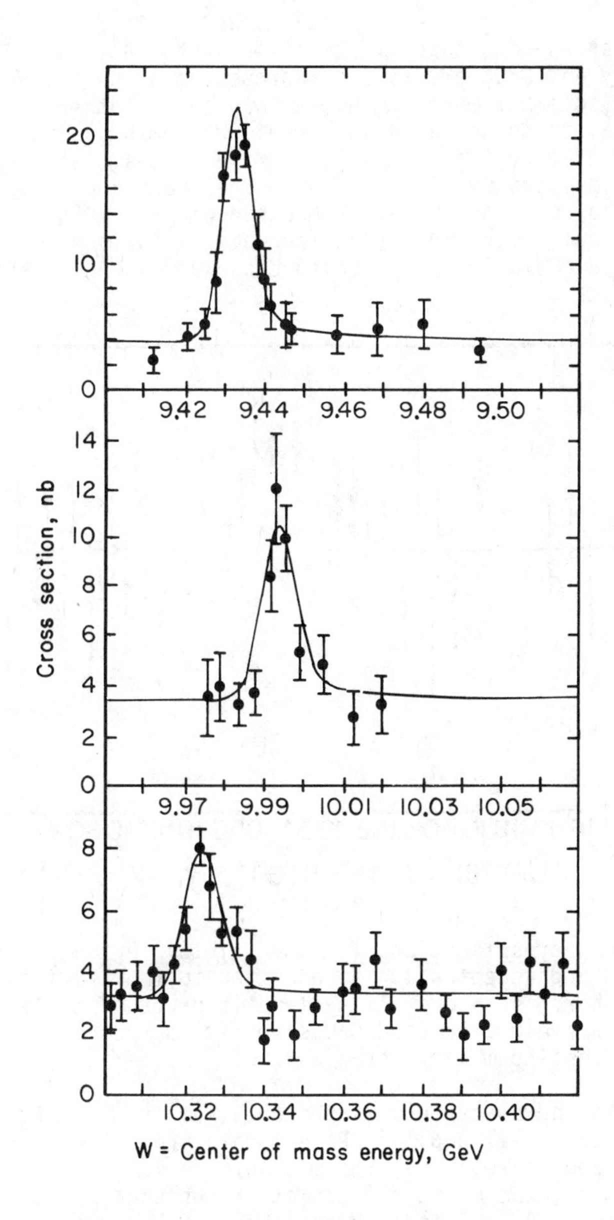

Fig. 9: Hadronic cross sections for the first three upsilon states, corrected for background and acceptance, but not for radiative effects. Errors shown are statistical only. An additional systematic normalization error of ±20% arises from uncertainties in detector efficiency and in the luminosity calibration. The absolute energy scale is uncertain to ±30 MeV. The curves show the fit described in the text.

Fig. 10 shows data taken in the 4S search. A total of 1090 nb^{-1} luminosity and 2398 events are represented. The cuts used in these data are somewhat more stringent than on the other resonances since the signal to noise ratio is worse. We demand a visible charged energy of more than 3 GeV to get rid of beam wall, beam gas and two photon physics events. No narrow resonances are observed; however, a broad, perhaps 20 MeV FWHM enhancement is visible. The fit curve uses a 1/s continuum (not the same as for the narrow upsilons) plus a radiatively corrected Gaussian resonance shape.

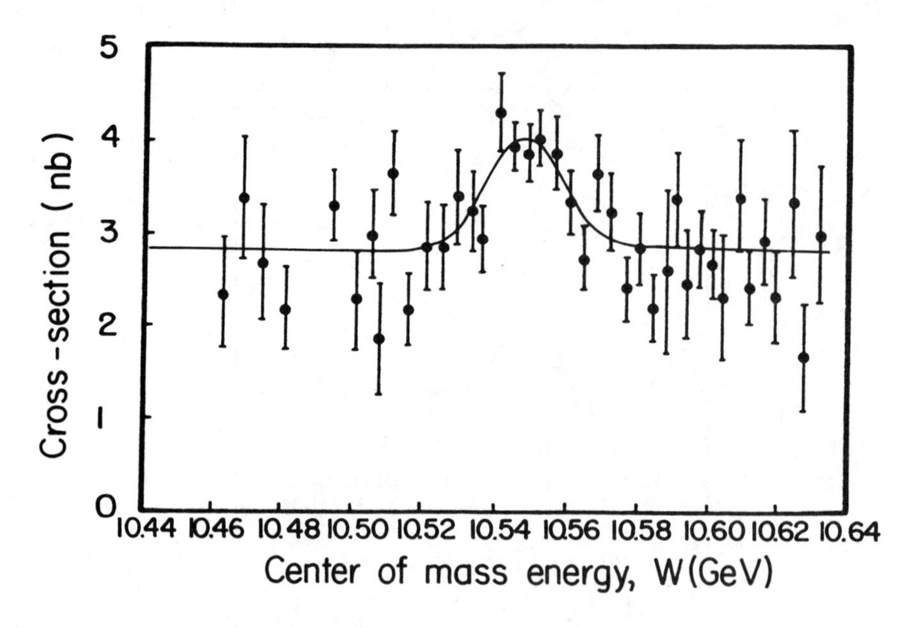

Fig. 10: Hadronic cross sections for the Υ(10.5) region, corrected for backgrounds and acceptance. In addition to the statistical errors shown, there is an overall systematic error of ±15%. The curve is a radiatively corrected Gaussian fit to the resonance above a smooth continuum varying as W^{-2}.

The masses and electronic widths obtained from the fits are summarized in Tables III and IV. Recent data from other detectors is also summarized there. Only the mass of the Υ(9.4) is given as an absolute number; the masses of the other resonances are expressed relative to the Υ(9.4). This is because the 30 MeV energy calibration uncertainty in CESR cancels in the difference. The CLEO electronic widths in Table IV were calculated assuming that the leptonic width is negligible compared to the hadronic width. The absolute leptonic width of the Υ(9.4), being sensitive to the absolute CLEO efficiency, is not quoted. The tables show a very satisfactory agreement among the four experiments.

TABLE III: Upsilon Masses
(First error is statistical, second systematic)
(Other detectors' data taken from presentations at the conference)

Resonance	CESR		DORIS	
	CLEO	CUSB	LENA	DASPII
Υ(9.4) (mass in GeV)	9.4331 ±.0003 ±.0300	9.4345 ±.0004 ±.0300	9.4616 ±.0005 ±.0100	9.4630 ±.0005 ±.0100
$\Delta M\Upsilon$(10.0) (mass relative to Υ(9.4)	0.5607 ±.0008 ±.0030	0.5590 ±.0010 ±.0030	0.5527 ±.0012 ±.0010	0.5537 ±.0017 ±.0010
$\Delta M\Upsilon$(10.3) (mass relative to Υ(9.4)	0.8911 ±.0007 ±.0050	0.8890 ±.0010 ±.0050		
$\Delta M\Upsilon$(10.5) (mass relative to Υ(9.4)	1.112 ±.0020 ±.0050	1.114 ±.0020 ±.0050		

TABLE IV: Upsilon Electronic Widths
Υ(9.4) width in KeV, others relative to Υ(9.4)
(Other detectors' data taken from presentations at the conference)

Resonance	CESR		DORIS	
	CLEO	CUSB	LENA	DASPII
Υ(9.4) (Γ in KeV)			1.23 ±.09	1.35 ±.11 ±.22
Υ(10.0) Γ/Γ(9.4)	0.44 ±.06 ±.04	0.39 ±.06	0.46 ±.07	0.48 ±.10
Υ(10.3) Γ/Γ(9.4)	0.35 ±.04 ±.03	0.32 ±.04		
Υ(10.5) Γ/Γ(9.4)	0.21 ±.06 ±.03	0.25 ±.07		

IS IT QUARKONIUM?

Are these resonances really the predicted family of S-wave bottomonium states? It is not yet possible to answer this question definitively, but there are some checks that can be made.

The mass spectrum predicted by most quarkonium models has level spacings of several hundred MeV, getting somewhat closer together for more highly excited bound states. This is qualitatively what is observed in Table III. None of the potential models has much difficulty matching the observed spectrum. In fact, the predicted spectra were used as a guide for the resonance searches. They were right on for the ϒ(10.3) state and about 50 MeV too high for the ϒ(10.5).

The leptonic widths of the four states in Table IV are of the same order of magnitude. If they were unrelated this would be a coincidence. If one of the states were a D-wave state, its leptonic width would be expected to be substantially smaller because of the small size of the D-wave function at the origin. (A definitive proof of the relation among the four states would be to observe cascade decays between them, e.g. ϒ(10.0) decaying to ϒ(9.4) plus two pions. Statistics are not yet sufficient for this test.)

The shapes of the events for the four resonances should be similar. For the continuum hadronic events, thought to be $e^+ + e^- \to q + \bar{q}$, events ought to be jetlike. The upsilon resonances decay via three gluons or (if above threshold) into B mesons nearly at rest. In either case, the events should be much more nearly isotropic.

We have chosen to use Fox-Wolfram moments[11] to analyze event shape. H_ℓ is basically a sum over all pairs of particles of the ℓth Legendre Polynomial of the angle between the particles, with a weight proportional to the particles' momenta:

$$H_\ell = \sum_{ij} \frac{|Pi|\ |Pj|}{s} P_\ell\ (\cos\theta_{ij})$$

The H_0 moment is 1 if all particles are detected; as suggested by Fox and Wolfram, it is used to normalize the higher moments. The H_1 moment should be zero by momentum conservation. Fig. 11 shows R_2 $(=H_2/H_0)$ for hadronic events on the ϒ(9.4) resonance and in the continuum near it. The on-resonance data show a clear excess of events with R_2 less than 0.3. The same effect is seen at the ϒ(10.5) resonance as shown in Fig. 12. The subtraction, depicted in the right half of that figure, shows that nearly all the increase in cross section comes from events with R_2 less than 0.3. This fact will be used later to enhance the signal-to-notice at the 4S resonance.

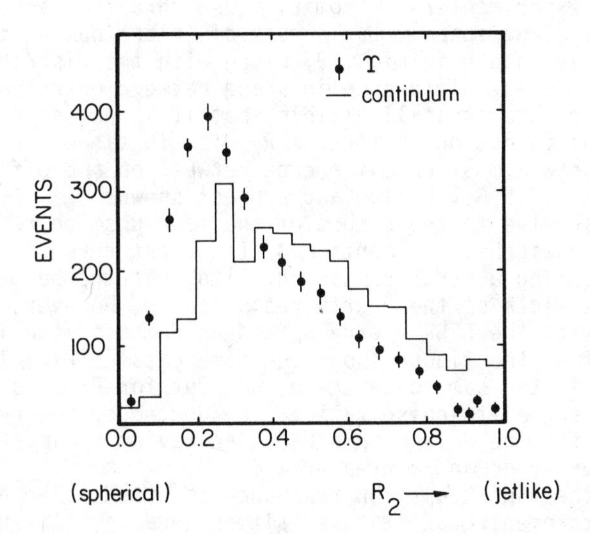

Fig. 11: Differential cross sections in $R_2=H_2/H_0$, not corrected for acceptance, for the Υ(9.4) resonance. The histogram represents data in the continuum regions above and below the resonance, normalized to the same number of events.

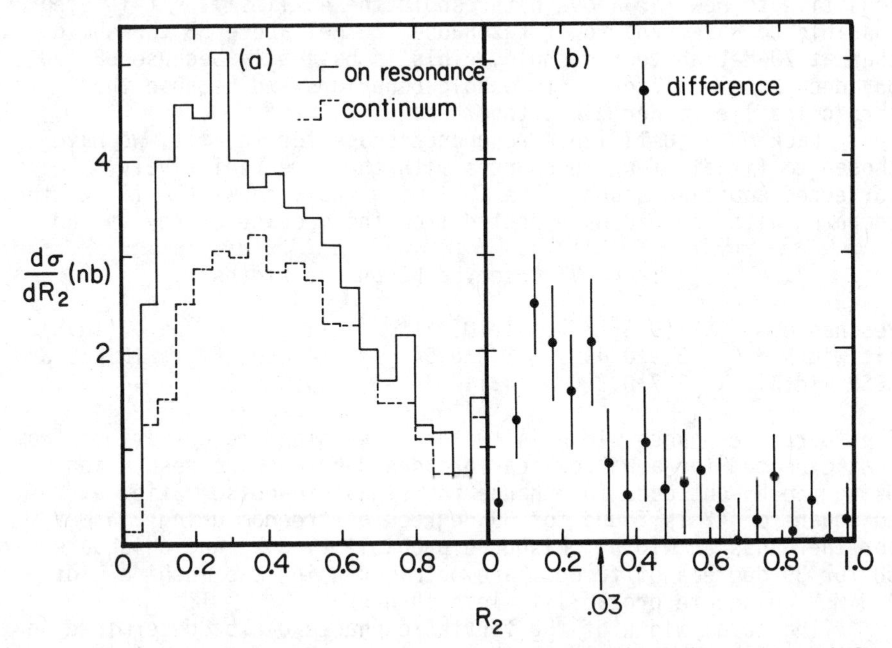

Fig. 12: Differential cross sections in $R_2=H_2/H_0$, not corrected for acceptance, for (a) the resonance region (10.536 GeV to 10.564 GeV) and continuum regions above and below the resonance; and (b) the resonance region with the continuum distribution subtracted.

Other experimenters customarily use thrust or sphericity to characterize event shape. Our thrust distributions at the Υ(9.4) and Υ(10.5) are shown in Fig. 13, along with the distribution off resonance. A clear difference in shape between on-resonance and off-resonance data is still visible, but it is not as pronounced nor as convenient to cut on as for the R_2 distributions. Plots in sphericity show almost no difference between on and off resonance shapes in the 10.5 GeV region and are not shown.

A final clue to the nature of the four upsilon resonances are the resonance widths. As mentioned, the first three resonances are well fit assuming a width consistent with just the beam energy spread. The width of the fourth resonance is, however, not consistent with the CESR energy spread as demonstrated in Fig. 14. The top half of the figure shows the same data as Fig. 10 and the bottom half is the same data again, but cut for R_2 less than 0.3 to enhance the signal-to-noise ratio. The dotted curves represent attempts to fit the data with a Gaussian having the machine width. The data clearly demand a greater width.

Thus the fourth upsilon resonance is broad. The attempt to make this statement quantitative falters because it is necessary to make some assumption about the resonance shape before a fit can be made. Theory gives little guideline for a sensible guess. Making the reasonable assumption that the resonance is wide because it decays into B and/or B* mesons leads to a shape that is very sensitive to how far above B threshold the 4S lies[12]. It is even possible to have a narrower resonance 100 MeV above $B\bar{B}$ threshold than at 70 MeV above threshold. This is basically because $B\bar{B}$ and BB* decays have different threshold behaviors and because these thresholds lie rather close together.

Lacking a justifiable resonance shape for the fit, we have chosen to fit all four resonances with the same radiatively corrected Gaussian shape. The following table shows the fit widths compared with the widths expected from the machine energy spread.

TABLE V: Observed Resonance Widths

Resonance	Υ(9.4)	Υ(10.0)	Υ(10.3)	Υ(10.5)
fit width	3.9±0.44	2.9±0.54	4.8±0.56	10.2±2.3
CESR width	3.7±0.3	4.1±0.3	4.4±0.3	4.6±0.3

The fourth resonance width is two or three standard deviations from that expected for a narrow peak broadened by machine resolution. Using the R_2-cut data to enhance the signal-to-noise ratio, a chisquare of 42 is found for 39 degrees of freedom using 9.6 MeV for the Gaussian width (chisquare probability 34%) and chisquare 58 for 39 degrees of freedom are obtained using the machine width 4.6 MeV(chisquare probability less than 3%).

The total width of the fourth resonance can be determined to be 21.5 ± 6 MeV by unfolding the machine energy spread. This evaluation is fraught with systematic error due to the unjustifiable assumption of a Gaussian resonance shape.

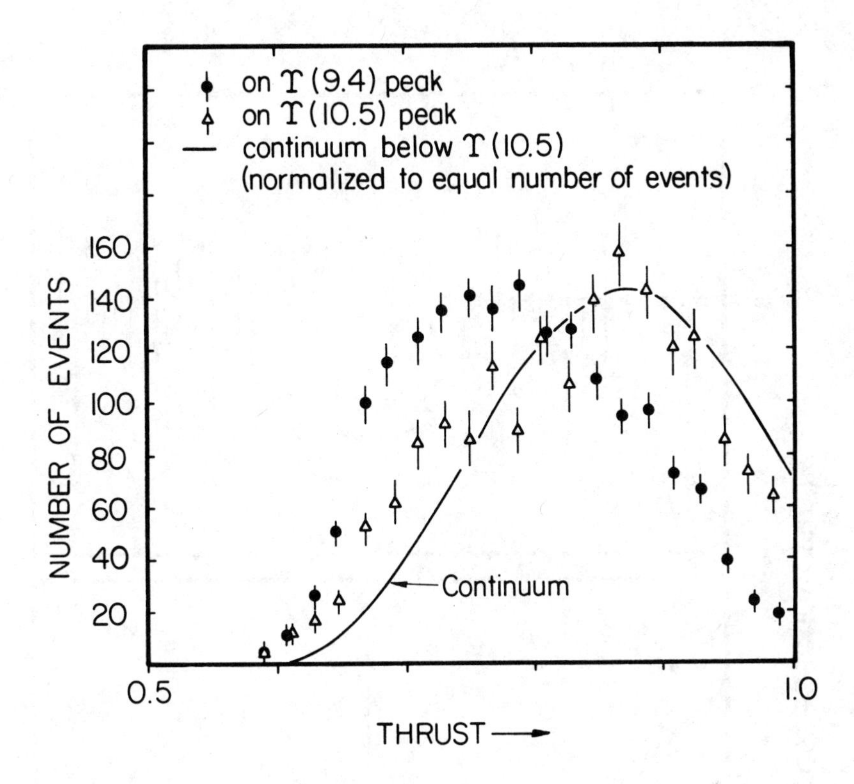

Fig. 13: Differential cross sections in thrust in the Υ(9.4), and Υ(10.5) peak regions. The curve represents the distribution in the continuum.

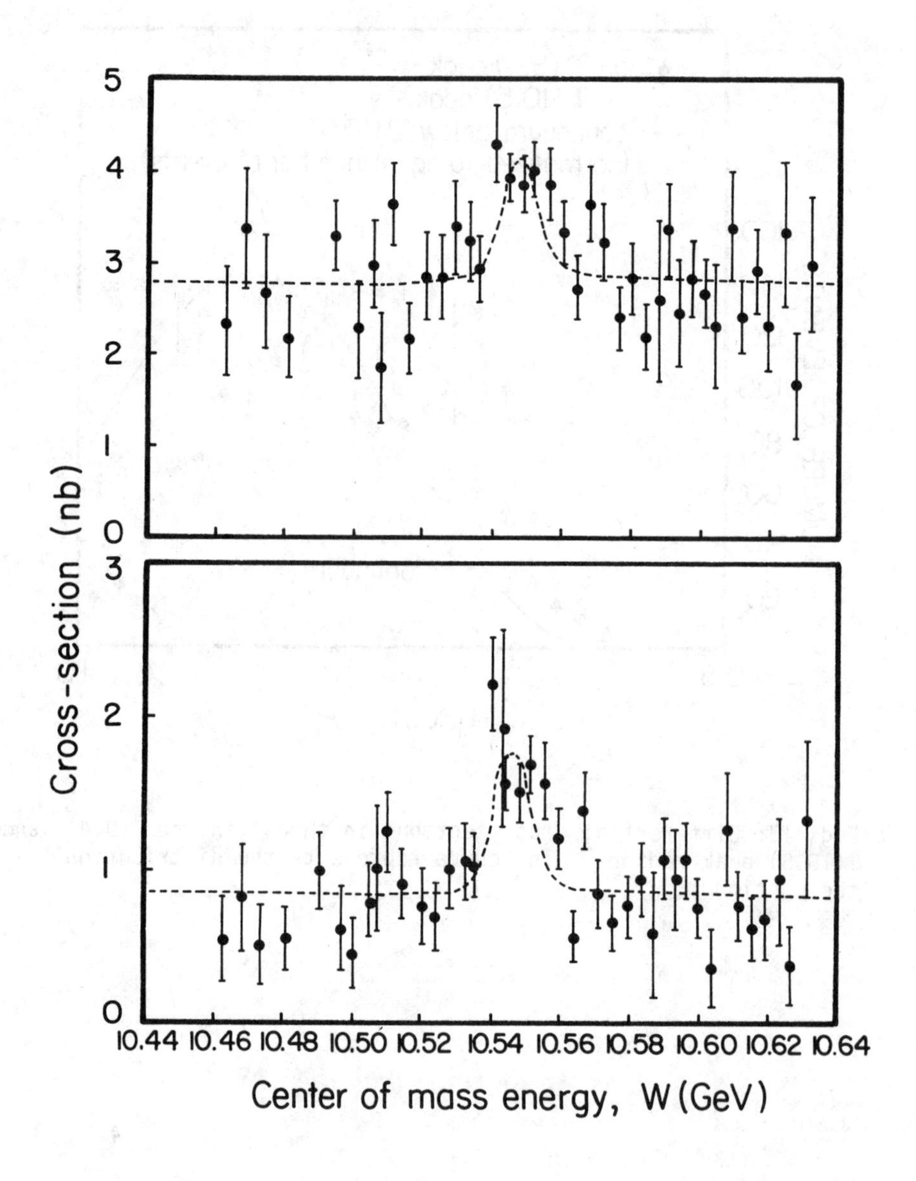

Fig. 14: Hadronic cross sections for the Υ(10.5) region as in Fig. 10. The dashed curves are attempts to fit the data with a Gaussian of the expected CESR width, showing that the enhancement is not narrow. (a) The total cross section. (b) Partial cross section for events with $H_2/H_o<0.3$.

In all likelihood, then, the fourth upsilon resonance is above threshold for production of B pairs. Because $\Upsilon(10.3)$ is narrow, it must be below or only very slightly above B pair threshold. Thus the B meson mass must be between 5.15 and 5.29 GeV, where the limits include systematic errors such as the uncertainty in the CESR energy calibration.

SUMMARY AND FUTURE PROSPECTS

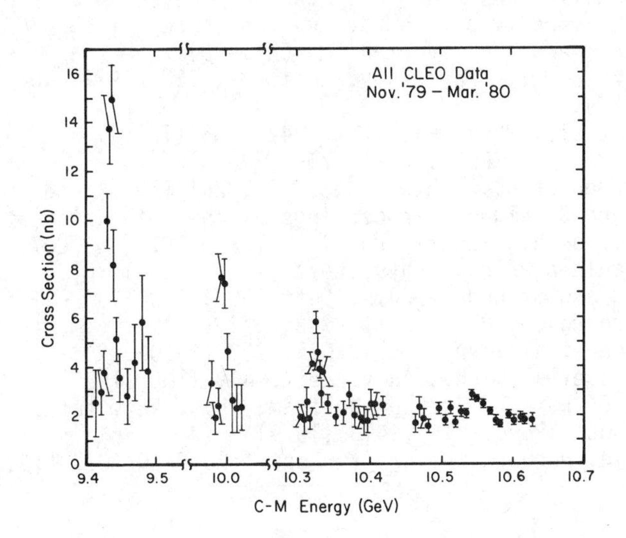

Fig. 15: Hadronic cross section as a function of energy for all CLEO data taken up to April 1980.

The total cross section versus energy for all data recorded so far with CLEO is shown in Figure 15. To summarize the results obtained so far:

1) The CESR machine is running reliably.
2) The CLEO detector is complete and also running well.
3) We are able to see the $\Upsilon(9.4)$ and $\Upsilon(10.0)$ resonances.
4) We confirm the existence of the $\Upsilon(10.3)$, determine its mass more precisely and show that it has a small width.
5) We have discovered a new, broad resonance which is presumably in the upsilon family. Assuming that it is wide because it decays into B or B* mesons, the mass of the B meson must lie between 5.15 and 5.29 GeV.

In the near future the CLEO collaboration has decided that the attempt to establish these resonances as bound states of a new b quark ought to have highest priority. We are attempting to observe an increase in the electron and muon production rates in hadronic events as evidence for a weak decay. A similar increase in the kaon production rate would be evidence for the preferred decay of b quarks to c quarks as opposed to u quarks. Although limited by the paucity of events, searches for specific B meson decays are

are being made. As part of the study of the continuum near the $\Upsilon(10.5)$, we are filling in the region between 10.3 and 10.5 GeV and searching for further narrow resonances. Later this month, we will study $\Upsilon(10.0)$ decays, and will likely be able to establish or set strong limits on its cascade decay to the $\Upsilon(9.4)$.

REFERENCES

1. A more detailed description of the CLEO detector may be found in Cornell University of Laboratory Studies report CBX-79-6, E. Nordberg and A. Silverman (March 1979).
2. S. W. Herb et al., Phys. Rev. Lett. 39, 252 (1977) and ibid. 1240 (1977).
3. K. Ueno et al., Phys. Rev. Lett. 42, 486 (1979).
4. C. Berger et al., Phys. Lett. 76B, 243 (1978).
5. C. W. Darden et al., Phys. Lett. 76B, 246 (1978) and 78B, 364 (1978), and G. Flugge, Proceedings of the 19th International Conference on High Energy Physics, Tokyo 1978, p. 807.
6. J. K. Bienlein et al., Phys. Lett. 78B, 360 (1978).
7. G. Bhanot and S. Rudaz, Phys. Lett. 78B, 119 (1978).
8. J. L. Richardson, Phys. Lett. 82B, 272 (1979).
9. E. Eichten et al., Phys. Rev. D21, 203 (1980).
10. D. Andrews et al., Phys. Rev. Lett. 44, 1108 (1980).
11. Geoffrey C. Fox and Stephen Wolfram, Phys. Rev. Lett. 41, 1581 (1978); Nucl. Phys. B149, 413 (1979).
12. E. Eichten, Harvard University Report HUTP-80/A027 (1980).

FUTURE ACCELERATOR PLANS AT DESY

B.H.Wiik
Deutsches Elektronen-Synchrotron DESY
Notkestrasse 85 - 2000 Hamburg 52 - Germany

INTRODUCTION

Experimental high energy physics has progressed by studying the collisions of what was thought to be the elementary constituents of matter at ever increasing energies. Now leptons and quarks are the basic constituents thus quark-quark, lepton-quark and lepton-lepton collisions are fundamental and should all be investigated. Indeed there is a proliferation of proposals to construct e^+e^- facilities to cover the region from PETRA and PEP energies up to and beyond the conjectured mass of the Z^o. There are three projects, the $p\bar{p}$ collider at CERN, TEVATRON at FNAL and ISABELLE at BNL aimed at a study of quark-quark collisions. In contrast only DESY has detailed plans[1] to construct a large electron-proton colliding beam facility.

Colliding electrons and protons in a colliding beam machine was first discussed[2] in connection with the ISR but abandoned because of the anticipated low counting rates. After the discovery of the deep inelastic pointlike ep cross section, the interest was revived in 1971 by an LBL/SLAC group[3]. In the meantime several possible electron-proton facilities[4-6] have been investigated.

At DESY the interest in electron-proton colliding beam machines dates back to 1971-1972. Detailed studies[4] were then carried out how to use DORIS as a model machine for a larger facility. These plans, abandoned after the J/ψ discovery, were revived in 1978 with the suggestion to install a superconducting magnet ring for protons in the PETRA tunnel With this option,called PROPER,it would be possible to collide 17.5 GeV electrons with 280 GeV protons in eight intersection regions.

The physics potential[7] of large electron-proton colliding rings was discussed in a meeting[8] organized jointly by DESY and ECFA in Hamburg in March of 1979. The study of neutral and charged weak currents is one of the prime motivations for constructing a large electron-proton colliding beam facility. Indeed the properties of charged currents at small distances can only be explored using colliding electron proton machines. It is crucial that a new accelerator is able to explore the region above 100 GeV in the center of mass system, the presumed mass scale of weak interactions, where new phenomena might be expected to occur. Although the kinematical region liberated by PROPER would greatly exceed that available at present fixed target machines the meeting concluded that the energy was to low to explore the region above 100 GeV with sufficient rate. DESY therefore investigated various possibilities aimed at raising the energy and found that to install a ep machine in a 6.5 km long tunnel ajoining the DESY site is the favoured option. This machine called HERA (Hadron-Elektron-Ring-Anlage) is designed to collide 820 GeV protons with 30 GeV electrons yielding 314 GeV in the center

ISSN:0094-243X/80/620333-35$1.50

of mass and a maximum momentum transfer squared of 98400 GeV^2. This is equivalent to an electron of 52000 GeV impinging on a proton at rest HERA thus explores a similar kinematical range as that accessible to LEP and various pp ($p\bar{p}$) colliders. In the interesting region above 10^4 GeV^2 HERA produces roughly 100 times more events than PROPER[5] for the same luminosity.

A feasibility study[1] of this project done in collaboration with ECFA has been completed. The project has received strong support from the German high energy community and a project definition endorsed by ECFA has been forwarded to the German Government. A first response is expected later this year and this might be followed by a decision in 1981.

A programme to develop superconducting magnets suitable for HERA is underway and the first prototypes are expected to be available in 1982. The construction time of the electron ring is estimated to be five years compared to some seven years for the proton ring. Since superconducting magnets presumably take longer to produce and test than to install, some of this time can be used to study e^+e^- collision between 40 GeV and 70 GeV in the center of mass using the HERA electron ring. Originally DESY has proposed[8] to upgrade PETRA to cover this energy range and a relatively modest improvement programme is now underway aimed at reaching about 45 GeV in the center of mass. Since superconducting RF cavities or a pulsed RF system are needed to extend this to higher energies it seems advantagous to use the larger radius electron ring of HERA to cover the energy region up to 70 GeV with a conventional RF system. Indeed this can be done at little extra cost with the RF system as foreseen for the ep option.

In this talk I'll describe the HERA project, first the physics and then the machine. The talk is based on the ECFA-DESY report listed in reference 1. A more complete list of references can be found in that report.

THE PHYSICS PROGRAMME AT HERA

Electron-Proton Collisions

It has been found experimentally that for large values of Q^2 the incident electron interact directly with one of the quarks in the nucleon as shown in Fig. 1. The interaction is mediated by spacelike currents charged or neutral and is characterized by the kinematical quantities Q^2, ν or the scaling variables x and y defined in Fig. 1.

A deep inelastic event has a rather striking final state topology which makes it easy to recognize among the much more numerous beam gas events. As indicated in Fig. 1 the scattered lepton appears at a large angle with respect to the beam axis with its transverse momentum balanced by the struck quark which fragments into a jet of hadron appearing on the opposite side of the beam axis. The remains of the proton give rise to a forward jet of hadrons with no net transverse momentum with respect to the beam axis. Because of the imbalance between the incident electron and proton momenta the particles will in general emerge in the forward hemisphere along the

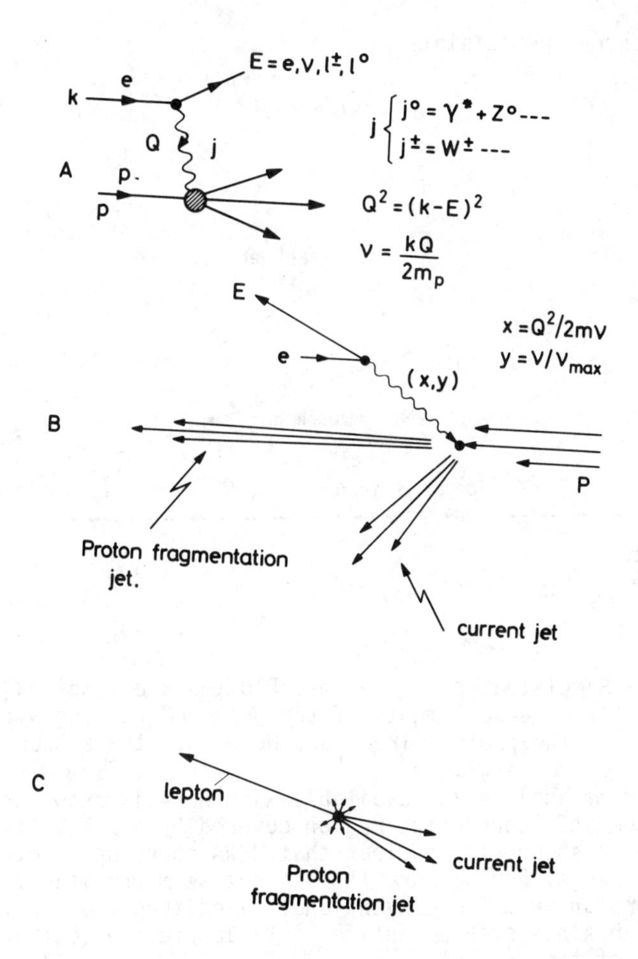

Fig. 1 - Kinematics of deep inelastic electron-proton collisions

proton direction. The proton jet, the quark jet and the lepton defines a plane with small momenta transverse to the plane and large momenta in the plane.

The Monte Carlo simulation of a deep inelastic charged current event is shown in Fig. 2, where the transverse and longitudinal momentum of each particle is plotted with respect to the beam axis. In this event the struck quark radiates a gluon leading to three well separated hadron jets, the outgoing ν is not shown. Because of their distinctive signature it will presumably be possible to identify deep inelastic events cleanly. Very low counting rate experiments are therefore feasible.

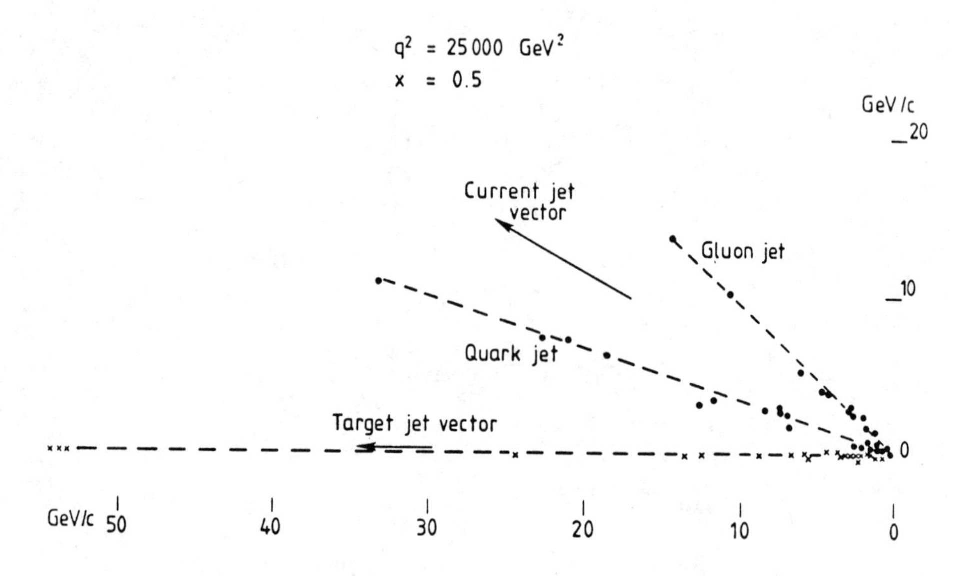

Fig. 2 - Simulated ep → νx event. Plotted are longitudinal and transverse moments of the produced hadrons with respect to the proton direction. Note that the struck quark radiates a gluon

The kinematical region available with HERA is shown in Fig. 3. In the bottom left corner the region covered by a 1 TeV fixed target machine is shown. It is clear that HERA opens up a totally new kinematical range. Furthermore it does not seem profitably to extract the proton beam for conventional fixed target experiments in particular since such a venture would double the cost of the proton part of the project.

The Q^2 value at which the electromagnetic and the weak interaction is expected to coalesce[9] is shown as the horizontal dotted line. A large kinematical area is available above this line and we will show below that the rate is sufficient to explore a part of this region.

A deep inelastic process is well defined and the data can be used to determine the properties of these currents - charged or neutral - at very small distances. Conversely once the currents are understood, measurements can explore the protons and its constituents down to distances of 10^{-17} cm. The spectrum of electron-like leptons or the spectrum of heavy quarks which couple to u or d quarks can be investigated up to a mass of 200 GeV with HERA.

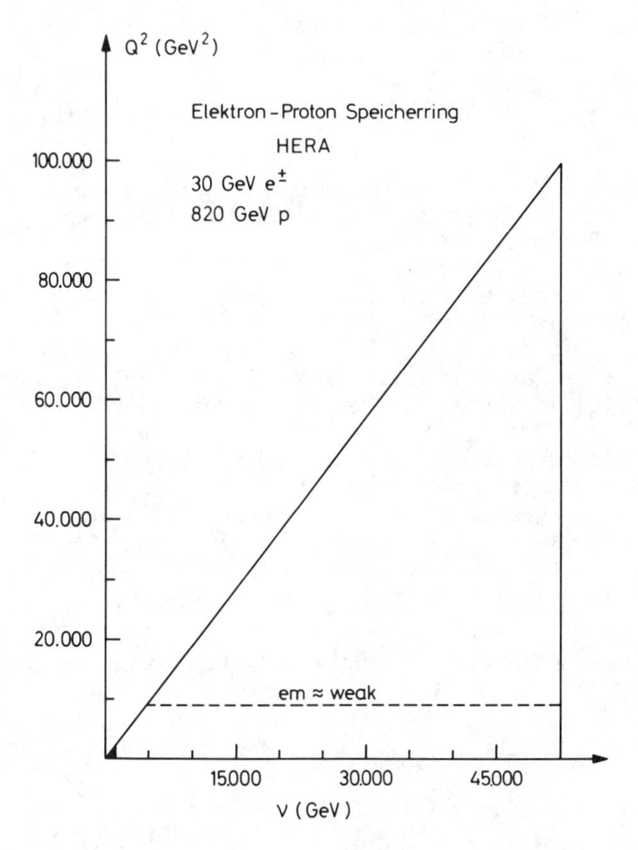

Fig. 3
Kinematical region in Q^2 and ν which can be explored with HERA

It is important to bear in mind that for small values of Q^2 ($Q^2 \lesssim m_\pi^2$) the photons are nearly real and hence the electron beam is equivalent to a well defined bremsstrahlungs beam with an end point energy of 52000 GeV. Therefore also classical hadron physics can be investigated with HERA up to very large energies.

Charged current events

To observe the charged weak current at small distances is one of the prime motivations for the construction of a large electron-proton colliding beam facility. Present data are all consistent with a lefthanded current which is mediated by a single charged vector boson with a mass around 80 GeV. However, present experiments can only probe the weak interaction at Q values which are small compared to the characteristic mass scale of the weak interactions. The observed simplicity of the charged current might reflect only the static limit studied so far and a rich structure with many vector bosons, some perhaps giving rise to righhanded currents, might appear at higher energies. Measurements at HERA will enable us to investigate the region well beyond 80 GeV and answer these questions.

From a purely experimental point HERA has some unique features compared to present fixed target experiments.

- Very high energy.
  The beam is equivalent of a monoenergetic neutrino beam with an energy up to 52 TeV
- Choice of helicity.
  It will presumably be possible to change the helicity of the incident lepton - i.e. the cross section for left and righthanded electrons (or positrons) can be measured directly.
- Visibility.
  The target is massless and can be surrounded by fine grained detectors including particle identification.
- Favourable kinematics.
  The lepton, the current jet and the target fragmentation jet are presumably well separated in space and the event is easily recognized.

The number of charged current events expected in a bin $dxdy = (0.2)^2$ after one month of data taking with an unpolarized 30 GeV electron beam colliding with protons of either 820 GeV or 200 GeV is shown in Fig. 4, assuming a luminosity of 10^{32} $cm^{-2}s^{-1}$. The rates were estimated in the standard model[9] with m_W = 78 GeV and formfactors parametrized according to Buras and Gaemers[10].

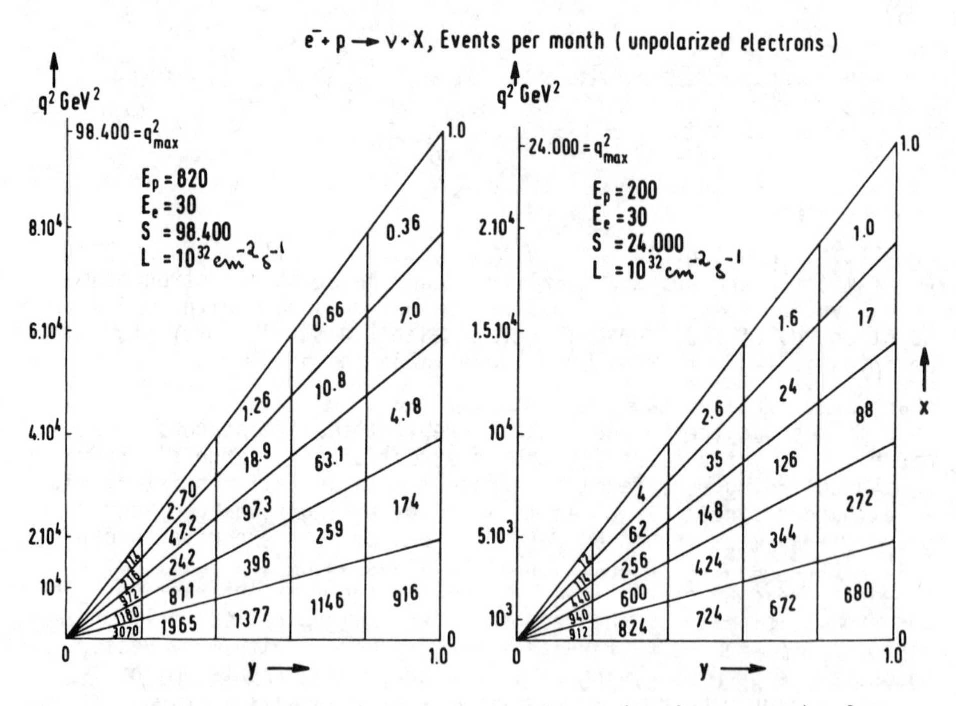

Fig. 4 - Number of charged current events produced per month of running time. In bins $dxdy = (0.2)^2$ assuming the standard model

Given the distinctive signature of a charged current event it seems possible to measure the cross section for $Q^2 \leq 4 \times 10^4$ GeV^2.

The expected counting rate for $e^- p \to \nu x$, evaluated with the assumptions listed above, is plotted in Fig. 5 versus Q^2 for various propagator masses. It is clear that HERA experiments can be used to determine the mass of the propagator as long as it is below say 300 - 400 GeV. The data can also be used to determine whether the charged current is dampted by a single vector boson as presently believed or by several. As an example, a model containing two vector bosons has been evaluated with the assumption that they have the same coupling constant at $Q^2 = 0$ and that one of the vector bosons has a mass of 78 GeV. The expected event rate for the two vector boson model, normalized to the standard model event rate, is plotted in Fig. 6 versus Q^2 for various mass values of the second vector boson. The effects are large.

The existance of righthanded currents can of course be deduced directly from a measurement of $\sigma(e^-_R p \to \nu x)$ or $\sigma(e^+_L p \to \bar{\nu} x)$.

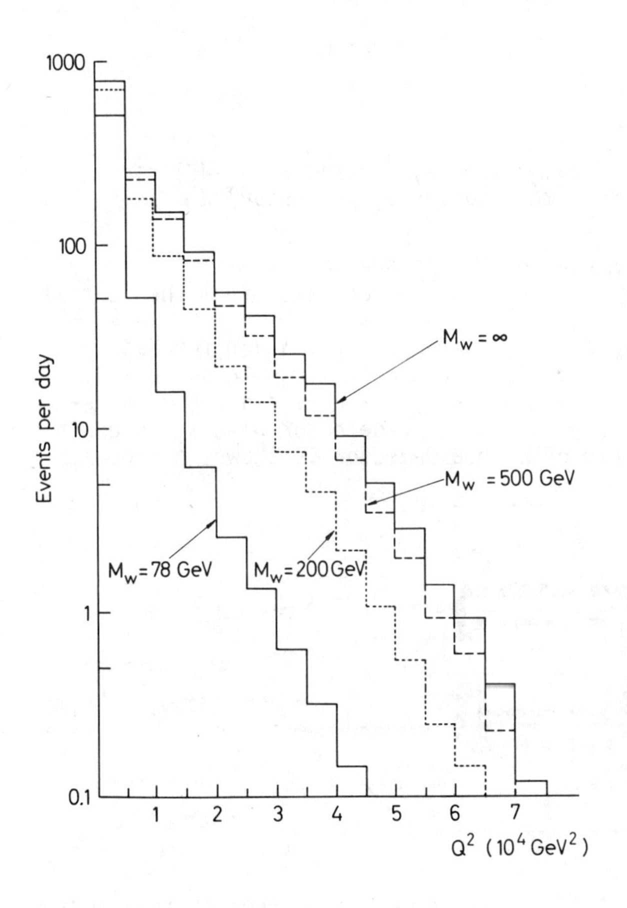

Fig. 5
Events per day for $e^- + p \to \nu + x$ in Q^2 bins of 5000 GeV^2 with the standard assumptions

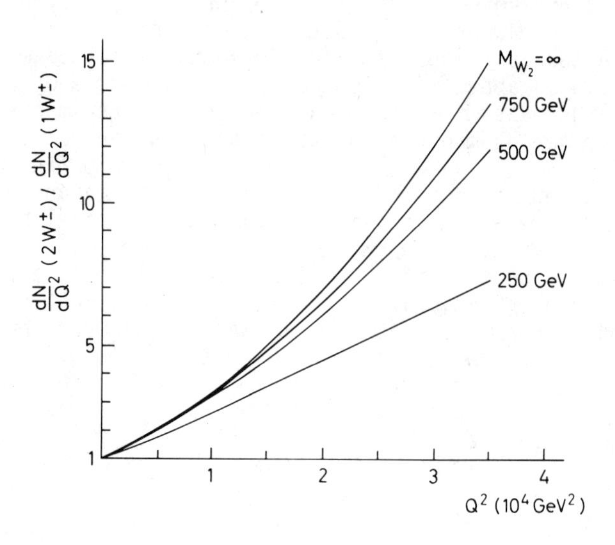

Fig. 6 - Ratio dN/dQ^2 (two $W^\pm$) / dN/dQ^2 (one $W^\pm_+$) for different mass values of the second $W^\pm$

Experimental remarks

The obvious experimental questions are:

1) Can charged current events $ep \to \nu x$ be separated from the neutral current events $ep \to ex$?
2) Can Q^2 and ν be measured with sufficient precision for each event and can the formfactors be extracted from these data ?

The answer to both questions is: yes - provided the measurements are done using a large solid angle detector without holes in the acceptance. An example of such a detector is shown in Fig. 7.

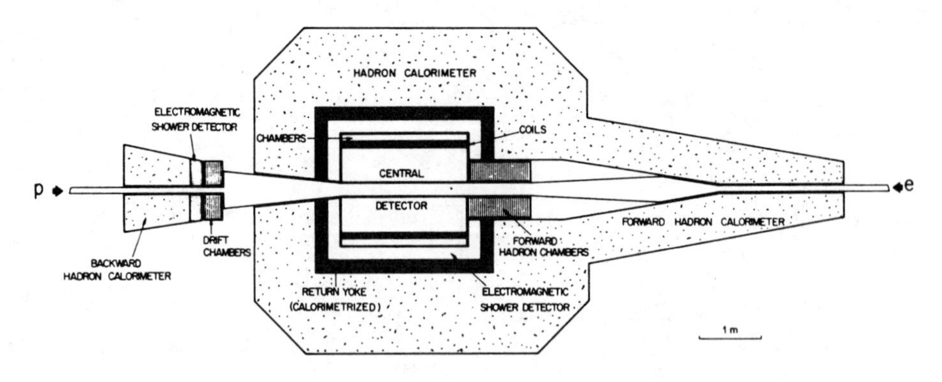

Fig. 7 - Schematic layout of a detector optimized for charged current events $ep \to \nu x$

Interesting deep inelastic events can be separated from beam-gas events by summing the transverse momenta of the final state hadrons with respect to the beam axis and demanding that this sum is above a few GeV. The direction of the final state lepton can be roughly determined from a measurement of the final state hadrons. One then only needs to examine the detector if it registered a single high energy electrón travelling within the solid angle defined by the hadrons or nothing. Since weak and the electromagnetic cross section are expected to be similar this should lead to a rather complete separation.

The resolution expected in Q^2 and $x = Q^2/2m\nu$ from a measurement of the hadron final state only using the calorimeter type detector depicted in Fig. 7 is shown in Fig. 8. The detector measures angles and momenta with the uncertainties listed in table 1.

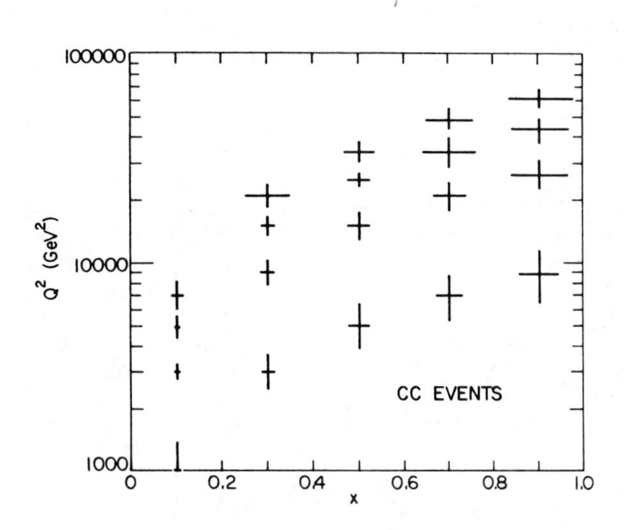

Fig. 8
Measurement errors in Q^2 and x for the resolution listed in table 1

Table 1 - Detector resolution

Charged particles	Δp	=	$0.3 \times \sqrt{E}$ or $0.005\ p^2$ (calorimeter or solenoid)
	$\Delta\phi$	=	$(2+3/p)$ mrad
Neutral particles	Δp	=	$0.3\ \sqrt{E}$
	$\Delta\phi$	=	30 mrad
π^0	Δp	=	$0.15\ \sqrt{E}$
	$\Delta\phi$	=	15 mrad

Furthermore also the formfactors can be determined with a good precision using such a detector. This is shown in Fig. 9, where the ratio of the measured formfactor to the input formfactor is plotted versus Q^2 for different values of x. For this computation it was assumed that all particles travelling within 30 mrad of the proton beam or within 45 mrad of the electron beam are lost. It is clear that the Q^2 region between a few hundred GeV^2 up to 40000 GeV^2 can be covered with good precision for proton energies between 820 GeV and 200 GeV and an electron energy of 30 GeV.

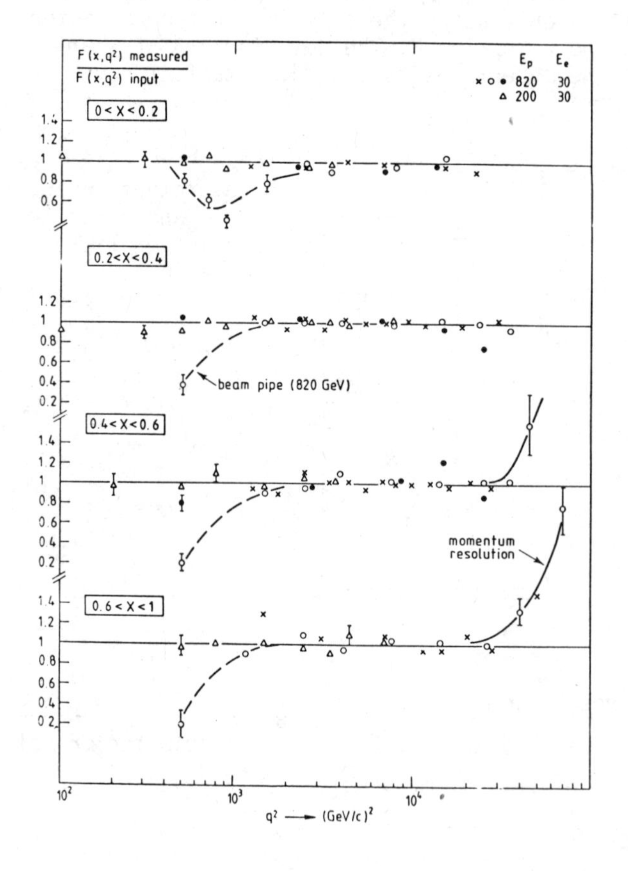

Fig. 9
Ratio of reconstructed to input structure functions for charged current events generated by three independent Monte Carlo programs

New Fermions

HERA is ideally suited to produce electronlike charged or neutral leptons and new heavy quarks which couple to u or d quarks in the proton. We know that such couplings are rather weak in the standard model, however, new currents may exist. Indeed the basic fermions must have excited states if they are not pointlike. The rate for producing a heavy quark from a light quark is plotted in Fig. 10 with the mass of the lepton at the upper vertex (Fig. 1) as a parameter. The rates were evaluated with the assumptions listed above plus the assumption that the new current couples with the same strength as the old one. Leptons and quarks with masses up to 150 - 200 GeV can be found in this way. The decay of these particles lead to rather spectacular signatures as shown in Fig. 11 for $ep \to L^o x$ with $L^o \to e^- Qq'$.

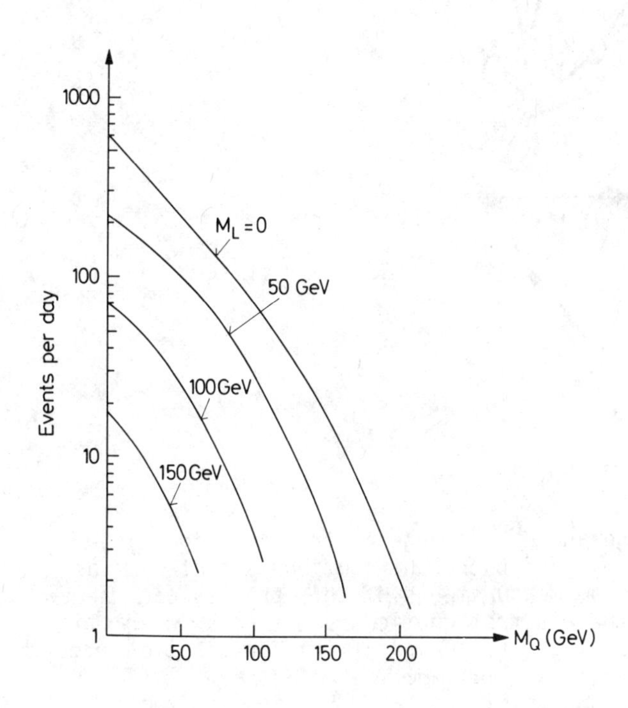

Fig. 10
Number of events per day for
$e^- + p \to L^o + Q + x$
at $s = 9.6 \times 10^4 \mathrm{GeV}^2$
assuming lefthanded couplings, unpolarized electrons,
$m_{W'} = 78$ GeV,
Buras-Gaemers QCD parametrization with $\Lambda = 0.5$ GeV and a luminosity of $10^{32}\ \mathrm{cm}^{-2}\mathrm{s}^{-1}$

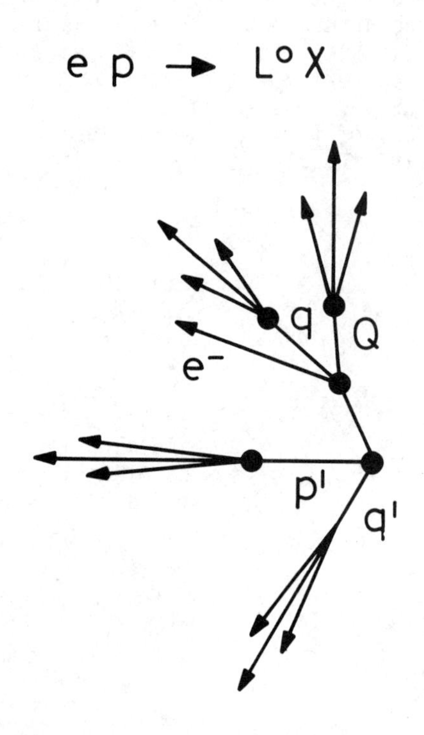

Fig. 11
Topology of
$e\ p \rightarrow L^{o}\ x$ with
$L^{o} \rightarrow e^{-}\ q\ Q$

Neutral Currents

One photon exchange and Z^{o} exchange contribute (Fig. 1) coherently to $e + p \rightarrow e' + x$ and both contributions are of similar strength at HERA energies. Measurements of this process can therefore decide if indeed the electromagnetic and the weak interactions are manifestations of a single force and if this unification occurs as conjectured in the Salam-Weinberg model[9] or if a more complicated mechanismn involving many Z^{o}'s is realized in nature. The number of neutral current events produced in a bin $dxdy = (0.2)^2$ per day by 30 GeV electrons colliding with 820 GeV protons and a luminosity of $10^{32}\ cm^{-2}s^{-1}$ is plotted in Fig. 12. Again due to the characteristic topology of deep inelastic events HERA can extend the Q^2 range from the present few hundred GeV^2 out to some 30000 GeV^2.

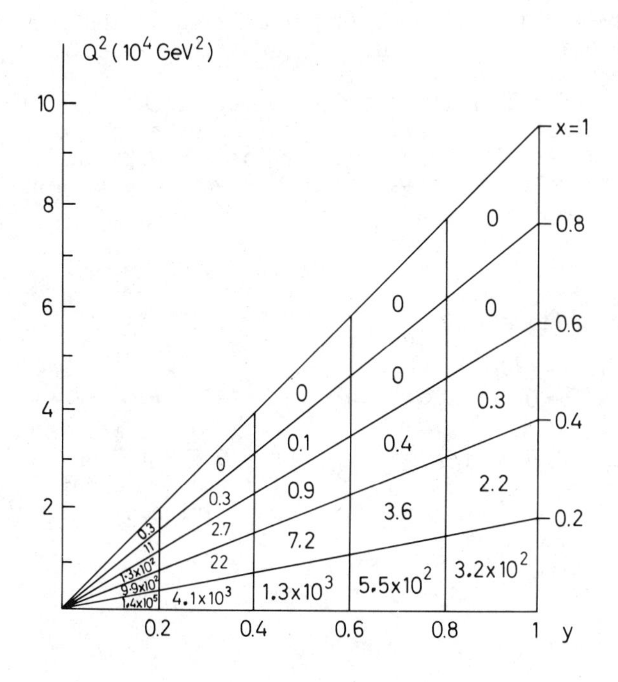

Fig. 12
Number of events per day for
$e^- + p \quad e^- + x$
at $s = 9.6 \times 10^4 GeV^2$
and the standard assumptions

The presence of a weak current in the amplitude has clear signatures.

1) Parity violation

$$\sigma(e_L^- \, p \to e^{-\prime} \, x) \neq \sigma(e_R^- \, p \to e^{-\prime} \, x)$$

$$\sigma(e_L^+ \, p \to e^{+\prime} \, x) \neq \sigma(e_R^+ \, p \to e^{+\prime} \, x)$$

This effect can only be caused by a neutral weak current.

2) Appearant C-violation

$$\sigma(e_L^- \, p \to e^{-\prime} \, x) \neq \sigma(e_L^+ \, p \to e^{+\prime} \, x)$$

$$\sigma(e_R^- \, p \to d^{-\prime} \, x) \neq \sigma(e_R^+ \, p \to e^{+\prime} \, x)$$

Two-photon exchange will also give rise to a charge asymmetry. This effect, however, is expected to be of order $\alpha/\pi \ln Q^2/m^2$ with $m \sim 300$ MeV. At large values of Q^2 this effect is small compared to the charge asymmetry caused by Z^0 exchange and it has furthermore a different Q^2 dependence. The two photon effects can be determined at relatively low values of Q^2 where Z^0 exchange has a small effect only.

3) The presence of a $1-(1-y)^2$ term which is not allowed in the one photon exchange approximation. This effect cannot be caused by two photon exchange.

The size of these effects in the standard model is shown in Fig. 13 where the ratio for the left and righthanded electrons and positrons is plotted as solid line versus y for x = 0.25 and $s = 9.8 \times 10^4\ \mathrm{GeV}^2$. Note that the rates are sufficient to determine these asymmetries in a few months of running.

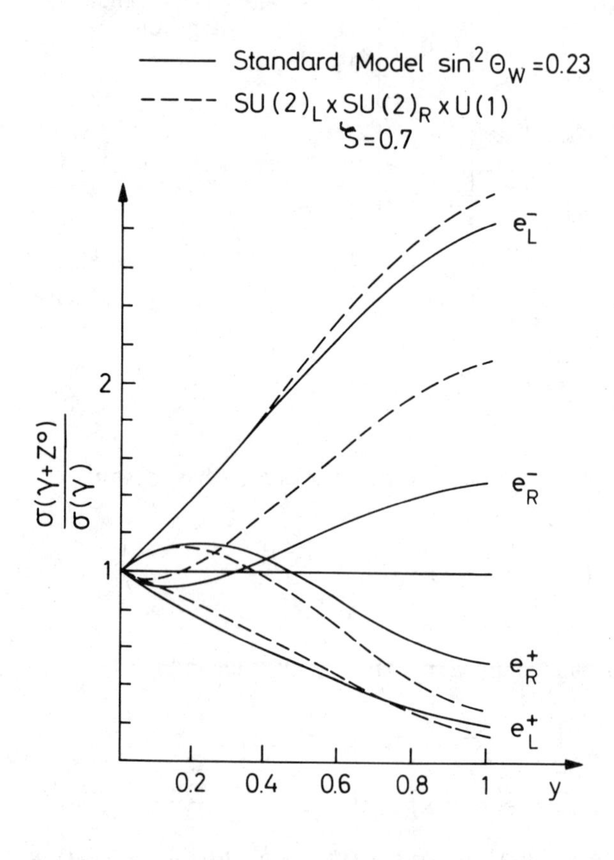

Fig. 13
The ratio
$\sigma(\gamma + Z^0)/\sigma(\gamma)$
at x = 0.25 and
$s = 9.6 \times 10^4\ \mathrm{GeV}^2$
for two different weak interaction models.

A measurement of these asymmetries can be used to pin down the properties of the neutral weak current. Suppose that $SU(2)_L \times SU_R(2) \times U(1)$ is realized in nature. The parameters of such a group can be adjusted to agree with all known data. The dotted lines in Fig. 13 show the cross sections expected in this model with the mass of the second Z^0 at 224 GeV. It is clear that the two models can be separated.

New flavour changing neutral currents might also appear. Such currents could lead to spectacular processes like $e^-d \rightarrow \tau^-b$.

Test of Strong Interactions

QCD[11] makes clear, unambigous predictions for deep inelastic processes.

Such a prediction for non-singlet moments is plotted in Fig. 14 versus Q^2. Note that the value of the moments in QCD is nearly constant for Q^2 above 1000 GeV^2. This is a very strong prediction unique to QCD. For example a simple power behaviour expected in other types of field theories can mimic the observed behaviour over the present available Q^2 range - however it will deviate from the QCD predictions at large values of Q^2. This constancy makes it also easy to observe threshold like color liberation if it should occur. Measurements of the final state hadrons will enable us to carry out detailed tests of QCD.

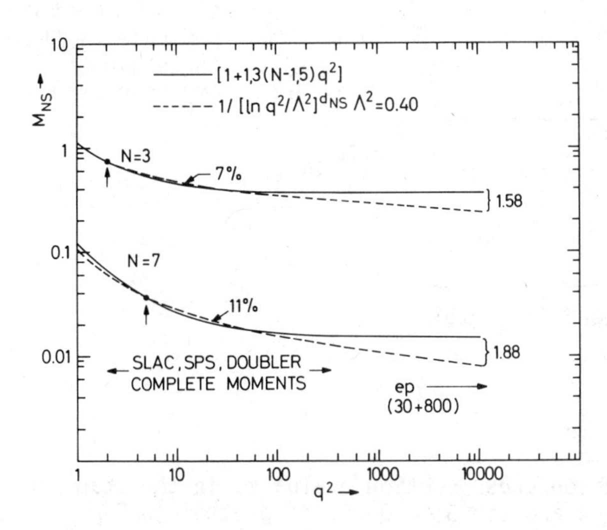

Fig. 14
A simple $1/Q^2$ power behavior compared with the QCD logarithmic behavior for non-singlet moments

Probing Quarks and Leptons

Faced with the large numbers of leptons and quarks many physicists find it natural that these entities are made up of new building blocks. With HERA we can probe the fermion structure down to distances of 10^{-17} cm corresponding to 10^{-4} of the size of the proton.

If the leptons have a size we would expect to observe a leptonic form factor and ultimately the production of excited leptons. The cross section would be modified by a form factor $F(Q^2) = 1/(1 + q^2/M^2)$ giving rise to a scaling violation which is very different from that expected in QCD. A 10% measurement at 4×10^4 GeV^2 would be sensitive to any mass of the order of 1 TeV.

An excited lepton could decay into $e + \gamma$, $e + Z^0$ and $e + W$ leading to peaks in the invariant spectrum.

The cross section would also be modified in the case of a quark structure in a similar manner - i.e. again one might probe down to distances of $(1\ \mathrm{TeV})^{-1}$. Again there might be excited quark states.

Another possibility is that the proton contains new gluon-like particles which interact neither weakly nor electromagnetically. These particles would show as a step in the momentum fraction of the protons carried by the quarks. Example of such particles are the spin 1/2 gluinos expected in supersymmetric theories. The expected behaviour of the momentum[12] sum rule in crossing the threshold $Q^2 = 4\,m_{\tilde{g}}^2$ is shown in Fig. 15.

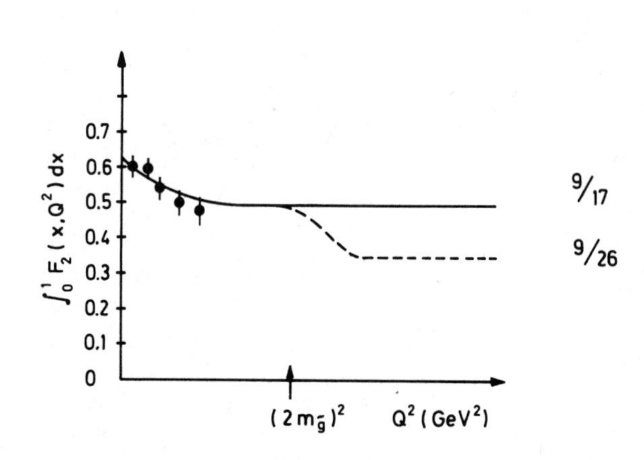

Fig. 15
Possible behavior of the momentum sum rule as the threshold $Q^2 = 4m_{\tilde{g}}^2$ is crossed

e^+e^- Physics below the Z^0

The e^+e^- annihilation cross section evaluated in the standard model with 3 generations and $\sin^2\theta_W = 0.25$, is plotted in Fig. 16 versus energy. The energy range covered by HERA is indicated. At the upper end of that range $R = \sigma(e^+e^- \to \text{hadrons}) / \sigma(e^+e^- \to \mu^+\mu^-)$ is about 30 compared to an R value of maybe 5 for purely one photon annihilation indicated by the dotted line. The cross section is thus dominated by the Z^0-pole and the same decays as on the Z^0 peak can be studied except for a 20 GeV difference in mass. About 3000 Z^0 decays per day are observed with a luminosity of $10^{32}\mathrm{cm}^{-2}\mathrm{s}^{-1}$.

The region covered by HERA is also well suited to determine the coupling constants of the neutral current to the various fermions. In Fig. 17 the forward backward asymmetry expected in the standard model for various types of fermions is plotted versus center of mass energy squared in units of the Z^0 mass squared. Essentially, this asymmetry measures the product of the axial couplings of the electron and the outgoing fermions to the neutral current. By colliding an unpolarized beam with a polarized beam one can measure the product of the vector and the axial couplings. It seems feasible to produce polarization at HERA whereas it seems to be more difficult at LEP.

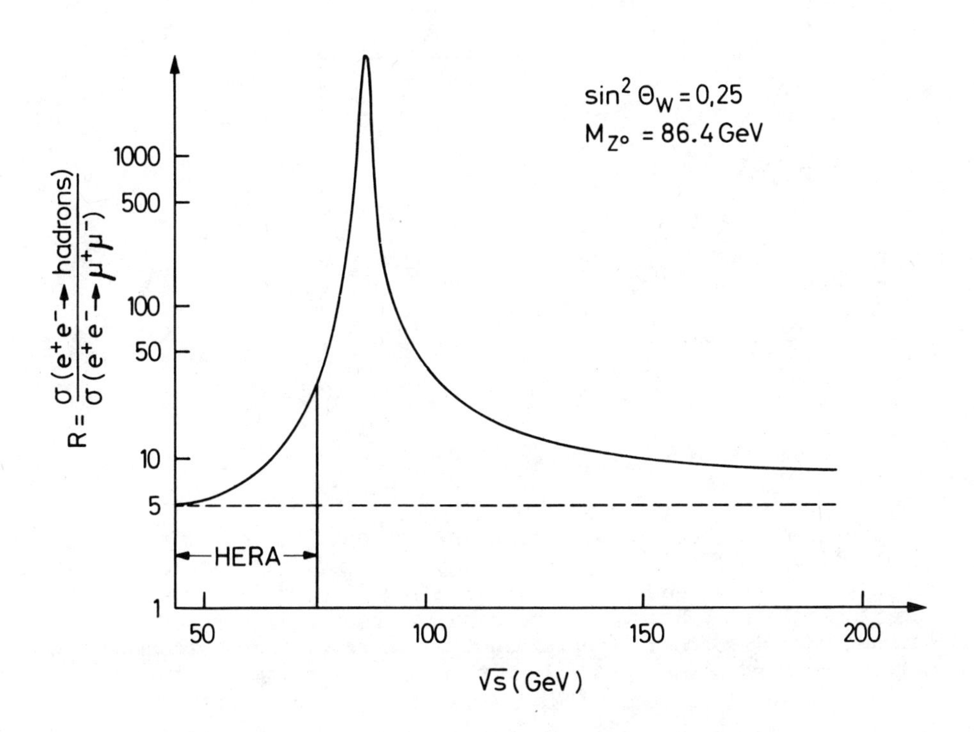

Fig. 16
Annihilation cross section for $e^+e^- \rightarrow$ hadrons evaluated in the standard model with 3 generations and $\sin^2\theta_W = 0.25$

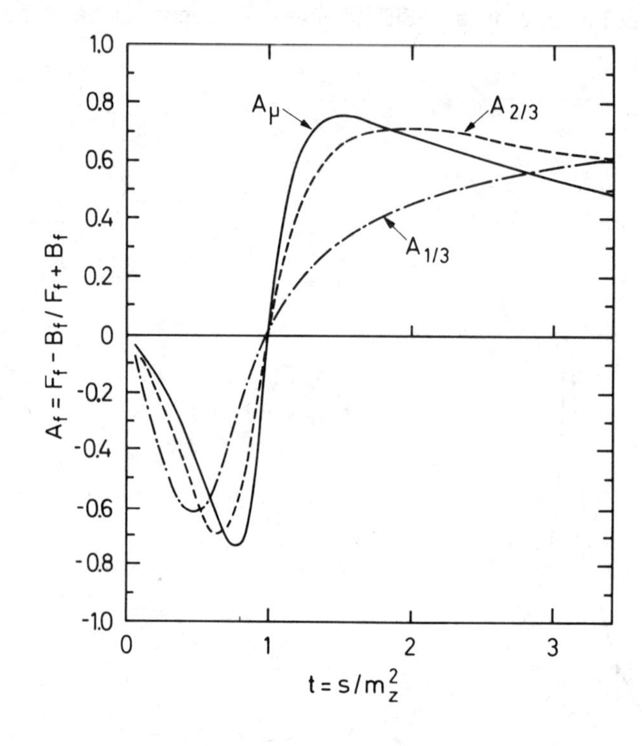

Fig. 17 Forward-Backward asymmetry expected in the standard model for various types of fermions versus s/m_Z^2 (from ref. 13).

However in order to measure coupling constants one must be able to identify the final state. For example a $c\bar{c}$ final state can presumably be identified by observing leading $D\bar{D}$ particles in the two back to back jets. However, demanding to observe leading charmed hadrons will lead to a strong reduction in rate. Such measurements will therefore only be possible with a machine optimized for this energy region. This is even more true for heavier quarks.

Toponium, the 1^{--} state made of $t\bar{t}$ quarks, is generally expected to occur below the Z^0 peak. The $t\bar{t}$-states and their various decay modes are the ideal laboratory for the study of strong interaction. For example the decays

$$T(1^{--}) \rightarrow 3\ g\ (g = \text{gluon}) \qquad \text{or}$$
$$T'(1^{--}) \rightarrow \gamma P_T(0^{++},\ 2^{++}) \rightarrow \gamma\ gg$$

yield clean gluon jets and can be used to measure the properties of the gluons, - i.e. are they coloured, flavour-neutral spin 1 particles with non-abelian coupings as in QCD.

The decay $T(1^{--}) \rightarrow H^0\gamma$ has a large branching ratio, a clean signature and a favourable ratio of signal to background.
Thus Higgs particles with masses rather close to the mass of the 1^{--} state can be found in these decays.

It might even be possible to determine the number of ν's in the world by a measurement of $T' \rightarrow \pi^+\pi^- T \rightarrow \pi^+\pi^- \nu\bar{\nu}$. This measurement however is rather difficult.

It seems rather obvious that the energy region between PETRA and PEP energies and the Z^0 mass is rich and contains information not obtainable from measurements of Z^0 decays.However, to exploit this physics a machine with the luminosity optimized for this energy region is needed.

HERA

Layout

The layout of HERA on a site ajoining DESY is shown in Fig. 18.

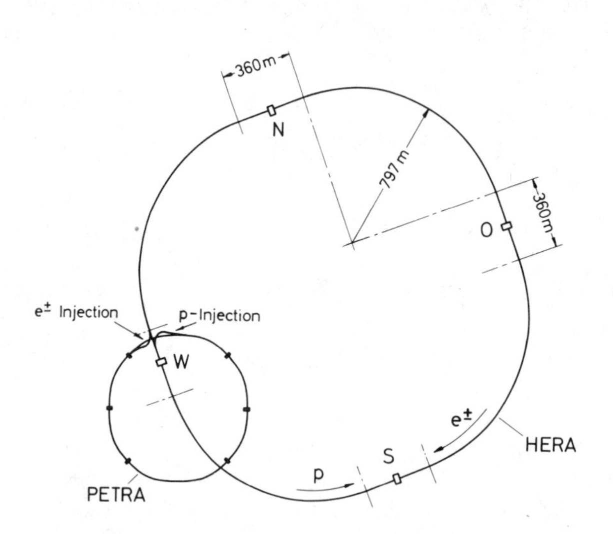

Fig. 18
Layout of HERA

The machine has a fourfold symmetry; four 360 m long straight sections are joined by four arcs with a geometric radius of 797.6 m yielding a total circumference of 6451.2 m. HERA consists of two rings, one for electrons (positrons), the other for protons and the rings cross the middle of the long straight sections. The machine and the four experimental areas will be burried some 10 - 20 m below the surface avoiding any disturbance to the surroundings. The tunnel traverses largely land belonging either to the Federal Government or to the City of Hamburg and it intersects the PETRA ring about 20 m below the surface. Thus the physical plant can be located on the present DESY site and only short injection paths are needed to connect PETRA and HERA.

The site, according to records made available by the Geologisches Landesamt in Hamburg is well suited for tunneling. The tunnel will be drilled using special boring machines equipped with driving shields. These machines, protected by the driving shield, can bore tunnels below the water table without the use of pressurized air. This method has been extensively used in Hamburg and is well adapted to the requirements posed by the HERA tunnel.

A cross section of the tunnel in the arcs with both accelerators and their utilities installed is shown in Fig. 19. To install all this in a tunnel with a diameter of only 3.2 m amenities like overhead cranes, survey monuments and air conditioning has been dispensed with.

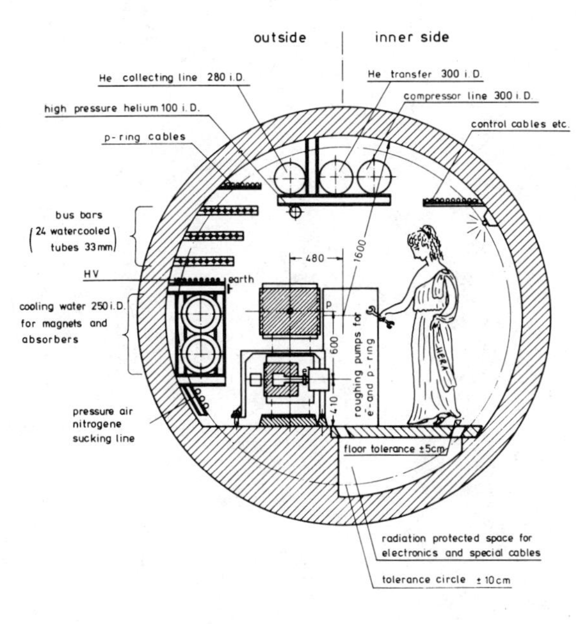

Fig. 19 Cross section of the tunnel in the arcs

The experimental halls are designed to accomodate a small, more specialized experiment, in addition to a large facility type experiment. Also machine components like klystrons, power supplies and compressors must be located in the halls. These requirements have lead to a hall 30 m long and 50 m wide with respect to the beam direction. Offices and controll rooms will be located in a multistory structure at the front face of the hall. The halls will be completely covered after construction and the only visible feature will be an access road leading to the elevator shaft which connects the halls with the surface.

Parameters

The general parameters of HERA are listed in table 2 and discussed below.

The maximum electron energy is a compromise between cost and the desire to reach high energies. We have chosen 30 GeV as the nominal upper limit. At this energy the transverse polarization builds up in 19.5 min compared to an expected beam life time of several hours. Note that the polarization time varies with energy as E^{-5}. The lower limit on the electron energy is determined by the upper limit on the electron damping time. Electrons of 5 GeV have been stored in PETRA, and if we take this to represent the upper limit on the damping time, electrons down to energies of 10 GeV can be stored in HERA.

Table 2 - Machine parameters

	p-ring		e-ring
Nominal energy (GeV)	820		30
$s = Q^2_{max}(GeV^2)$		98400	
Luminosity ($cm^{-2}sec^{-1}$)		0.35×10^{32}	
Polarisation time (min)			19.5
Number of interaction points		4	
Length of straight sections (m)		360	
Free space for experiments (m)		10	
Circumference (m)		6451.2	
Bending radius (m)	579.436		550.0395
Magnetic field (Tesla)	4.725		0.1819
Total number of particles	$6.72 \cdot 10^{13}$		$0.78 \cdot 10^{13}$
Circulating current (mA)	500		58
Energy range (GeV)	100 → 820		10 → 33
Emittance ($\varepsilon_x/\varepsilon_z$) ($\times 10^{-8}$m)	0.47/0.23		3.2/0.8
Beta function β^*_x/β^*_z (m)	7/0.6		5/0.15
Dispersion function D^*_x/D^*_z (m)	0.5		0
Beam-beam tune shift $\Delta Q_x/\Delta Q_z$	0.0007/0.0007		0.014/0.01
Beam size at crossing σ^*_x(mm)	0.18(0.97)**		0.40
Beam size at crossing σ^*_z(mm)	0.038		0.035
Number of bunches		224	
Bunch length (cm)	9.5		1.1
RF frequency (Mhz)	208.189		499.665
Maximum circumferential voltage (MV)	100***		288
Total RF power (MWatt)	2		13.2
Filling time (min)	8.5		5
Injection energy (GeV)	40.0		14.0
Energy loss / turn (MeV)			140.2
Critical energy (keV)			109

* At the interaction point

** Including the bunch length

*** 25 MW is foreseen initially

The proton energy for a given geometry is determined by the maximum field strength and the dipole packing fraction in the arcs. Several programs aimed at producing superconducting magnets with niobium-titanum coils have been underway for many years mainly in the US, but also in Europe. The maximum field of the HERA magnets is chosen to be 4.725 Tesla compared to 4.3 Tesla for the FNAL TEVATRON[14] and 5.0 Tesla, the design value for ISABELLE[15]. With a dipole packing fraction of 0.727 in the arcs this yield an maximum proton energy of 820 GeV. The lower limit on the proton energy for long term storage is determined by the effect of persistent currents in the superconducting coils. The relative importance of these currents increases with decreasing energy as they cause constant higher multipole fields to disturb the dipole guide field. In particular there is a large sextupole component which leads to a large chromaticity at injection. An estimate of these effects, based on the FNAL magnets, shows that it should nevertheless be possible to inject protons at 40 GeV and to store them down to energies of about 100 GeV. Note that the lower end of the Q^2-range available with 200 GeV protons in HERA overlaps with the top Q^2-range explored at FNAL and CERN.

In experiments[16] with a stored bunched beam in the SPS it has been found that the bunch length should be no more than about 30% of the bucket length. If this condition is violated, then RF-noise will lead to a loss of beam. We have chosen 208 MHz as the RF frequency for the proton beam. At this frequency the bunch will be stable with an RF voltage of 25 MV, a voltage which can be provided with high Q cavities at a modest power consumption. A lower frequency would be better but the hardware then becomes cumbersome and expensive.

Several estimates have shown that the cost of an RF system for an electron ring varies little for a choice of frequency between 300 MHz and 700 MHz. We have therefore chosen 500 MHz, the frequency adopted for the other DESY machines. This choice allows us to exploit fully both the expertise and the hardware available at DESY and makes it attractive to construct the RF system in stages. The final stage employs 192 cavities and a total RF power of 13.2 MW, sufficient to reach 35 GeV electron energy with 0 current.

The electron and the proton beam cross in the horizontal plane at an angle of ± 5 mrad in the middle of the four long straight sections. A horizontal crossing was chosen since the beam size in the horizontal plane is larger than the beam size in the vertical plane. A system of vertical bending magnets turns the transverse polarization of the electrons in the arcs into a longitudinal polarization in the interaction point. A free space of ± 5 m is foreseen for the experiments. This free space can be made longer at the cost of a loss in luminosity.

The luminosity of an electron-proton colliding ring is given by

$$L = \frac{f_o \cdot n_b \cdot N_e \cdot N_p}{2\pi(\sigma^2_{xp,eff} + \sigma^2_{xe})^{1/2}\,(\sigma^2_{zp} + \sigma^2_{ze})^{1/2}} \qquad (1)$$

In this formula f_o is the revolution frequency, n_b the number of bunches in each ring, N_e and N_p the number of electrons and protons per bunch respectively, $\sigma_{xp,eff} = (\sigma^2_{xp} + (\sigma_{sp} \cdot \phi)^2)^{1/2}$ with σ_{sp} denoting the proton bunch length and ϕ the crossing angle of ± 5 mrad, σ_{xe} is the width of the electron beam and σ_{zp} and σ_{ze} the heights of the proton and the electron beam respectively. The beam sizes are all defined in the interaction point and are calculated from the emittances of the two beams. In the case of the proton beam this is determined by the injectors and here we use the standard CERN value of

$$\varepsilon_x = \frac{20\pi \cdot 10^{-6}}{\beta\gamma} \text{ rad m} \quad \text{and} \quad \varepsilon_z = \frac{10\pi \cdot 10^{-6}}{\beta\gamma} \text{ rad m}$$

for the normalized emittances. The emittance of the electron beam is not a constant of motion but determined by the focussing strength of the quadrupoles and the synchrotron radiation.

One potential limit to the luminosity is the maximum values of the beam-beam tune shifts. The tune shifts, which must not exceed this maximum, can be expressed as:

$$\Delta Q_{ze} = \frac{\beta_{ze} \cdot r_e \cdot N_p}{2\pi\ \gamma_e(\sigma_{xo,eff} + \sigma_{zp}) \cdot \sigma_{zp}} \tag{2}$$

$$\Delta Q_{zp} \quad \frac{\beta_{zp} \cdot r_p \cdot N_e}{2\pi\ \gamma_p(\sigma_{xe} + \sigma_{ze}) \cdot \sigma_{ze}} \tag{3}$$

Here γ_e and γ_p are the electron and the proton energy respectively measured in units of the rest energy and β_e and β_p the values of the amplitude function at the origin. ΔQ is quoted per interaction region.

For a fixed value of the tune shift the number of particles which may be stored and hence the luminosity improves with decreasing values of β at the interaction point. The minimum value of β_e is determined by the condition that β_e must be larger than length (σ_s) of 9.5 cm. For the protons β_p is given by the distance between the quadrupoles and the interaction point and the maximum value allowed for the field in the superconducting coils of the quadrupole. It was chosen to be 60 cm. This results in a maximum value of β elsewhere in the proton ring of about 1000 m.

The luminosity for a given value of the tune shift also increases in proportion to the number of bunches. We have chosen 224 equidistant bunches for each ring corresponding to a distance of 96 nsec between adjacent bunches. With $3 \cdot 10^{11}$ protons per bunch at 820 GeV and $0.35 \cdot 10^{11}$ electrons per bunch at 30 GeV we obtain a luminosity of $3.5 \cdot 10^{31}$ $cm^{-2} sec^{-1}$. The tune shifts are $7 \cdot 10^{-4}$ and $1.4 \cdot 10^{-2}$ per crossing respectively for protons and electrons.

These values for the Q shift are not as high as the limits assumed by proponents of other machines. In fact they are not the primary limits to performance, although rather close to the tune shift at which beam-beam interaction affects luminosity. HERA's luminosity is limited rather by the currents in the two rings. The electron current is fixed by available RF power while the proton current is limited by single particle instabilities in HERA and its injectors to $3 \cdot 10^{11}$ protons per bunch or a circulating current of 500 mAmp.

The luminosity is plotted in Fig. 20 versus the proton energy for electron energies of 10, 20, 28 and 30 GeV. The curves are evaluated with the following assumptions: 13.2 MW is the available RF power for the electron beam, $\Delta Q_p < 0.0025$ (ΔQ_e always remains well below 0.03) and the proton bunch length $\sigma_s = 9.5$ cm. In the electron ring the phase advance per cell $\Delta\phi$ is adjusted in discrete steps of 30^0, 45^0, 60^0, and 90^0 to optimize the luminosity.

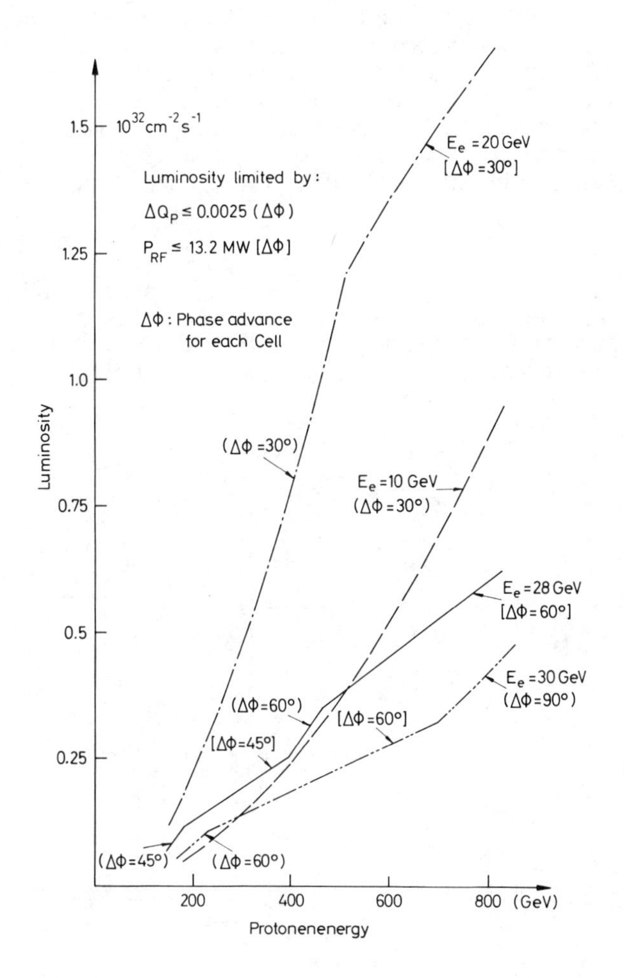

Fig. 20
Luminosity versus
proton energy

At lower proton energies each curve is limited by the tune shift and this is shown by curved parentheses around the $\Delta\phi$ parameter. At higher energies the RF power limits and this is shown by rectangular parentheses.

The injection system is capable of filling the electron ring of HERA with 224 bunches of 14 GeV electrons (positrons) with a maximum intensity of 1.3×10^{11} particles per bunch in a time which is short compared to the expected lifetime of the beams. The proposed injection system is based on Linac II, the DESY synchrotron and PETRA and leads to filling times of the order of 10 - 20 min for the peak current.

To a large extent also the proton injection scheme is based on existing accelerators. Protons from a 50 MeV linear accelerator are injected into the DESY synchrotron, accelerated to 7.5 GeV and transferred to PETRA where they are accelerated to 40 GeV and injected into HERA. The injection time is of the order of 10 min. To implement the injection scheme DESY must construct a 50 MeV linear accelerator with properties similar to the one used at CERN and equip the DESY synchrotron and PETRA for proton acceleration.

Magnet Layout

Lattice

Both rings have a periodic FODO cell structure consisting of equidistant focusing and defocusing quadrupoles which alternate in sign. The magnet structure of the standard cells is depicted in Fig. 21. As much of the interveening space as possible is filled

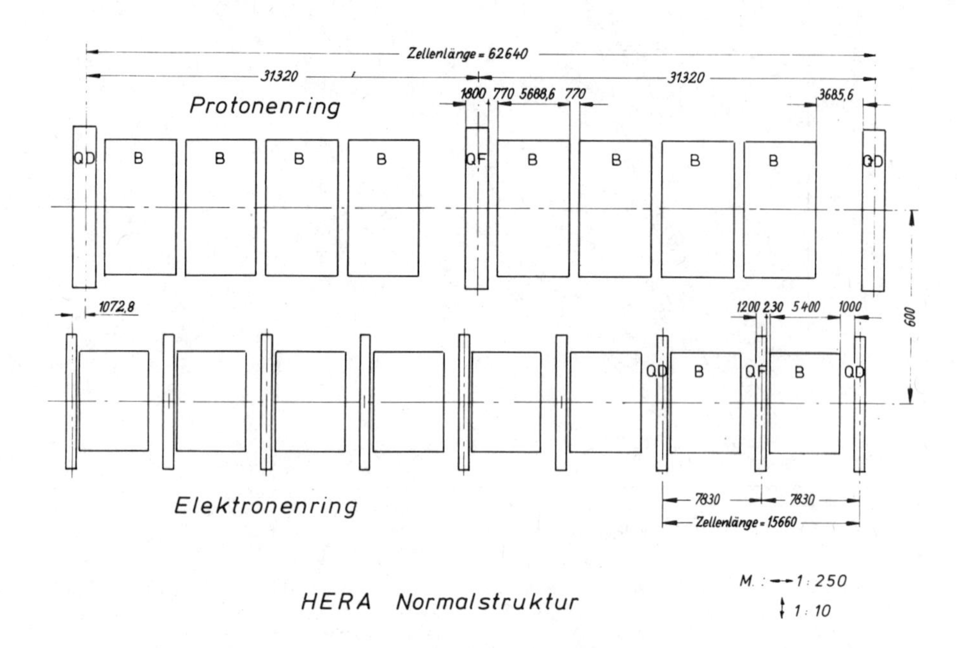

Fig. 21 - Magnet structure in the arcs.

with bending magnets in order to reach the highest proton energy and in case of the electron machine, to minimize synchrotron radiation. A short straight section placed to one side of each quadrupole provides space for sextupole magnets, orbit detecting pickups, correction dipoles and vacuum equipment together with other beam detection equipment and correction windings.

The length of the proton cell is determined by the need to make the phase advance per cell 90° while at the same time arriving at a Q value sufficiently high to minimize the effects of non-linearities. There are 20 proton cells per arc. In contrast there are 80 electron cells per arc reflecting the need to have a dense focusing structure for the electron ring whose design is dominated by synchrotron radiation effects. Such a dense focusing structure leads to a small beam emittance, low synchrotron frequency and low RF voltage. The number of electron periods is so chosen that with 45° phase advance per cell the electron beam size matches that of the protons. The lattice can also be run with stronger focusing and 90° phase advance per cell when the highest electron energy is of interest.

The electron lattice is made of 640 5.4 m long dipole magnets, 660 1.2 m long standard quadrupoles and 56 1.6 m long matching quadrupoles and a total of 604 0.30 m long sectupoles.

The proton lattice consists of 640 5.69 m long dipoles and 164 1.8 m long quadrupoles.

Interaction region

Measurements with electrons and positrons in well defined helicity states are needed to untangle electromagnetic and weak effects and to determine the properties of the charged current. Only lefthanded electrons and righthanded positrons have been found to interact via the charged weak current - i.e. all four interaction regions are designed to produce these helicity states.

The insertion is shown in Fig. 22. A side view of the rings is shown in the upper part of the figure. The rings make an S bend antisymmetric around the interaction point. The vertical bends in the first part of the straight section cause the polarization vector, which is vertical in the arcs of the machine, to precess until it is antiparallel to the beam in the interaction point. The antisymmetry of the configuration restores the vector to be vertical on reentering the arcs. Each insertion is identical producing identical helicity states in all four interaction points. The precession angle θ of the spin is related to the electron energy E and the bending angle α by $\theta = 2.273 \cdot E \cdot \alpha$ - i.e. the precession angle varies with energy. We have chosen $\alpha_o = 23.04$ mrad which precesses the spin of a 30 GeV electron by 90°.

Whereas vertical bends are necessary to produce electrons with well defined helicities, the choice of a horizontal crossing angle or head on collisions is a matter of design. Both possibilities can be realized with the present design - however we have first investigated the geometry shown in the lower part of Fig. 22 with the beams crossing at an angle of 10 mrad. The reasons to choose a finite crossing angle can be listed as follows:

1) It is possible to use a continous proton beam. This is advantagous at lower energies where it otherwise might be different to synchronize the beams.
2) The proton quadrupoles can be mounted closer to the interaction point yielding a higher luminosity for the same tune shift.

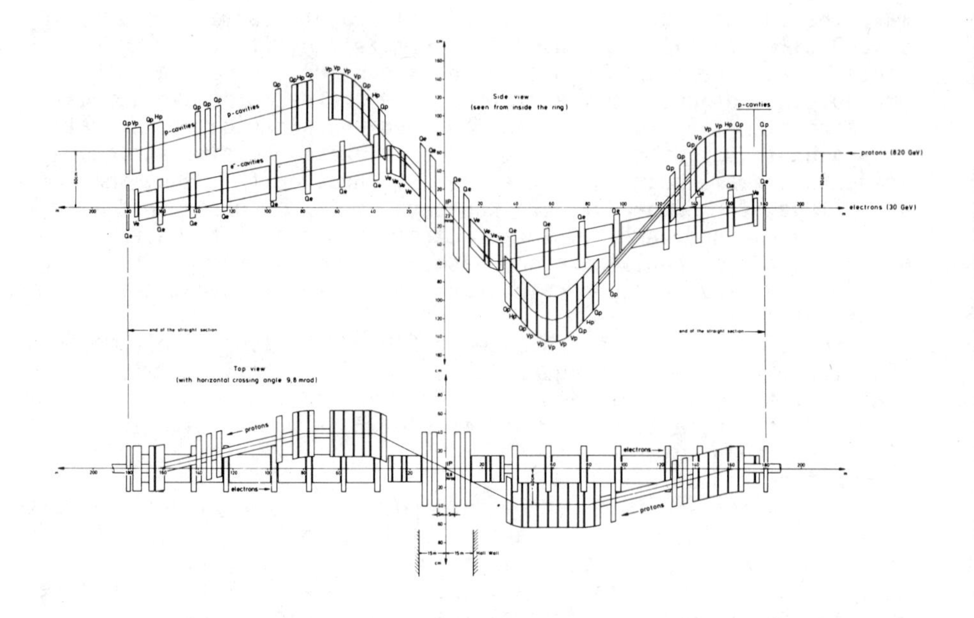

Fig. 22 - The insertion

3) The electron and the proton beams have only a few elements in common. Thus the beam energies can be chosen independently.
4) The synchrotron radiation does not hit the superconducting coils. However the allowable tune shift might be smaller[17] for a finite crossing angle.

The design shown above minimizes the amount of synchrotron radiation produced adjacent to the interaction point and maximizes the space available to install RF cavities.

Electronring

The electron ring of HERA is similar to the PETRA machine except that the circumference is larger by a factor of 2.8. Although the PETRA component in principle could be used directly, some changes, based on PETRA experience, are made to simplify the design and to reduce the cost.

The RF system is similar to the one used at PETRA except that the cavity will be designed with seven cells instead of five and a reduced hole size between adjacent cells. These design changes should increase the shunt impedance per meter by 50% to $R = 18\ M\Omega/m$ and a corresponding reduction in power W for a given voltage $U (W = U^2/2R)$.

The magnets and the vacuumsystem should be considered as a single system. The circulating electron beam is an intense source of synchrotron radiation and instead of trying to contain more than 99% of the radiation in the beam pipe we propose to use the magnets as absorbers. The beam can be contained both horizontally and vertically in a rather small vacuum chamber. However a small horizontal vacuum chamber makes it difficult to use integrated pumps at the lowest electron energies. We have therefore chosen to use discrete pumps spaced about 2 m apart. With this arrangement the pumping speed will be independent of energy. A cross section of a dipole magnet with vacuumchamber and a pump is shown in Fig. 23. The vacuumchamber is made of a 4 mm thick copper pipe.

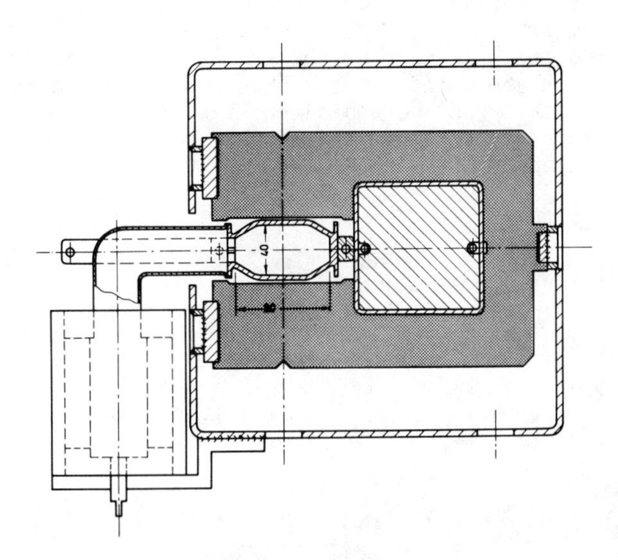

Fig. 23
Cross section of a dipole magnet for the electron ring with vacuumchamber and pump

At 30 GeV about 90% of the synchrotron intensity is contained in the beam pipe and the remaining 10% ($\sim$ 200 Watt/m) is absorbed in the magnets. During the anticipated lifetime of HERA this is equivalent to some 10^{10} rad deposited in the magnet coils. For the simple coil design shown in Fig. 23 it is very easy to insulate the coil using radiation resistant material. This would still be true if, for convenience, we would add a return conductor mounted on the front of the magnet. The coils of the quadrupolmagnets can be shielded by placing lead in front of the coils.

Proton Ring

The proton ring of HERA will be constructed from superconducting magnets to allow the acceleration and storage of protons at energies up to 820 GeV. At this time no superconducting accelerator or storage ring of this size has been completed although a rather similar machine, the FNAL TEVATRON[14] is now under construction and a 400 GeV proton storage ring complex ISABELLE[15] is being built at Brookhaven. Altogether 716 dipoles and 236 quadrupoles are needed to guide the protons around the ring and bring them into collisions at the four interaction points.

Field strengths between 4 and 5 Tesla are currently produced in magnets which are wound from a niobium titanium superconductor imbedded in a copper matrix. Both the FNAL and the BNL magnets are based on this conductor and we have made the same choice. We have also decided not to cool down the magnet yoke but rather leave it at room temperature as done in the FNAL design. A short cool down time seems to us to be persuasive in view of the large number of magnets to be tested despite strong argements which favour cold iron. Contrary to the FNAL design, however, we decided to use a warm bore, as in the ISABELLE design to avoid any excessive load on the refrigerator system from higher order mode losses. These losses are proportional to the peak current squared and the design peak current in HERA is much above that in the TEVATRON. A warm bore also allows room temperature corrective and diagnostic equipment to be installed at frequent intervalls in the ring.

A vertical cut through a dipole magnet is shown in Fig. 24.

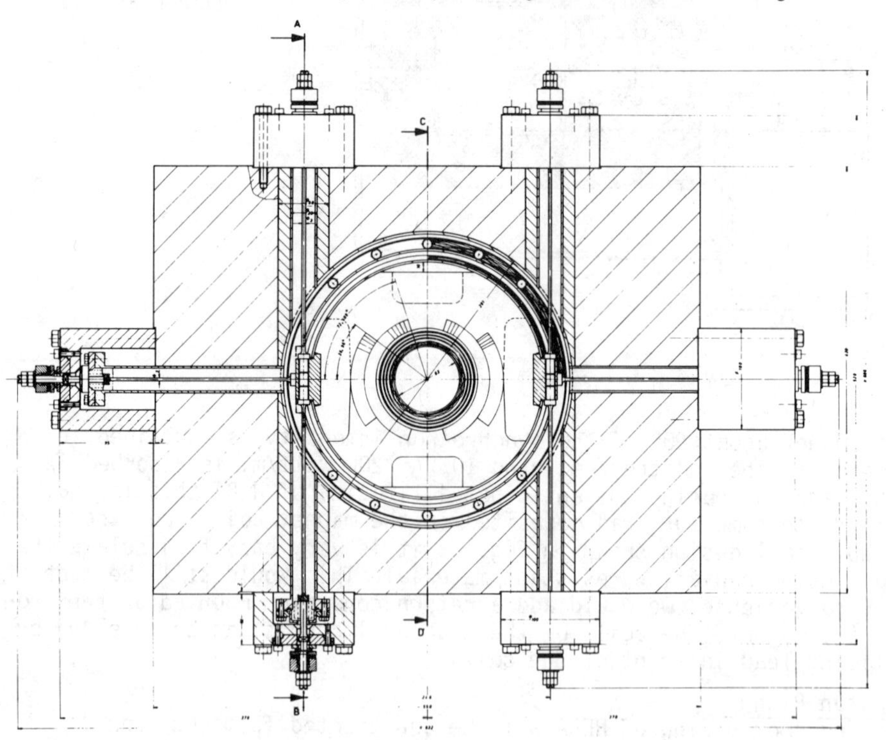

Fig. 24 - Cross section of a superconducting dipole magnet

The cryostat is mounted inside a 550 mm wide and 420 mm high iron yoke using four sets of six tie rods each. The heat loss through the tie rods is small and this system allows us to adjust the coil within the iron following magnetic measurement without warming up the magnet. In the present design the dipole field is approximated by a two shell conductor arrangement fixed by precision stamped stainless steel collars as in the FNAL design.

The coil is immersed in a liquid helium bath at 4 K. The heat shield is maintained at 50 K by passing cold Helium gas through the outer cryostat.

The parameters of the dipole and quadrupole magnets are listed in table 3.

Table 3 - Superconducting Magnet Parameters

Parameter	Dipole	Quadrupole
Magnetic length (m)	5.686	1.80
Induction (T)	4.725	-
Gradient (T/m)	-	74.4
Bore (cm)	6.3	6.3
Current (A)	6348	6348
Critical short sample current in cable at B = 5.5 T and the maximum tolerable operating temperature of t = 4.6 K	$I_{cr.s}(4.6\ K) \geq 8250\ A$	
Stored energy (kJ)	914	126
Mass (kg)	8514	934

An important parameter which is yet to be determined is the required purity of the magnetic field - i.e. the fraction of higher multipole field which might be present in the dipole field without severely restricting the operation of the machine. In the present design we assumed the values given in the FNAL design report[13].

An overview of the refrigeration system is given in Fig. 25. Rather than one central plant, the refrigeration system is subdivided into 4 units each cooling two octants consisting of 99 dipoles and 38 quadrupoles plus one experiment.The total heat load is 16.3 kW at 4 K plus 52 g liquid He/s. The load can be handeled by 3 of the 4 units. The cryogenic system can thus be maintained and repaired without interrupting operations. The compressors, located on the DESY site, are feeding compressed helium in a warm circuit in the tunnel to the refrigerators which work with expansion turbines and Joule-Thompsen effect. A string of magnets is cooled by supercritical helium (i.e. phase one) at a temperature between 3.8 K and 4.2 K and a pressure greater than 2.3 bar. Because of the limited specific heat of phase one helium, it is recooled by the returning two phase helium between every cell (four bending magnets and one quadrupole). The return stream of low pressure helium is brought back to the compressor through a circuit in the tunnel. To minimize the cross section for this return line the helium from the remote octants is first passed through auxiliary compressors.

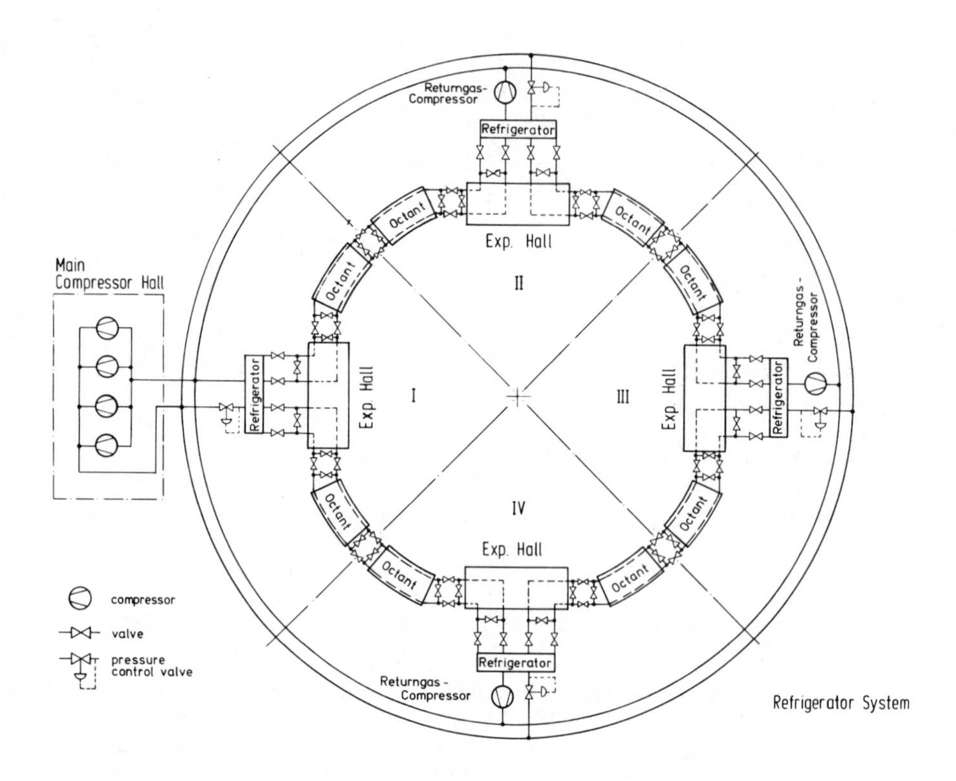

Fig. 25 - Overview of the refrigeration system

Special safety measures are needed in the superconducting system for the case that one or more magnets quench. Without these precautions all the stored energy of the magnet ring might be deposited in a single superconducting magnet which would then be destroyed. As soon as a quench is detected heating strips are fired ensuring that all the magnets in a cell quenches. The magnets will of course be so designed that they can absorb their own stored energy. At the same time the cell is short circuit passing the energy of the remaining cells around the quench region. Furthermore the liquid helium in the cell is vented into the return line and the helium flow is passed around the quenched cell to the neighbouring cells.

DESY is continuing the machine studies with the aim of submitting a detailed proposal in 1981. In parallel a vigorous effort to design and build prototype superconducting magnets is underway. This work is done in collaboration with both Saclay and industrie and is expected to yield the first dipole and quadrupole prototypes in the middle of 1982. The preparation for the civil engineering work is progressing well and the necessary construction permits might be obtained towards the end of 1981 such that construction would start early 1982. Once the project has been approved the construction work will be completed in four years such that the electron ring could be operating in the fifth year and the proton ring in the seventh year of construction. The two years between the completion of the electron ring and the turn on of the proton ring will be used to install the proton magnets and to commission the electron ring - i.e. learn how to operate a high current multibunch electron machine and how to produce polarized electrons. However, we also hope that some time will be available for e^+e^- physics using the central part of the ep-detectors.

The HERA program is very ambitious, it is based on new technologies and it tries to collide electrons and protons which has never been done before. However, the scientific rewards should be great and HERA would ensure DESY of an exciting future.

References

1. Report of the Electron Proton Working Group of ECFA
   Study of the Proton-Electron Storage Ring Project HERA.
   This study was organized by U.Amaldi with the following convenors:
   Theory: G.Altarelli, J.Ellis, G.Kramer and C.H.Llewellyn-Smith
   Machine: B.H.Wiik and E.J.N.Wilson
   Experiments: D.Dalpiaz, W.Hoogland, H.E.Montgomery, D.H.Perkins, P.Söding, K.Tittel and R.Turlay
   Superconducting magnets: G.Horlitz
2. H.G.Hereward, K.Johnsen, A.Schock and C.J.Zilverschoon
   Proc. 3rd International Conference on High-Energy Accelerators, Brookhaven, 1971 p. 265
   L.Goldzahl and E.G.Michaelis, CERN 66-12 (1966)
3. C.Pellegrini, J.Rees, B.Richter, M.Schwartz, D.Möhl and A.Sessler
   Proc. 8th International Conf.on High-Energy Accelerators
   CERN, 1971 p. 153
4. H.Gerke, H.Wiedemann, B.H.Wiik and G.Wolf
   DESY H-72/22 (1972)
   M.Tigner, DESY Techn.Notiz 2/73
5. E.Dasskowski, R.D.Kohaupt, K.Steffen, G.A.Voss
   DESY 78-02 (1978)
6. R.Chasman and G.A.Voss, IEEE Trans.Nucleus Science NS-20, No. 3 (1973) 777
   EPIC Machine Design Study Group, 9th International Conf. on High Energy Accelerators, Stanford, 1974, p. 548
   T.L.Collins, D.A.Edwards, J.Ingebretsen, D.E.Johnsen, S.Ohnuma, A.G.Ruggiero and L.C.Teng
   IEEE Trans.Nucleus Sci. NS-22, No. 3 (1971) 1411
   G.E.Fisher, D.Bleckschmidt, A.Hofman, H.Hoffmann and B.W.Montague pg 161
   B.H.Wiik pg 220 in CERN 76/12 (1976)
   CHEEP - An e p facility in the SPS
   J.Ellis, B.H.Wiik and K.Hübner (editors)
   TRISTAN PROJECT
   T.Nishikawa, in proceedings of the 9th International Conference on High Energy Accelerators, SLAC (1974) p. 584
   for an update see
   Y.Kimura et a., KEK report
7. C.H.Llewellyn-Smith and B.H.Wiik - DESY 77/38 (1977)
8. PETRA, updated version of the PETRA Proposal, DESY Hamburg 1976
9. S.L.Glashow, Nucl.Phys. 22 (1961), 579
   S.Weinberg, Phys.Rev.Lett. 19,(1967), 1267
   A.Salam, Proc. 8th National Symposium, Stockholm, Almquist and Wiksells, Stockholm 1968 p. 363
10. A.J.Buras and K.J.F.Gaemers, Nucl.Phys. B 132 (1978), 249
11. For reviews see: H.D.Politzer, Phys.Rep. 146 (1974), 129
   M.Marciano, H.Pagels, Phys.Rep. 366 (1978), 137
   J.Ellis, CERN-preprint TH 2744 to appear in the Proceedings of the 1979 Int.Symposium on Lepton and Photon Interactions, FNAL, Batavia 1979

12. For a review and references, see G.Barbiellini et al., DESY 79/67 (1979)
13. K.Winter, LEP Summer Study / 1-4 (January 1979)
14. TEVATRON - Superconducting Accelerator Design Report (1979) Fermi National Accelerator Laboratory
15. ISABELLE - a 400 x 400 GeV Proton-Proton Colliding Beam Facility - Brookhaven National Laboratory, BNL 50718 (1978)
16. D.Boussard et al., IEEE Transactions on Nuclear Science, NS-26, No. 3 (1979)
17. A.Piwinski - DESY 77/18.

PRESENT STATUS OF THE LEP PROJECT

Eberhard Keil
CERN, Geneva, Switzerland

ABSTRACT

This paper is devoted to LEP, the CERN project for an e^+e^- storage ring of about 30 km circumference, covering an energy range from 22 to 130 GeV. The first chapter contains the main considerations upon which the design and the performance estimates are based. The second chapter describes those engineering aspects in which LEP differs most from smaller machines.

1. LEP PERFORMANCE

Design concepts[1] for a large electron-positron storage ring (LEP) have been under study at CERN since early 1976. In the first study, 50 km circumference and 100 GeV per beam - to be obtained with a conventional RF system - were chosen. This study[2] was terminated with several problems still unsolved, including high sensitivity of orbit stability to closed-orbit distortions, operation in collision mode with electrostatically separated beams and technical problems due to the low magnetic field at injection. In addition, the estimated cost was considered high. A fresh start was made in the second half of 1977. In order to explore the variation of difficulties and cost with machine size and in an attempt to arrive at a solid base for an entirely feasible machine, it was decided to reduce the nominal energy to 70 GeV while retaining the target for maximum luminosity at $10^{32}cm^{-2}s^{-1}$. The optimum radius for this design - later confirmed by the outcome of the study - is 3.5 km. This phase of the study ended in August 1978 with the completion of a detailed Design Report[3] including a cost estimate. The conclusions are that such a machine is not only feasible but that it can be developed to reach 100 GeV per beam when suitable superconducting cavities become available. In fact, the design is made such that this extension of energy requires no major change other than the substitution of cavities.

Following strong encouragement from ECFA, the European Committee for Future Accelerators, a larger machine - LEP Version 8 - has been under study since the autumn of 1978. A report on this machine, called the Pink Book[4], was published last summer. This machine is composed of similar building blocks to those of the previous machine. However, further reductions in cost have been sought by improvements of details and the application of novel solutions to some components.

1.1 MAIN LEP PARAMETERS

The nominal design of LEP includes a maximum luminosity of $10^{32}cm^{-2}s^{-1}$ at about 86 GeV, to be obtained with copper RF cavities. The design of at least the magnet and vacuum systems permits

ISSN:0094-243X/80/620369-14$1.50

extension to 130 GeV by means of superconducting RF cavities. The optimum value of the circumference is about 30 km. Its exact value has been chosen so as to permit e-p collisions[5] in a bypass[6] to the CERN SPS, respecting the different radio frequencies and bunch numbers in the SPS and LEP.

The machine is almost tangential to the SPS between SPS straight sections 5 and 6 as required for the e-p bypass. The machine is situated underground. The main tunnel will be bored by methods similar to those used for the SPS tunnel and it will have the same width, 4 m. In order not to disturb the landscape outside the immediate vicinity of the eight experimental areas, it is planned to feed the input power, the primary cooling water and all controls connections through the main tunnel. General machine parameters are shown in Table 1.

TABLE 1. LEP PARAMETERS AT 86 GeV

Machine circumference			30.6	km
Number of interaction points			8	
Number of bunches			4	
Horizontal tune			70.3	
Vertical tune			74.5	
Length of regular cell			79	m
Beam-beam bremsstrahlung lifetime			8.2	h
Maximum luminosity/$10^{32}cm^{-2}s^{-1}$		1	0.5	
Horizontal ampl. fct.	β_x^*	1.6	3.2	m
Vertical ampl. fct.	β_y^*	0.1	0.2	m
Free space	ℓ_x	± 5	± 10	m
Beam-beam tune shift	ΔQ	± 0.06	0.06	

1.2. EXPERIMENTAL AREAS

Eight experimental areas are possible. However, in the initial Phase 1 of LEP construction it is foreseen to construct only four experimental halls. These interaction regions are designed for a luminosity of about $10^{32}cm^{-2}s^{-1}$ with a free space between the nearest quadrupoles of ± 5 m. Detailed studies[7] have shown that this is adequate for most foreseeable experiments. In this context, it should be noted that the solenoidal fields which often form a vital part of the experiments can be compensated by skew quadrupoles outside the central region of the insertion.

The four interaction regions which will not be equipped in Phase 1 could have a free space of ± 10 m, and half the nominal luminosity. In the present design, the two types of interaction region alternate around the circumference. Once LEP has been in operation for some time, it might be possible to modify individual interaction regions and to adapt them to the requirements of specific experiments.

In Phase 1, two experimental areas are situated in the flank of a mountain range and will be accessible via individual, roughly horizontal access tunnels, as shown in Fig. 1. The other two experimental areas will be accessible via vertical shafts. This type of underground experimental area is shown in Fig. 2. It is very similar to one of the colliding-beam halls now being constructed for the $p\bar{p}$ project at the SPS. Of the four experimental areas to be constructed later, one will be accessible by an almost horizontal tunnel, and the other three by vertical shafts.

The size of the experimental hall is determined by the experiments to be performed and the manner in which they are to be installed, operated and eventually changed. The size of the expected experiments is therefore only one of the necessary ingredients. It can be estimated - with considerable uncertainty - from the present generation of experiments at PETRA and PEP. The resulting central detector is roughly a cube with sides 10 m long. Forward detectors for certain experiments, two-photon for example, are expected to extend ± 15 m along the beams.

On the other hand, space requirements for installation and access can be rather safely predicted as they are defined more by the size of the basic components of an experiment which can be conveniently transported and handled.

It is also important that the installation and exchange of experiments in one area does not require long shutdowns of the whole machine. For LEP, where apart from two-photon experiments the detectors are relatively compact, this is most easily provided for by mounting experiments on rails perpendicular to the beam. A rapid exchange can then be carried out if an installed experiment can be rolled out to one side and a previously fully mounted experiment rolled in from the other. This push-pull mode of operation with experiments of the size discussed requires a total hall length transverse to the beams[7] of about 70 m as shown in Fig. 2.

1.3 LUMINOSITY VARIATION WITH ENERGY

The machine has been designed to reach, with full confidence, an energy at which the best present estimates predict W pair production at a good rate. In its nominal form, an energy of 86 GeV and a luminosity of $10^{32}\text{cm}^{-2}\text{s}^{-1}$ will be obtained by means of an RF system using copper cavities, the only solution which can be safely proposed at present, and an RF power of 96 MW. However, since LEP is expected to be the major European high-energy physics facility in the late 1980's, it must not only offer a high peak luminosity at a nominal energy, but also adequate luminosity over a wide range of lower energies and the potential of extension to considerably higher energies. In our study, particular attention was paid to this point. Between 22 and 86 GeV, the luminosity is proportional to E^2. This variation is obtained by a system of wiggler magnets. Once LEP is in operation, the low-energy luminosity might be increased somewhat by filling more of the aperture with beam. With an RF system using copper cavities, the LEP performance at high energies is limited by the RF power dissipation in the cavities. There is good reason to believe, however, that superconducting cavities, whose basic development is being actively pursued in several laboratories, will become operational with adequate

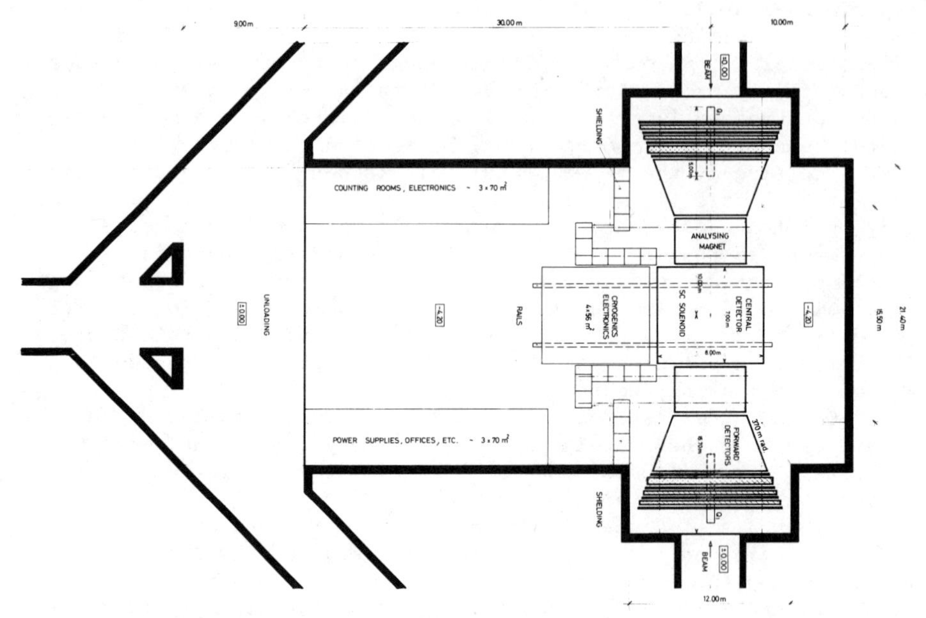

Fig. 1 : Deep underground experimental hall with two-photon experiment.

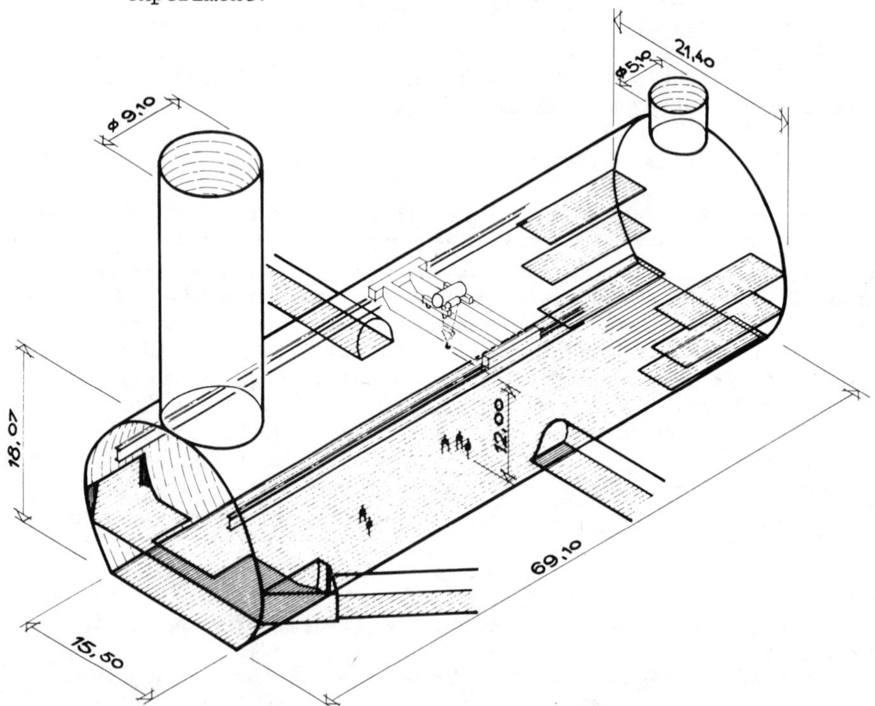

Fig. 2 : Schematic view of underground experimental area.

performance and at reasonable cost during the lifetime of LEP. Therefore, another phase using superconducting RF cavities is foreseen for energies much above 86 GeV. The circulating current is assumed to be limited by collective phenomena at injection and hence kept constant up to an energy where most of the available RF power is converted into synchrotron radiation. Above that energy, about 117 GeV, the synchrotron radiation power is kept constant. The maximum energy at full luminosity, 130 GeV, is determined by the installed RF power. The accelerating gradient in the cavities reaches 5 MV/m at this energy. The magnet and vacuum systems will be constructed so as to permit this extension to 130 GeV with only minor modifications, as soon as a suitable RF system becomes available.

In planning the LEP construction, much emphasis is put into starting colliding-beam physics at the earliest possible date, albeit at reduced energy. The most economic and practical phased construction method consists of installing a fraction of the RF system, but the complete magnet and vacuum systems. In Phase 1 of the LEP project - previously called Stage $^1/_6$ - it is proposed to construct four sets of klystron galleries, i.e. half the full complement, and to install 128 RF cavities and 16 klystrons, i.e. $^1/_6$ of the full complement. In this Phase 1, the peak luminosity of $0.4 \times 10^{32} cm^{-2} s^{-1}$ is obtained at about 50 GeV, which should be adequate for Z^0 physics. The option is left open of continuing by either installing more room-temperature cavities and klystrons until the available klystron galleries are filled, or, possibly, going to superconducting cavities at once. It should be possible to reach about 76 GeV in the first case, with $^2/_3$ of the nominal RF system installed and about 86 GeV in the second case at a voltage gradient of 3 MV/m. Reaching even higher energies would require constructing

TABLE 2 : PHASES OF LEP CONSTRUCTION

Fraction of RF installed	$^1/_6$	$^1/_3$	1	Superconducting RF
Design energy	49.4	62.3	86.1	130 GeV
Luminosity	0.385	0.616	1.07	$1.04 \times 10^{32} cm^{-2} s^{-1}$
Current	5.71	7.20	9.15	6.16 mA
RF power	16	32	96	96 mW
Length of RF	272	543	1629	1629 m
Number of five-cell room-temperature cavities	128	256	768	-

the missing four sets of klystron galleries and installing the extra klystrons and cavities. In this way, an energy of about 86 GeV at full luminosity could be achieved with the nominal complement of room-temperature cavities and 96 MW of RF power, while with superconducting cavities 130 GeV could be obtained at a voltage gradient of 5 MV/m. Work is under way now to increase all these limits by a few GeV by increasing the tunes of the machine.

The projected luminosity variation with energy in these phases of LEP construction is shown in Fig. 3. A summary of the most important parameters of these phases is given in Table 2.

1.4 COLLECTIVE PHENOMENA

The LEP performance is strongly influenced by collective phenomena, which occur both for single beams and for two colliding beams.

The single-beam collective phenomena determine the current which can be stored in LEP at the injection energy where they are most pronounced. The actual injection energy, 22 GeV, has been chosen as a compromise between the cost of the injection system and the severity of single-beam phenomena, at the currents shown in Table 2. If higher currents were desired the choice of the injection energy would have to be reassessed.

The dominant beam-beam collective phenomenon is the beam-beam limit. It describes the maximum permissible change in the tune of the machine, i.e. the number of betatron oscillations in one turn, by the focusing effect of the electromagnetic field due to one bunch on a counter-rotating bunch, for one beam-beam collision, and for particles with betatron amplitudes small compared to the beam size. Since the charge density distribution inside the beam is not uniform, but rather Gaussian, the forces are non-linear. It is their non-linearity which is believed to be at the origin of the beam-beam limit, and the linear beam-beam limit ΔQ is just a convenient measure of the strength of the non-linear forces of the beam-beam interaction. If the forces were linear, the tune-shift would be the same for particles of all amplitudes, and its effects could easily be compensated by a slight retuning of the low-β insertions.

The performance described in the last section is obtained by assuming that the linear beam-beam limit is at $\Delta Q = 0.06$. In this assumption, we follow the practice in the most recent machine designs, PEP[8], PETRA[9] and CESR[10]. This figure corresponds to the highest values experimentally observed in ADONE[11], SPEAR[12] and VEPP-2M[13]. More recent data from DCI[14] and SPEAR[15], and also from PETRA[16], have led to some doubts as to whether a tune shift $\Delta Q = 0.06$ can actually be achieved in LEP.

From the data already available, a tune shift $\Delta Q = 0.03$ is a safe lower limit. In principle, it would be possible to obtain the same luminosity at half the beam-beam tune shift just by doubling the current. But there are several serious drawbacks in doing this, such as the longer filling time for positrons, stronger collective instabilities in particular at injection, as mentioned already, and higher RF power. When discussing the consequences of a smaller value of the beam-beam limit, e.g. $\Delta Q = 0.03$, it is therefore more reasonable to keep the circulating current constant, and hence a reduction in luminosity is inevitable. However, there is sufficient flexibility in the LEP lattice and its aperture is large enough for increasing the beam size as required to keep the luminosity proportional to ΔQ. Hence, the overall result of reducing ΔQ from 0.06 to 0.03 is a loss in luminosity by about a factor of two over the whole energy range. Having established this lower bound for the luminosity, we believe that it would be premature to change the design value of ΔQ at this moment, since this change would have little consequence on the lattice design.

2. LEP ENGINEERING

This chapter of the LEP description is devoted to engineering aspects in which LEP differs most from other electron-positron storage rings.

2.1 MAGNET SYSTEM

The dipoles are designed as C-magnets. Their lengths are about 6 m for reasons of mechanical rigidity. They will be made of precisely punched steel laminations and concrete[17], the field distribution in the gap being determined by the steel profile. The unusually low field, 0.123 T at 130 GeV, permits a reduction of the steel filling factor to less than 0.3 without leading to saturation. To this end, spacers are pressed into the laminations by the punching die, so that the laminations of 1.55 mm thickness will be spaced at 5.5 mm pitch once they are stacked in a jig. Then the assembly is placed in a mould and the space between laminations is filled with a low shrinkage, corrosion-resistant mortar composed of cement and finegrain silica. Longitudinal tie rods, passing through punched holes near the outer edges of the laminations, will precompress the mortar. As the price of the mortar is very much lower than that of punched laminations the cost of these steel-concrete cores will be substantially reduced - by about a factor of two compared with a conventional core. Other advantages of the steel-concrete cores are their much-improved mechanical rigidity and their reduced weight (also by a factor of two).

So far, two full-size dipoles have been completed (Fig. 4), and the first has undergone extensive mechanical and magnetic measurements. The rigidity in torsion and flexion is comparable to that of a concrete block of the same dimensions, and improved by factors of 20 and 6, respectively, compared to a typical laminated magnet. The field uniformity is the same as for a conventional core. The small increase in the remanent field and of the excitation current at the highest energy do not affect the machine performance and have a negligible effect on the overall cost of the magnet system.

The low field required also permits excitation by simple aluminium-bar conductors instead of the usual multiturn coils.

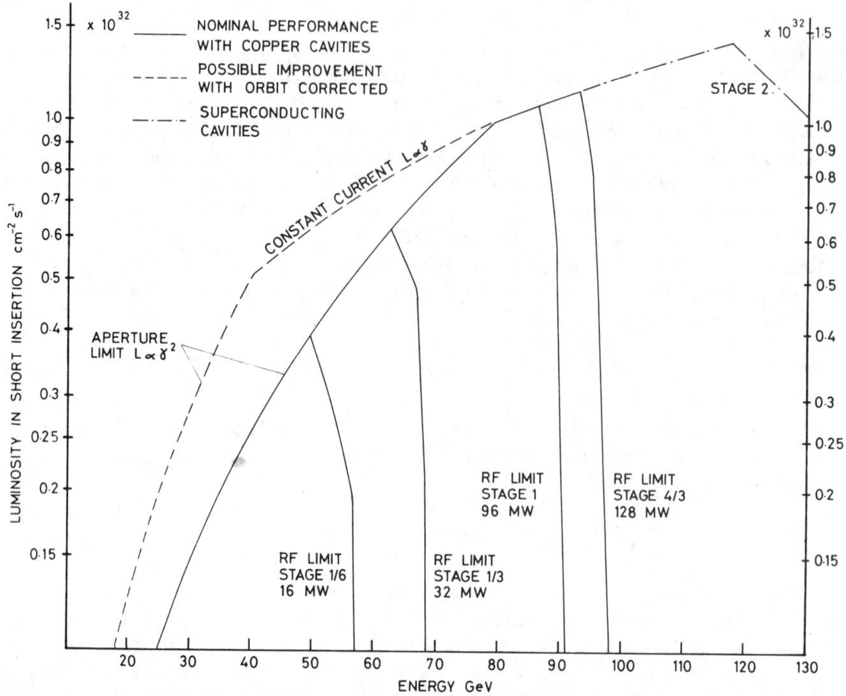

Fig. 3 : Projected luminosity variation with energy. The rising curve is determined by the aperture and the circulating current, the falling curves by the RF power.

Fig. 4 : Full-size prototype of a steel-concrete magnet with vacuum chamber.

Several cores can be placed end-to-end with little or no space lost, and excited by one set of bars. A regular half-cell contains six dipoles arranged in this way. All dipoles are connected in series.

The strength of the lattice quadrupoles is entirely governed by 130 GeV operation, requiring an increase in gradient from 4.1 T/m at 86 GeV to at least 10T/m at 130 GeV. Their cores will be of conventional construction, i.e. they will be made of densely stacked punched laminations. However, the excitation coils will be fabricated from anodized aluminium strip. Up to 80 or 90 GeV, all insertion quadrupoles can be of conventional copper-steel construction, but the strongest ones then become rather large, with a length of about 5 m and a diameter of about 1.1 m. It may be advantageous, and it will be necessary above 80 or 90 GeV, to build the strongest ones as superconducting magnets. A superconducting LEP insertion quadrupole would have half the transverse size of a copper-steel one and present much less interference with the experiments.

2.2 VACUUM SYSTEM

The linear density of synchrotron radiation hitting the dipole chamber is 1.1 kW/m at 86 GeV and 3.9 kW/m at 130 GeV. These figures are of the same magnitude as those for PEP[8] or PETRA[9]. The critical energies are 400 keV and 1.4 MeV, respectively. They are about an order of magnitude higher than in PETRA or PEP. This fact and the lower magnetic field in the dipoles pose new problems for LEP although the basic vacuum system design involves a water-cooled chamber made of extruded aluminium and distributed sputter ion pumps immersed in the dipole field, similar to those of SPEAR[18], PETRA and PEP.

Considerable thickness of lead shielding is required to prevent an excessive amount of radiation in the tunnel, where corrosive and toxic chemicals would be generated from air and humidity. The lead shield required up to 80 or 90 GeV, 8 mm at the sides of the chamber and 3 mm opposite the magnet poles, will be bonded to the aluminium chamber by a continuous process of melting and extrusion. For higher energies, additional shielding will be installed between the magnet poles. The high quantum energy occurring then will give rise to noticeable neutron production in the chamber walls and in the cooling water. However, this does not seem to present a serious problem[19].

The distributed outgassing load due to radiative and thermal desorption will be absorbed by a linear, distributed sputter-ion pump situated in the field of the main dipole magnets. The pole width, determined by requirements of field uniformity, accommodates pump cells of 50 mm diameter. The discharge in these cells can be maintained down to about 180 Gauss, well below the injection field of LEP. Cells of smaller diameter will be inserted to improve the pumping speed at higher fields. Nevertheless, in order to obtain the required low pressure, in situ bakeout (by means of electric heaters attached to the chamber) and in situ glow-discharge (by means of the pump electrodes) are foreseen. In order to arrive at an acceptable cost for the 22 km of distributed pumps the pump anodes will be made of superimposed layers of thin stainless steel strips[20] which will be fabricated in a continuous process. Fig. 5 shows a cross-section of the vacuum chamber.

2.3 RF SYSTEM

At the nominal parameters, E = 86 GeV and L = $10^{32}cm^{-2}s^{-1}$, the RF system has to make up for 1.37 GeV of energy loss per turn and 25.1 MW of power loss due to synchrotron radiation. With the parameters finally chosen a peak RF voltage of 1.95 GV is required to cover also the parasitic electromagnetic energy losses (110 MeV per turn) and to provide over-voltage for sufficient quantum lifetime.

Following well-established practice, we propose to supply this voltage and power by an accelerating structure consisting of five-cell slot-coupled cavities fed by high-power CW klystrons. There will be 768 cavities fabricated from copper and operating at 353 MHz. They occupy 1630 m of active length, the total length being divided into 16 equal stations, located on either side of all interaction areas. To this conventional system we propose to add a device that decreases the power dissipation in the cavities by a factor 1.5 by modulation. The method consists of coupling a low-loss, H-mode, storage resonator to the accelerating cavity and exciting the coupled system, with CW power sources, at both its resonant frequencies. This makes the stored energy oscillate between the two resonators, spending on average half the time in the low-loss environment. The coupling is adjusted to make peaks of the accelerating field coincide with the passage of a pair of e^+e^- bunches. One common storage resonator is sufficient for each five-cell accelerating cavity. Fig. 6 shows a conceptual design, employing a spherical storage resonator (H_{110}-mode in spherical coordinates) formed from copper sheet. With this system, a total RF generator power of 96 MW enables 86.1 GeV to be reached at full luminosity, about 6 GeV more than the same system would reach without storage cavities.

The frequency of 353 MHz has been chosen in the region of an economic optimum involving effective shunt impedance, fabrication cost of cavities and cost of RF power. The strong longitudinal focusing resulting from this high frequency leads to some beam-dynamical problems. To alleviate these, a third-harmonic RF system is foreseen.

At the present time klystrons are considered the most advantageous power sources. A power conversion efficiency of 70% has been demonstrated. It is planned to install 96 klystrons - six per RF station - of 1 MW each and to branch out power to the cavities via a chain of hybrid power dividers. Alternative power sources, such as tetrodes, are being studied, however, with the aim of achieving even higher efficiencies or a simplification of power distribution to the cavities[21].

The performance of room-temperature accelerating structures might be further improved by more intense modulation or pulsing. However, only superconducting accelerating cavities will permit the achievement of the full performance potential of LEP. An active

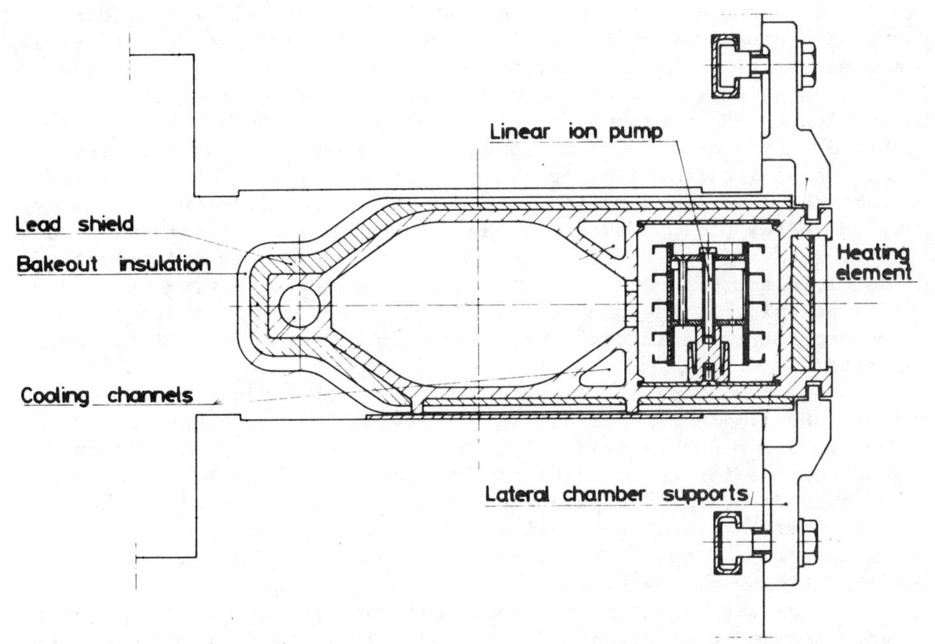

Fig. 5 : Vacuum chamber cross-section for dipole magnets showing the arrangement of cooling channels, the chamber supports and the integrated ion pump.

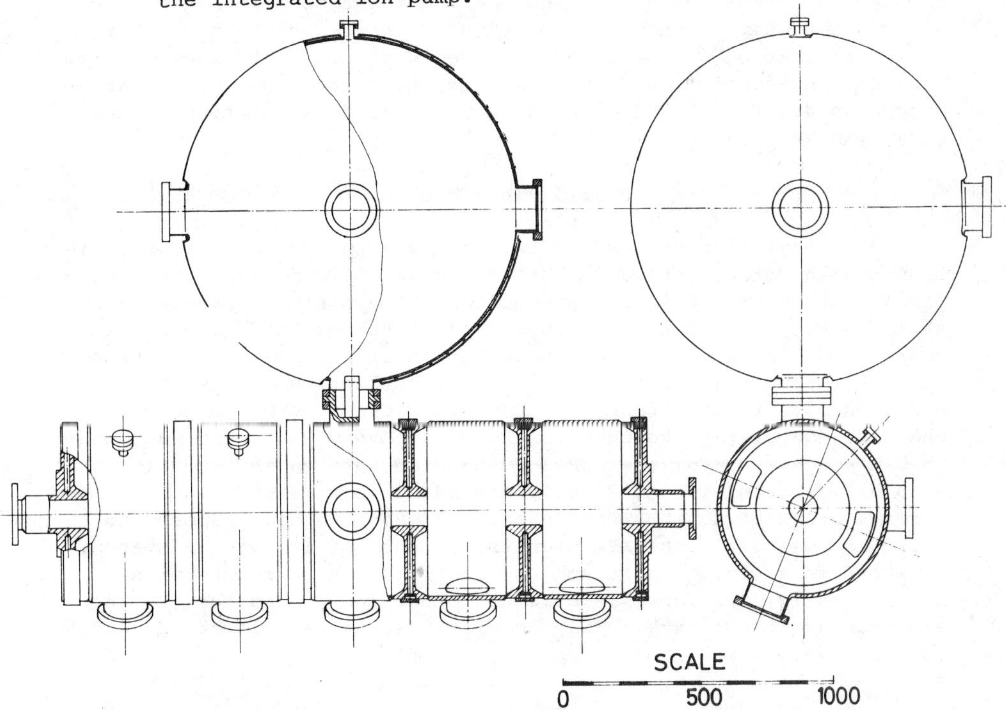

Fig. 6 : RF cavity with spherical storage cavity (scale in millimeters).

development programme is under way and no fundamental limitations known at present appear to preclude accelerating fields four to five times above the values currently used in room-temperature cavities for storage rings. LEP has been designed so that a progressive conversion to RF superconductivity is possible whenever the new technology is ready. We anticipate that superconducting cavities can be economically built for the same frequency as copper cavities so that the power sources can be retained in a conversion. A maximum energy of 130 GeV at $10^{32}cm^{-2}s^{-1}$ nominal luminosity will be reached when all the RF power is converted to beam power (including parasitic electromagnetic losses). If the superconducting structure achieving this has the same active length, 1630 m, as foreseen for room-temperature cavities, an accelerating field of 5 MV/m will be required.

Superconducting RF cavities are being developed at Karlsruhe[22] and at CERN[23]. In both cases a frequency of 500 MHz compatible with large storage rings is used. The aim of the work at Karlsruhe is a single-cell cavity to be installed and tested with realistic beam conditions in DORIS. The shape of the cavity is essentially cylindrical. An accelerating field of 3 MV/m was achieved in a continuous run of one week. Very little electron loading and no deterioration were observed.

In the CERN work, nearly spherical cavities are used because of the absence of electron multipactoring indicated by measurements of the Genoa group[24] at 4.5 GHz, and by the CERN group at 3 GHz, and predicted by computer simulation at Wuppertal[25]. In a single-cell cavity manufactured from niobium sheet metal by a spinning method and treated only by chemical polishing without a high-temperature annealing in a UHV furnace, an unloaded Q of 2.1×10^9 and a maximum accelerating field of 4.6 MV/m were achieved at an operating temperature of 4.2 K. This limitation was due to local heating of the cavity, shown in Fig. 7.

2.4 INJECTION

The injection system for LEP starts with a high-current electron gun injecting into a 200 MeV linac which is followed by a conversion target for positron production. The positrons are collected at a few MeV, accelerated in a positron linac of 600 MeV nominal energy, and injected into an accumulation ring (ACR). This system runs at a repetition rate of 100 Hz. The electron gun is pulsed such that it produces four electron bunch trains corresponding to the four equidistant bunches in the ACR. Every 2.5 s the four bunches are ejected from the ACR and injected into an injector synchrotron ISY, accelerated to 22 GeV, and added to the four bunches already circulating in LEP. In the Pink Book we had proposed to build ISY using all the ISR magnets and much of its vacuum system, arranged in a single racetrack machine of 1.7 km circumference. More recently, the idea has come up to use the PS and the SPS as electron synchrotrons[26]. This looks attractive and is being studied very actively now.

Fig. 7 : 500 MHz superconducting cavity developed at CERN.

Acknowledgements

The LEP description sums up the work of a Study Group based at CERN with wide outside-CERN participation. A list of contributors is given in the Pink Book[4].

References

1. R. Billinge et al., IEEE Trans. Nucl. Sci. NS-24, No. 3, 1857 (1977).
2. J.R.J. Bennett et al., CERN 77-14 (1977).
3. The LEP Study Group, CERN Int. rep. ISR-LEP/78-17 (1978).
4. The LEP Study Group, CERN Int. rep. ISR-LEP/79-33 (1979).
5. A. Hutton, IEEE Trans. Nucl. Sci. NS-26, 3520 (1979).
6. Y. Baconnier and W. Scandale, CERN Int. rep. SPS/AOP/DV/Note/78-23 (1978).
7. Proc. LEP Summer Study, CERN 79-01 (1979).
8. PEP Proposal, LBL-2688/SLAC-171 (1974). PEP Conceptual Design Report, LBL-4288/SLAC-189 (1976).
9. PETRA proposal (DESY, Hamburg, 1974). Updated version of the PETRA proposal (DESY, Hamburg, 1976).
10. B.D. McDaniel, Electron-Positron Colliding Beam Facility at Cornell, Cornell Univ. (1975).
11. F. Amman et al., Proc. 8th Internat. Conf. on High Energy Accelerators, Geneva, 1971 (CERN, Geneva 1971), p. 132.
12. SPEAR Group, IEEE Trans. Nucl. Sci. NS-20, 838 (1973).
13. I.B. Vasserman et al., Proc. 5th All-Union Conf. on Particle Accelerators, Dubna, 1976 (Izdatelstvo Nauka, Moscow, 1977), Vol. 1, p. 252 (in Russian), Eng. trans. as SLAC TRANS-177 (1977).
14. The Orsay Storage Ring Group, LAL/RT/79-01 (1979).
15. M. Cornacchia, PEP Note 275 (1978).
16. A. Piwinski, IEEE Trans. Nucl. Sci. NS-26, 4268 (1979).
17. J.-P. Gourber and L. Resegotti, IEEE Trans. Nucl. Sci. NS-26, 3185 (1979).
18. U. Cummings et al., SLAC-PUB 797 (1970).
19. W.R. Nelson and J.W.N. Tuyn, CERN Int. Rep. HS-RP/037 (1979).
20. J.-M. Laurent and O. Gröbner, IEEE Trans. Nucl. Sci. NS 26, 3997 (1979).
21. C. Zettler, CERN Int. Rep. SPS/ARF/79-12 (1979).
22. W. Bauer and J. Halbritter, Primärbericht 08.02.03p01J, Kernforschungszentrum Karlsruhe (1980).
23. P. Bernard, H. Heinrichs, H. Lengeler, E. Picasso and H. Piel, private communication (LEP Note 209, 1980).
24. V. Lagomarsino et al., IEEE Trans. Mag. 15 (1979).
25. U. Klein and D. Proch, Gesamthochschule Wuppertal, Int. Rep. WU 13-78-31 (1978).
26. Y. Baconnier, O. Gröbner and K. Hübner, private communication (LEP Note 212, 1980).

SOME TOPICS IN e^+e^- INSTRUMENTATION

David Hitlin
California Institute of Technology
Pasadena, California 91125

ABSTRACT

Progress in charged particle tracking, electromagnetic shower detection and particle identification in current e^+e^- colliding beam detectors is reviewed.

INTRODUCTION

The last few years have seen the development of a large number of general purpose e^+e^- colliding beam detectors for use at DCI, SPEAR, DORIS, CESR, PEP and PETRA. Many of these devices are now working; others will be installed within the next year. This, then, would seem to be an appropriate time to review some of the approaches taken to particular problems in these types of detectors in the context of the demands of the physics and to evaluate their success. It must be remembered that the center-of-mass energy spanned by these detectors covers more than an order of magnitude, and that different event characteristics and different physics demands require a different emphasis within the achievable performance of particular devices. The areas examined will be charged particle tracking, electromagnetic shower detection and particle identification. Some discussion of new ideas under development in each of these areas will be included.

CHARGED PARTICLE TRACKING

With the exception of the Time Projection Chamber for PEP, tracking is accomplished in this generation of detectors by large cylindrical drift chambers. Designs consist either of concentric planes of wires, (Mark II or Mark III), concentric low mass cylinders (TASSO, CELLO) or individual pie-shaped modules (JADE). Concentric wire planes provide the lowest mass design where multiple Coulomb scattering is important, but have the disadvantage that a minimum distance must be maintained between planes so that electric field configurations within a drift cell are not disturbed. Determination of position in a direction parallel to the wire (z position) is done by small angle stereo with a precision equal to the resolution divided by the sine of the stereo angle or by charge division. A further disadvantage of this technique is that a broken sense wire can disable many planes.

When low mass cylinders are used to separate drift chamber planes, the z position may be determined by measurement of the induced charge on cathode strips. This technique provides good resolution, and, if crossed cathode strips are used, resolves multitrack ambiguities.

ISSN:0094-243X/80/620383-18$1.50

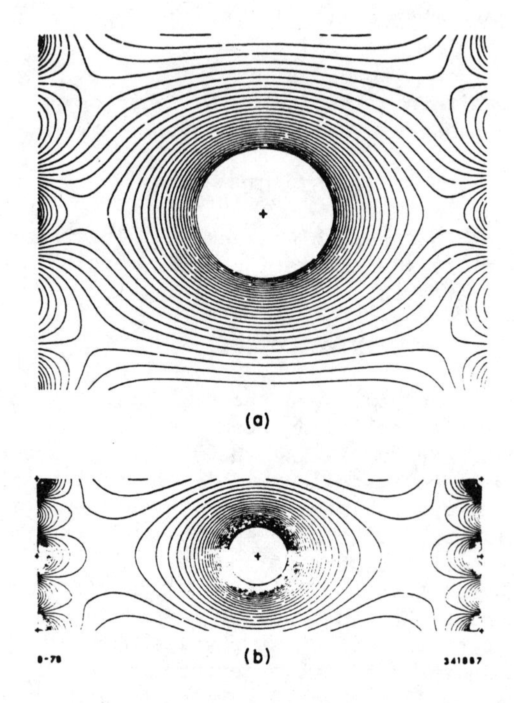
(a)
(b)
8-79
341887

Fig. 1

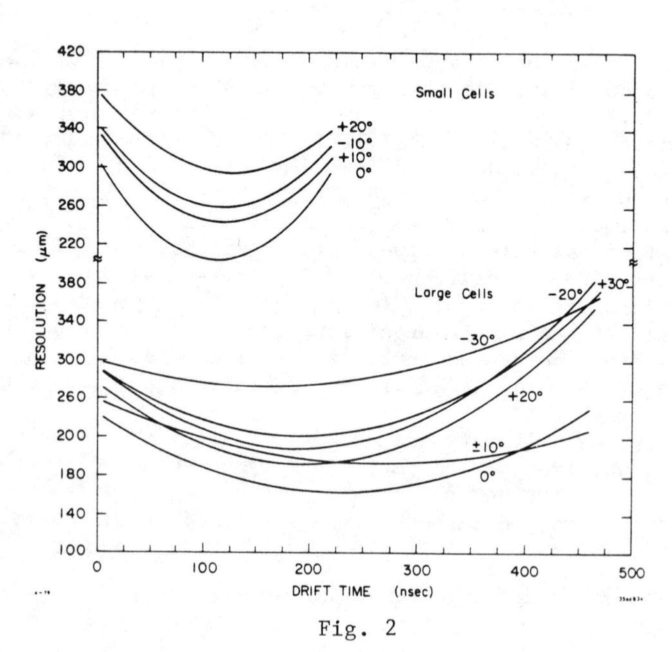
Small Cells
+20°
-10°
+10°
0°
Large Cells
-20°
+30°
-30°
+20°
±10°
0°
RESOLUTION (μm)
DRIFT TIME (nsec)
420
380
340
300
260
220
380
340
300
260
200
180
140
100
0
100
200
300
400
500

Fig. 2

Many detectors employ the "industry standard" cell configuration originally developed for the Mark II. This is shown in Figure 1. This design, while simple, does not have a particularly linear time-distance relation. Nonetheless, it is possible to achieve resolution of better than 200 μm (σ) with this cell. Figure 2 shows the spatial resolution as a function of angle of incidence for two different sizes of cell in the Mark II drift chamber.[1]

Other detectors have employed a more elaborate cell structure. As an example, Figure 3 shows the wire arrangement of the Mark III drift chamber. This design, which requires multi-hit electronics, has several advantages over the industry standard: the time-distance relation is more linear, the guard wires above and below the sense wires provide some degree of isolation so that ionization produced outside the cell does not reach the sense wire, and the left-right track ambiguity is resolved within each cell. Figure 4 shows the deviation from a linear time-distance relation in this cell for several different angles of inclination of the track. It will be seen that for angles less than 30°, the deviations are small and easily calibrated. Figure 5 demonstrates the resolution of the left-right ambiguity. The plot shows the difference between the average of the times for the first and third sense wire and the second, central wire. While no correction has been made in this plot for the proper time-distance function, the two peaks, corresponding to tracks originating in the left and right halves of the cell, are clearly resolved. Since the separation of the wires is accurately known, the time difference between the two peaks gives a measure of the average drift velocity in each cell on actual data. Thus changes in gas composition or purity can be followed and the determination of chamber constants is greatly simplified. Another feature of this design is that the direction of track segments within a cell is determined to about 10 mrad. This provides an advantage in pattern recognition, in that track segments can be projected into adjoining layers using this measured, rather than a calculated, direction.[2]

The motivation for obtaining high spatial resolution in tracking chambers is, of course the attainment of good charged particle momentum resolution. Typical detectors have achieved a precision corresponding to 1 to 2 % x p (in GeV/c) momentum resolution. The contribution of multiple Coulomb scattering which adds a term independent of momentum to the resolution, can be quite important, even at PEP/PETRA energies. Figure 6 shows the contribution to momentum resolution of measurement error and multiple scattering for regimes typical of current detectors. Note that since even for $\sqrt{s} = 30$, 40% of charged particles have momentum less than 1 GeV/c, momentum resolution can be multiple scattering limited for a substantial portion of the spectrum. At SPEAR/DORIS energies, multiple scattering is an even more important factor, and reduction of the amount of material in the path of charged particles can result in real improvement in resolution. Even within the realm of conventional detectors, careful attention to this point can achieve substantial reduction in material. For example, the Mark III detector has only 1/4 the number of radiation lengths in the charged particle path, compared to the Mark II. This is done by use of a berylium beam pipe, a drift chamber, rather than a scintillator for triggering and a thin main drift chamber entrance window.

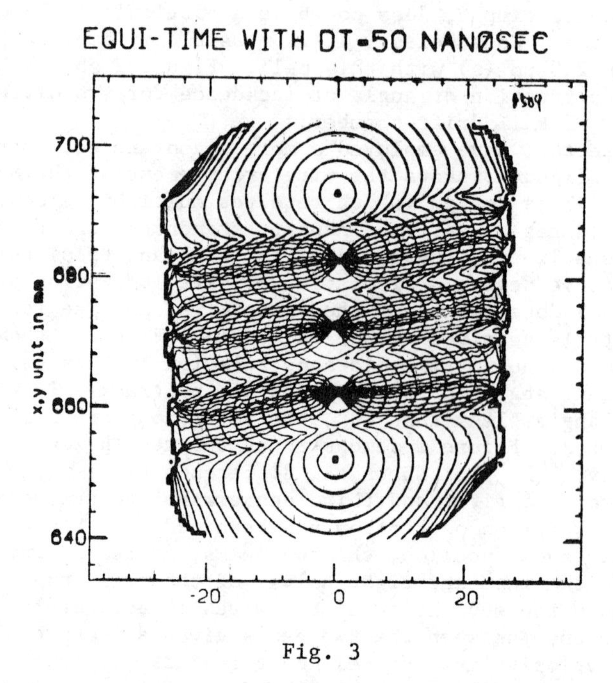
EQUI-TIME WITH DT=50 NANØSEC
x,y unit in mm
700
680
660
640
-20
0
20

Fig. 3

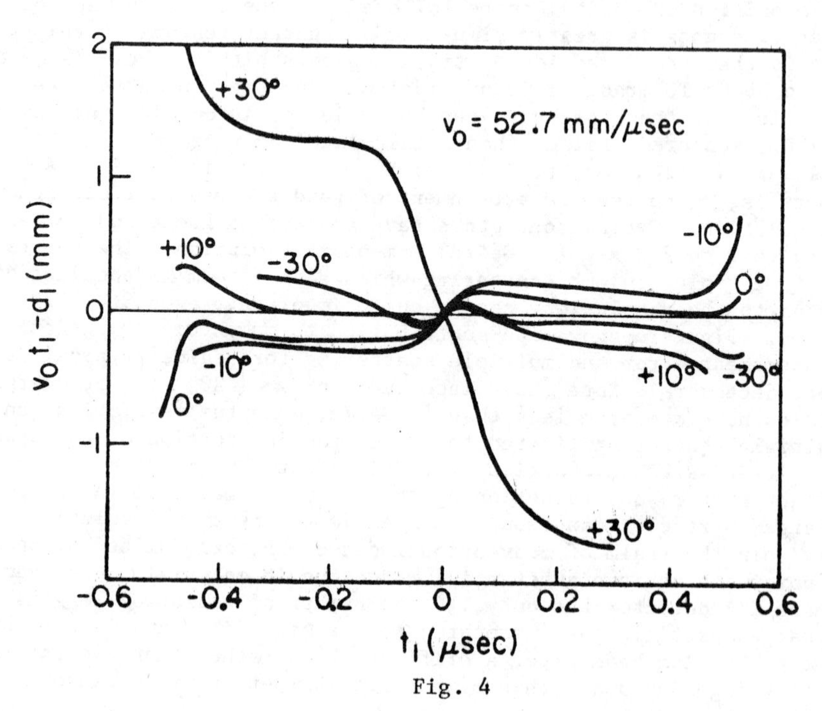
+30°
v_0 = 52.7 mm/μsec
+10°
-30°
-10°
0°
-10°
0°
+10°
-30°
+30°
$v_0 t_1 - d_1$ (mm)
t_1 (μsec)
2
1
0
-1
-0.6
-0.4
-0.2
0
0.2
0.4
0.6

Fig. 4

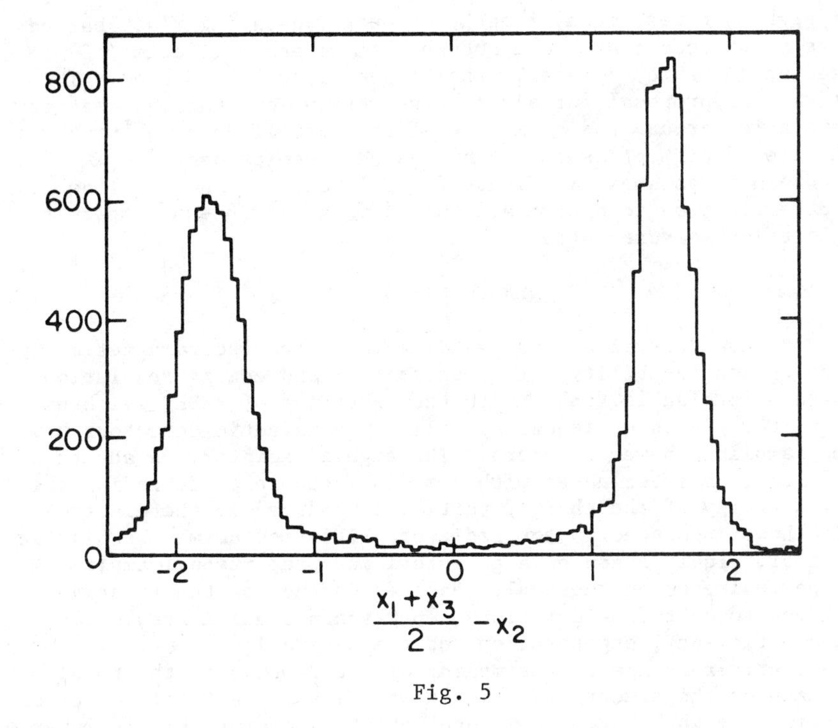

800
600
400
200
0
-2
-1
0
1
2
$\frac{x_1 + x_3}{2} - x_2$

Fig. 5

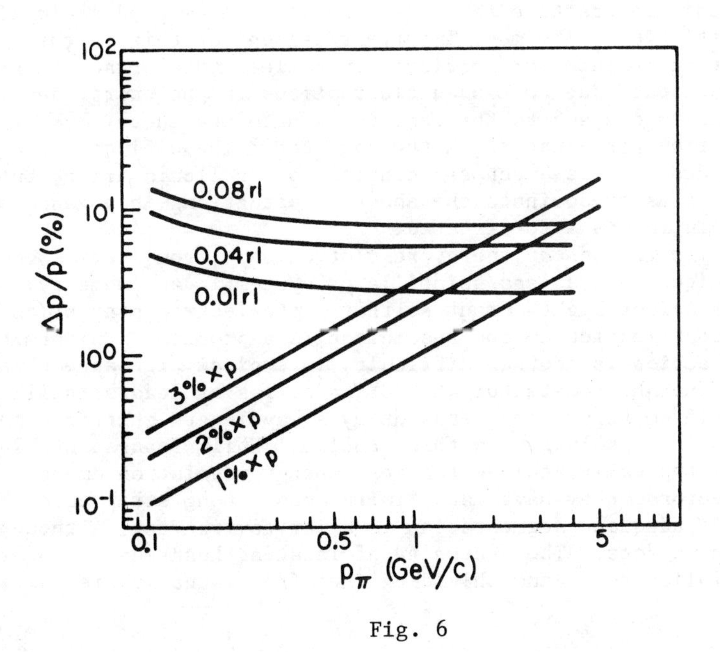

10^2
10^1
10^0
10^{-1}
$\Delta p/p$ (%)
0.08 rl
0.04 rl
0.01 rl
3% × p
2% × p
1% × p
0.1
0.5
1
5
p_π (GeV/c)

Fig. 6

At high energies, decay lengths of weakly decaying τ's, charmed particles and B mesons are long enough (50 μm for a lifetime of 5×10^{-13} sec at 15 GeV/c) to enable them to be directly measured. This has motivated a proposal for a very high resolution tracking chamber to be installed around the beam pipe of the Mark II detector at PEP.[3] With the use of high pressure organic gases, resolutions of ~40 μm can be obtained, as shown in Figure 7. This chamber should allow vertex determination to a precision of <100 μm, which should permit direct lifetime measurements.

SHOWER COUNTERS

Providing a general purpose 4π detector with electromagnetic shower detection capability with good spatial and energy resolution has proved a formidable task. With the exception of JADE, which uses more than 3000 pieces of leadglass,[4] all modern magnetic detectors have employed sampling shower counters. The high z radiator, in sheets of ~ .2- .5 r.l., is interleaved with a medium capable of detecting the ionization energy of the charged particles produced in the electromagnetic shower produced in the radiator. This medium may be plastic scintillator, liquid argon or a gas providing proportional amplification. The choice among these alternatives depends on the relative weight given to cost considerations, energy and spatial resolution, detection efficiency, segmentation and π/e separation.

The energy response of a sampling device depends on the total track length of the electromagnetic shower, i.e., the total number of gap crossings by the charged component of the shower. This is inversely proportional to the thickness, t, of a single radiator plate, and proportional to the incident photon energy E. Thus the energy resolution due to shower statistics improves as $\sqrt{t}$ and as $1/\sqrt{E}$. It has proven difficult in practice to achieve resolution as good as implied by shower statistics. The most important reason for this is the difficulty in maintaining proportionality between total track length and detector output, due to Landau fluctuations in the energy deposition by shower tracks and to the contribution to the energy sum of knock-on electrons originating in the radiator.[5] These fluctuations are small in dense media, such as scintillator or liquid argon, but may be so large as to dominate the shower fluctuations in a gaseous medium. Resolution is also affected.

Further limitations on energy resolution come from areas specific to each technique. In a lead/scintillator system in 4π geometry, achieving sufficient light output so that photoelectron statistics is not a major contribution to the resolution is a problem.[6] Uniformity of light collection is another difficulty in typical high aspect ratio designs. In the ARGUS detector at DESY, a series of lead/scintillator towers, 15 x 15 cm across and read out by a wavelength shifter along one side, provide a solution to this problem.[7] While lead/scintillator detectors can typically achieve the best energy resolution among sampling detectors, they have many limitations. Long strip counters do not provide adequate segmentation in a jet environment, although the tower scheme does. The necessity of locating hundreds or thousands of photomultiplier tubes and shielding them from magnetic fields is a

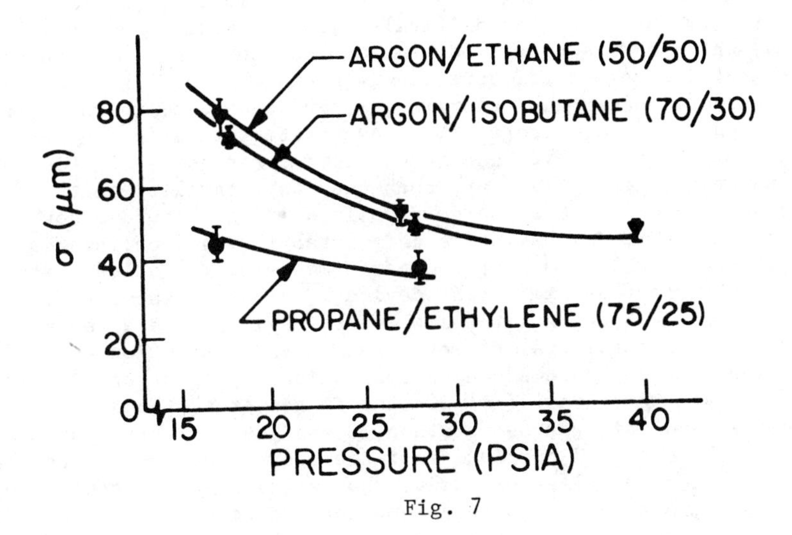
ARGON/ETHANE (50/50)
ARGON/ISOBUTANE (70/30)
PROPANE/ETHYLENE (75/25)
σ (μm)
80
60
40
20
0
15
20
25
30
35
40
PRESSURE (PSIA)

Fig. 7

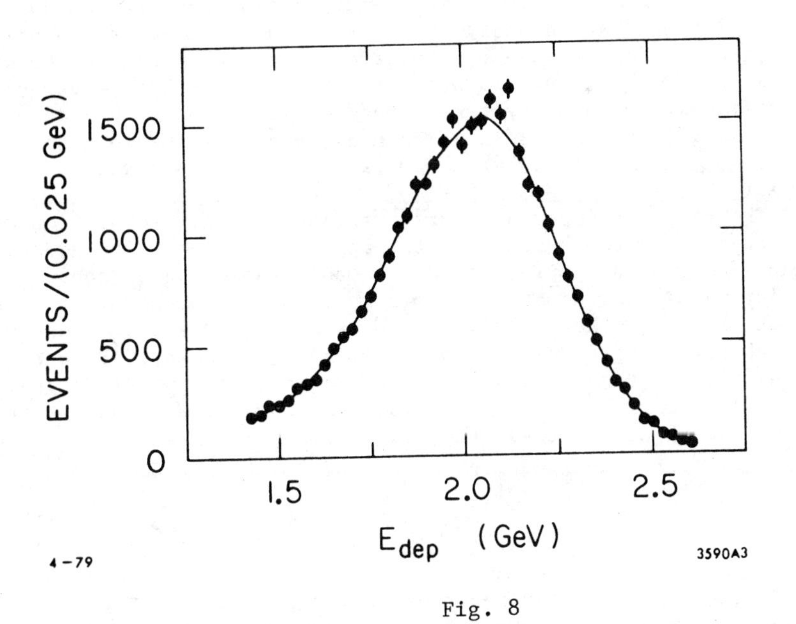
EVENTS/(0.025 GeV)
1500
1000
500
0
1.5
2.0
2.5
Edep (GeV)
4-79
3590A3

Fig. 8

problem, as is providing a calibration and monitoring system. A severe limitation is the difficulty in achieving good spatial resolution, which can be very important in some instances, and in having longitudinal segmentation, which is necessary for π/e discrimination.

Lead/liquid argon calorimeters have been used in the Mark II,[8] CELLO and TASSO detectors. In these systems a series of strip and/or tower electrodes, which can be configured for good longitudinal and transverse segmentation, are connected as a parallel plate ionization chamber to collect the charge deposited in the liquid argon by shower tracks. The medium having no gain, stability and calibration are relatively simple, so that the performance of installed systems closely approaches that of small test devices. Figure 8 shows a Mark II Bhabha peak. Note that the low energy asymmetry is a radiative tail and not an instrumental effect. Low energy resolution and efficiency of liquid argon systems can be compromised by electronic noise, since signals are very small ($\sim$ 1/10 that in gas proportional chamber systems). Spatial resolution of better than 10 mrad can be obtained in such systems. The cost and complexity of these systems is a limitation, for although they are less expensive than a lead glass installation, they are more expensive than other sampling techniques.

Gas sampling detectors are employed by MAC, CLEO and Mark III. The energy resolution of these devices is a good deal worse than that implied by sampling statistics. For example, for .5 r.l. sampling, used in MAC and Mark III, sampling statistics would yield a resolution of $\sigma \sim 12\%/\sqrt{E}$, whereas $18\%/\sqrt{E}$ and $16\%/\sqrt{E}$ have been achieved, respectively. The CLEO detector uses $\sim$.2 r.l. sampling but has a resolution similar to the others, since energy loss fluctuations dominate the resolution. An advantage of this technique is that z position determination can be done by charge division, eliminating the need for ambiguity resolving coordinates. Calibration and gain monitoring are difficult in these systems. Good segmentation with concomitant improvement in spatial resolution and π/e discrimination can be obtained, at the expense of a large number of electronic channels (12K for the Mark III).

The relative importance of energy and spatial resolution must be decided upon in the context of the physics problem to be addressed. Figure 9 shows the dependence of π° mass resolution on energy and angular resolution separately as a function of momentum. The resolution of a given device will be the sum in quadrature of one member of each family of curves. Another example is shown in Figure 10, which compares the χ state mass spectra obtained by the Crystal Ball NaI (Tℓ) detector with the Mark II liquid argon device. The spectra are the result of constrained kenematic fits to the reaction

$$\psi' \to \gamma\, \chi$$
$$\hookrightarrow \gamma\, \psi$$
$$\hookrightarrow \ell^{+}\ell^{-}\,.$$

The Crystal Ball has far superior energy resolution, the Mark II superior angular resolution, yet the mass resolutions obtained are comparable. This result is, of course, particular to the χ state regime and the comparison will be different for the states intermediate to the Υ and Υ'.

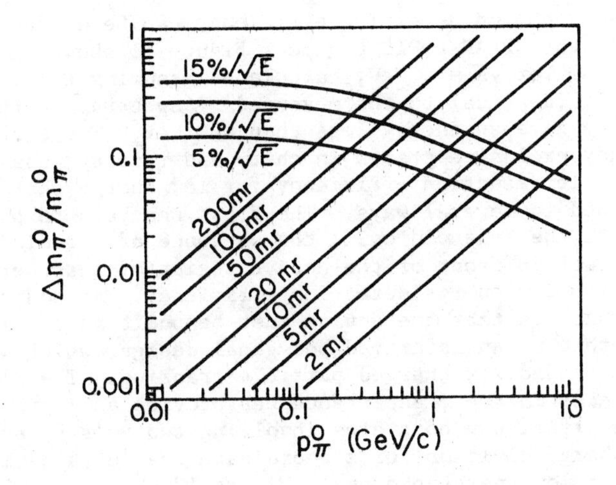
15%/√E
10%/√E
5%/√E
200mr
100mr
50mr
20 mr
10 mr
5mr
2 mr
Δm$_\pi^0$/m$_\pi^0$
p$_\pi^0$ (GeV/c)
1
0.1
0.01
0.001
0.01
0.1
1
10

Fig. 9

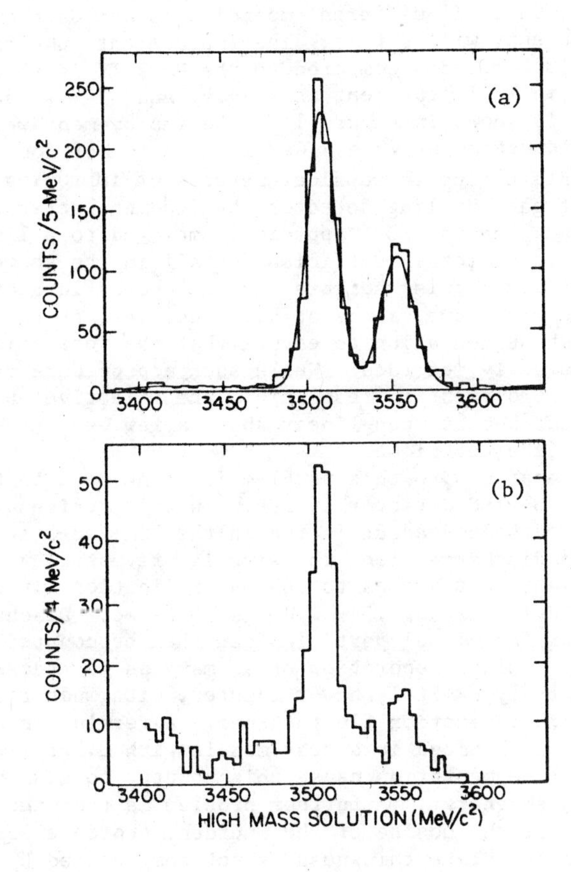
(a)
COUNTS/5 MeV/c^2
250
200
150
100
50
0
3400
3450
3500
3550
3600
(b)
COUNTS/4 MeV/c^2
50
40
30
20
10
0
3400
3450
3500
3550
3600
HIGH MASS SOLUTION (MeV/c^2)

Fig. 10

Low energy photon detection efficiency can be a crucial parameter, especially in the SPEAR/DORIS regime. Figure 11 shows an integral photon spectrum for $\sqrt{s}$ = 4 GeV, calculated assuming the inclusive π° spectrum to be identical to the measured $\pi^\pm$ spectra. Note that 90% of the photons have an energy less than 400 MeV. The Mark III detector, built to study exclusive states in charm and τ decay, has aimed at the highest possible detection efficiency for low energy photons. This has been approached in several ways. The first has been to place the shower counter inside the solenoid coil; the presence of a radiation length or more of material in front of the active region of a shower counter severely limits low energy detection efficiency. An additional benefit of this approach is that the device must be built as a cylindrical annulus rather than as an inscribed polygonal design, which wastes precious radial space needed for charged particle tracking. The resulting elimination of cracks in the ϕ acceptance can provide an improvement in reconstruction efficiency on states involving two π°'s by as much as 30%. The use of charge division for z coordinate readout also improves the efficiency for low energy photons. Figure 12 shows an EGS Monte Carlo simulation of seven 100 MeV photon showers in the twenty four .5 r.l. segments of the Mark III counter. The disjoint character of the track segments is clear. If different coordinates are determined at different depths, efficiency will suffer. The improvement in efficiency of the Mark III at low energies compared to the Mark II is shown in Figure 13. This results in an improvement in $\pi^\circ \to \gamma\gamma$ and $\eta^\circ \to \gamma\gamma$ detection efficiency which is shown in Figure 14. The improvement weighted by the inclusive π° spectrum at $\sqrt{s}$ = 4 GeV is a factor of four.

It is interesting to consider methods of improving the energy resolution of gas sampling detectors by finding better ways to measure the total track length. One approach, employed to a limited extent in the Mark III is to retain sufficient detail in the shower to allow truncation of large pulse heights due to Landau fluctuations. This is a delicate procedure, since at high energies, it becomes difficult to have enough segmentation to ensure that the ionization of a single track is separately recorded. While such a procedure can result in a few percent improvement in resolution, the effective energy response of the detector becomes non-linear above a few GeV, thus limiting the usefulness of the technique.

Another approach to this problem is being used in the shower counters for the TPC detector at PEP.[9] In this device a multiwire chamber with cathode readout is run in the limited-Geiger mode. Propagation of the discharge along the wire is prevented by stretching nylon monofilament at right angles to the multiplication wires. In this way, an effective cell size of .5 x 1 cm is obtained. Discharges due to the passage of individual particles can then be counted without active amplification. Clear separation of as many as 10 tracks is seen, resulting in a truly digital shower counter, with much reduced sensitivity to track ionization fluctuations. A resolution of 11.5%/$\sqrt{E}$ up to 4 GeV was obtained in a test module with 0.2 r.l. Pb plates. Limitations of segmentation cause this counter to also have non-linear response at high energy. A further problem is that the counter response varies at the cosine of the incident photon energy, since greater effective plate thickness is not compensated by larger pulse height per track as in conventional devices.

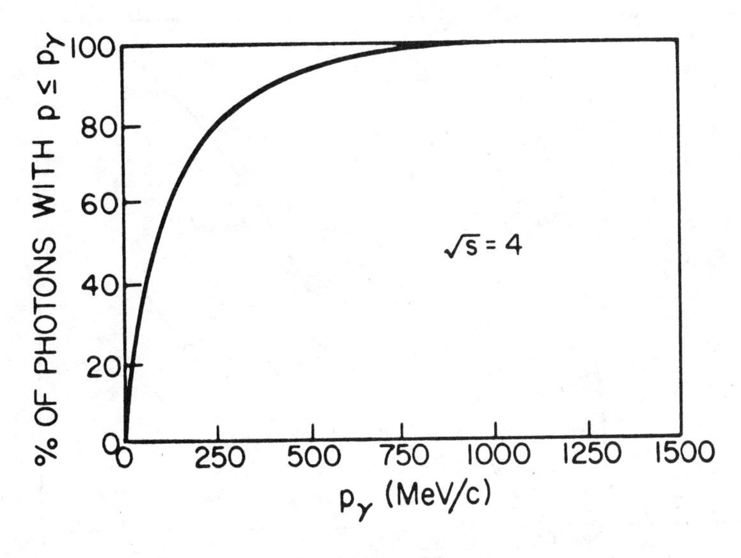
% OF PHOTONS WITH p ≤ pγ
100
80
60
40
20
0
0
250
500
750
1000
1250
1500
pγ (MeV/c)
√s = 4

Fig. 11

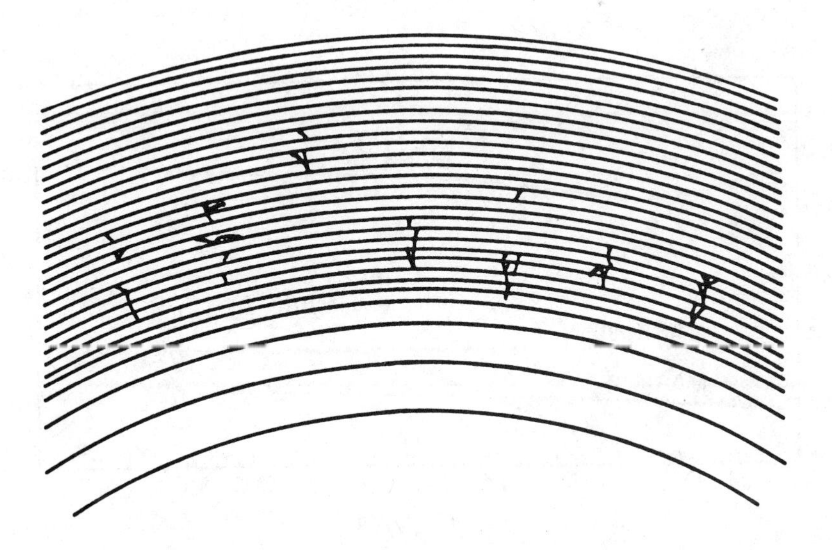
Eγ = 100.0 MeV

Fig. 12

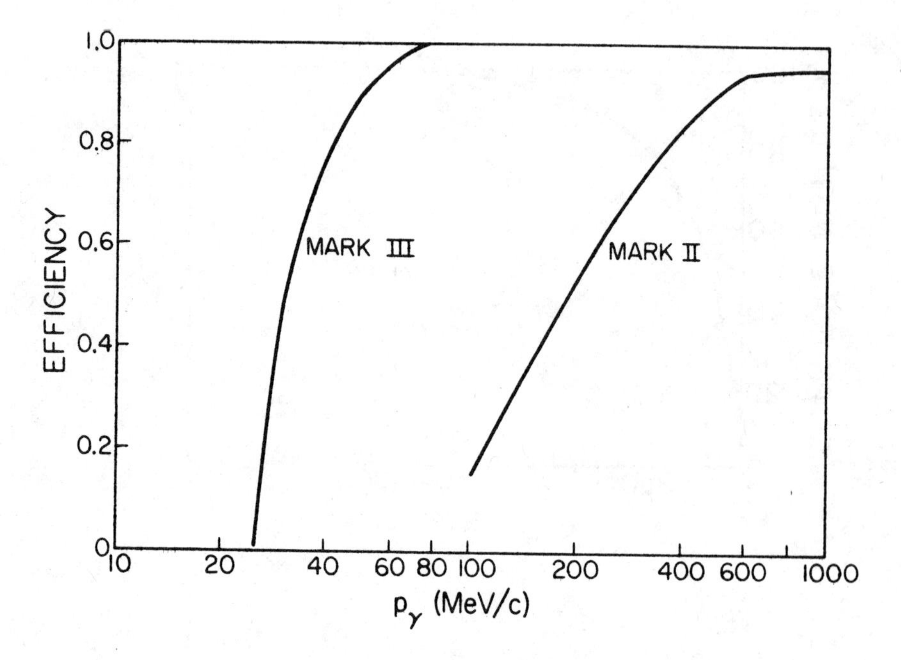
EFFICIENCY
1.0
0.8
0.6
0.4
0.2
0
10
20
40
60
80
100
200
400
600
1000
p_γ (MeV/c)
MARK III
MARK II

Fig. 13

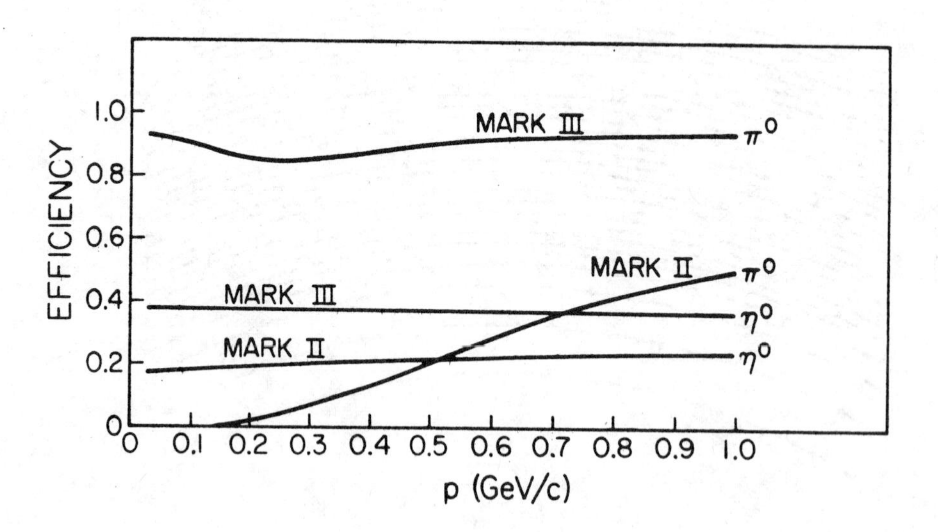
EFFICIENCY
1.0
0.8
0.6
0.4
0.2
0
0.1
0.2
0.3
0.4
0.5
0.6
0.7
0.8
0.9
1.0
p (GeV/c)
MARK III
π^0
MARK II
π^0
MARK III
η^0
MARK II
η^0

Fig. 14

PARTICLE IDENTIFICATION

Identification of hadron species and separation of leptons from hadrons are crucial requirements for many studies in e^+e^- annihilation. No single technique suffices to achieve all these aims. Rather, a combination of time-of-flight measurement, dE/dx, shower development, penetration and Cerenkov counters is necessary to obtain complete identification.

Time-of-flight measurement is perhaps the most straightforward approach to hadron identification. Time resolution of better than 300 psec has been realized for the 48 scintillators of the Mark II detector. With a flight path of something more than 1.5 meters, this resolution provides π-K separation at the 2σ level to more than 1 GeV/c (see Figure 15). The advent of practical large size gas avalanche (Pestov) counters, now under development,[10] promises time resolution in the 50-80 psec range, which could extend 2σ π/K by nearly a factor of two.

Particle identification by dE/dx has been incorporated in the JADE, CLEO, TPC, ARGUS and Mark III detectors. The Mark III, concerned with low momentum π/K separation, employs twelve samples in 1 atmosphere Argon/Propane. This relatively simple system is expected to achieve 2σ π/K separation to 700 MeV/c over 95% of 4π, as shown in Figure 16. Note also that useful π/e discrimination is also obtained over a substantial portion of important momenta.

Identification of higher momentum hadrons by dE/dx requires sensitivity to the smaller differences in ionization in the region of the relativistic rise. This makes it necessary to employ high pressures (4 atmospheres for JADE, 10 atmospheres for the TPC) and many samples (48 for JADE, 192 for the TPC). Achieving the design resolution demands careful calibration, gases of high purity and accurately known composition and great attention to systematic changes in pulse height connected with large drift distances. The JADE detector has measured a resolution of 13% to this point, whereas the design resolution is 9%.[11] It is expected that further work will allow them to more closely approximate their design goal. An example of preliminary results obtained in the dE/dx system of CLEO is shown in Figure 17. This system, when fully implemented, will consist of eight modules placed outside the solenoid coil. Each module contains $>10^4$ wires, run at 2 atmospheres. A particle produces 117 samples along its path.[12]

The use of Cerenkov counters for particle identification presents difficulties, because no single material provides thresholds which allow unique separation of hadron species over a large momentum range. The shielding of photomultiplier tubes from magnetic fields in a general purpose detector is also a difficult problem. Nonetheless, Cerenkov counters have been incorporated in the TASSO, CLEO and DELCO detectors.

The DELCO approach has been to build a very open magnet frame, and using a small central tracking package, surround the entire azimuth with a multicell Cerenkov counter. This approach has been used successfully at SPEAR; the detector has now been modified to improve segmentation and the optics and has been installed at PEP. The main task of this device is to distinguish pions from electrons. Figure 18 shows the π/e rejection obtained as a function of momentum.

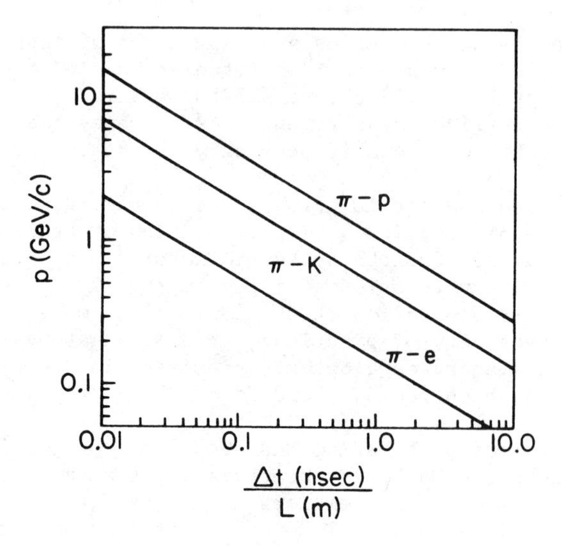
p (GeV/c)
10
1
0.1
0.01
0.1
1.0
10.0
Δt (nsec) / L (m)
π − p
π − K
π − e

Fig. 15

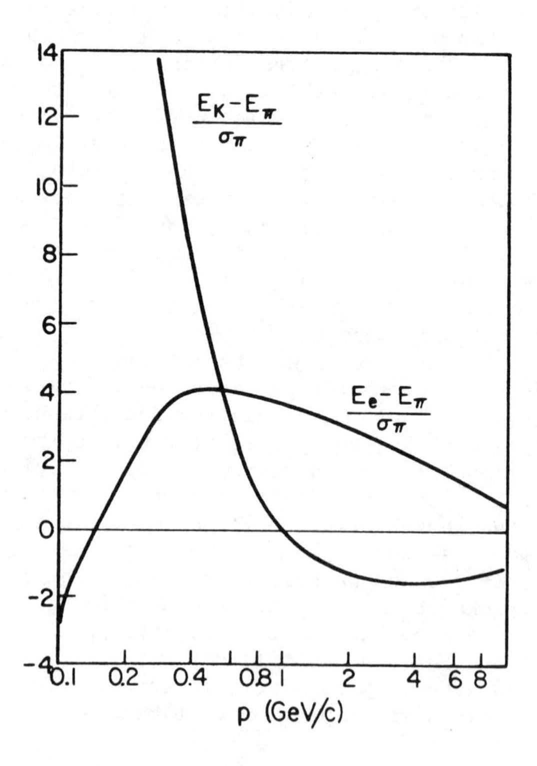
14
12
10
8
6
4
2
0
-2
-4
0.1
0.2
0.4
0.8
1
2
4
6
8
p (GeV/c)
$\frac{E_K - E_\pi}{\sigma_\pi}$
$\frac{E_e - E_\pi}{\sigma_\pi}$

Fig. 16

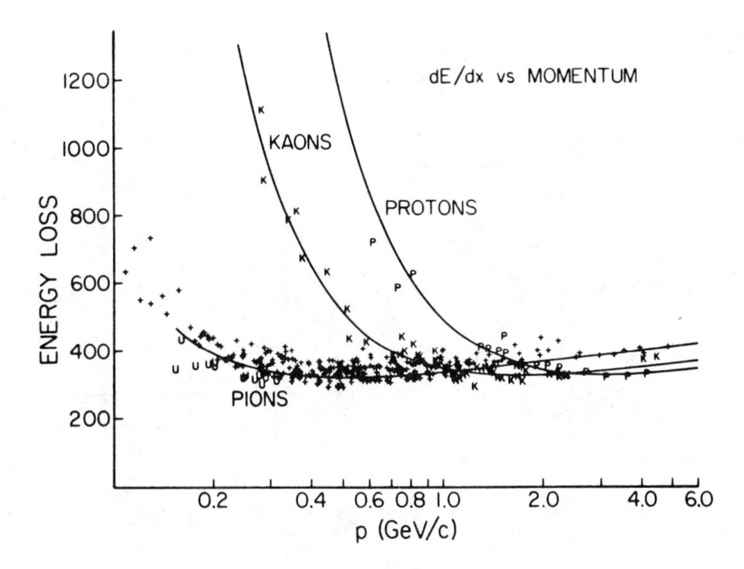
dE/dx vs MOMENTUM
ENERGY LOSS
1200
1000
800
600
400
200
KAONS
PROTONS
PIONS
0.2
0.4
0.6
0.8
1.0
2.0
4.0
6.0
p (GeV/c)

Fig. 17

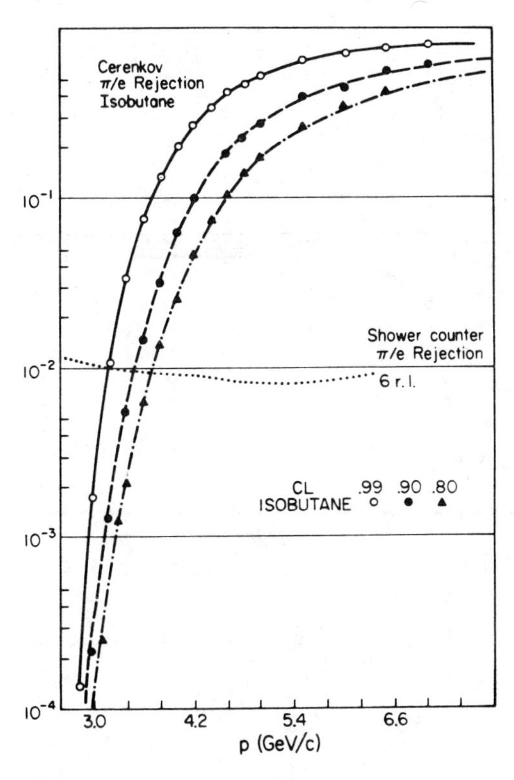
Cerenkov
π/e Rejection
Isobutane
Shower counter
π/e Rejection
6 r.l.
CL .99 .90 .80
ISOBUTANE
10⁻¹
10⁻²
10⁻³
10⁻⁴
3.0
4.2
5.4
6.6
p (GeV/c)

Fig. 18

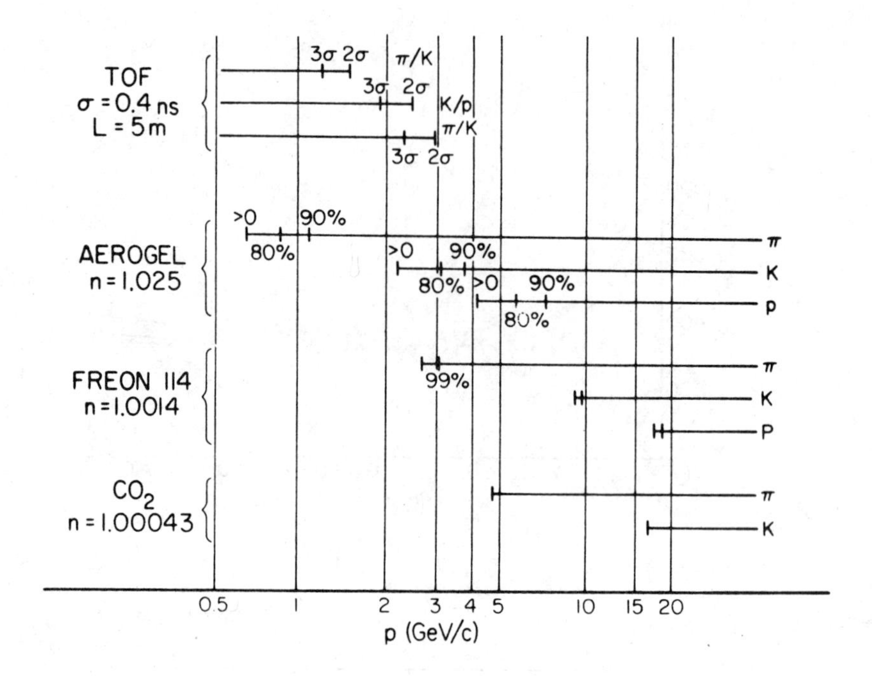
TOF
σ = 0.4 ns
L = 5 m
3σ 2σ
π/K
3σ 2σ
K/p
π/K
3σ 2σ
AEROGEL
n = 1.025
>0
90%
80%
π
>0
90%
K
80%
>0
90%
p
80%
FREON 114
n = 1.0014
99%
π
K
P
CO2
n = 1.00043
π
K
0.5
1
2
3
4
5
10
15
20
p (GeV/c)

Fig. 19

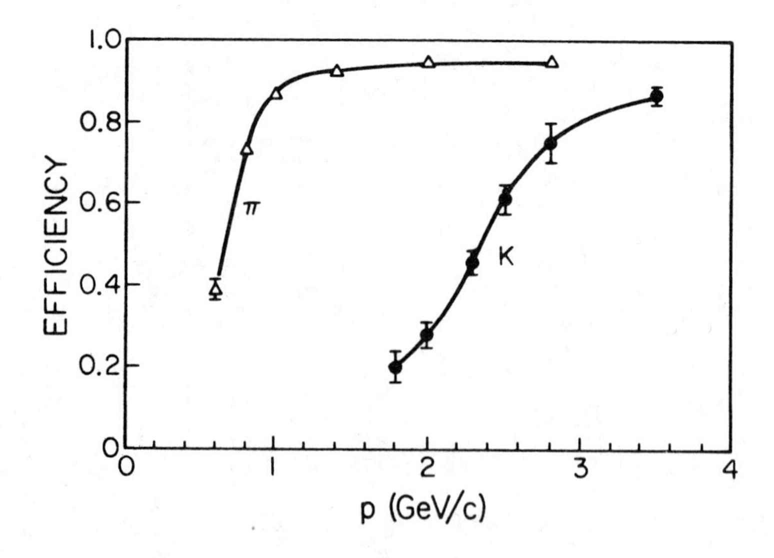
EFFICIENCY
1.0
0.8
0.6
0.4
0.2
0
π
K
0
1
2
3
4
p (GeV/c)

Fig. 20

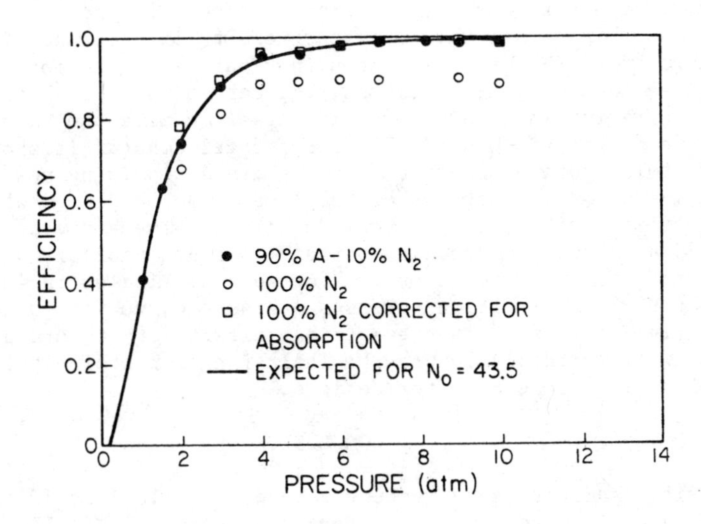
1.0
0.8
0.6
0.4
0.2
0
EFFICIENCY
0
2
4
6
8
10
12
14
PRESSURE (atm)
90% A-10% N2
100% N2
100% N2 CORRECTED FOR ABSORPTION
EXPECTED FOR N0 = 43.5

Fig. 21

TASSO has attempted to obtain good hadron identification in the PETRA regime by a combination of TOF and Cerenkov techniques. The Cerenkov counters consist of aerogel, Freon 114 and CO_2 types in two arms. Combining all these devices results in coverage of the entire spectrum with few gaps, as shown in Figure 19. The most innovative of these counters, is of course, the aerogel device which provides π/K separation in an otherwise inaccessible momentum range. The efficiency for π's and K's for the aerogel counter is shown in Figure 20.

Photoionization Cerenkov counters are a promising new development which, since no phototubes are employed, can be placed within the magnetic fields of general purpose detectors. These devices employ a proportional chamber containing benzene as a detector of UV photons. Such a counter is under development for possible inclusion in the HRS detector at PEP.[13] An efficiency versus pressure curve for 10 GeV/c pions is shown in Figure 21. The geometrical flexibility afforded by this technique will certainly insure the inclusion of photoionization Cerenkov counters in future detector designs.

CONCLUSION

Drift chambers, shower counters and particle identification technique have been discussed in the context of current e^+e^- colliding beam detectors. Some recent results and new ideas have been presented.

ACKNOWLEDGEMENTS

I would like to thank J. Jaros, W. C. Carithers, D. Meyer, S. L. Wu, J. Kirkby, D. Kreinick, H. Lynch, W. B. Atwood, and J. Heintze for providing information on the design and performance of their detector systems.

CONFERENCE PROGRAM

Thursday, May 1

Greeting by Provost W. Holladay

Theory (I) - Chairman: R. Mickens (Fisk)

QED Tests and Neutral Current Structure - J. J. Sakurai (UCLA)

Experiments (PETRA) - Chairman: B. Wiik (DESY)

CELLO - Design and Performance - V. Schröder (DESY)

Recent Results from JADE - R. D. Heuer (DESY)

New Accelerator Plans (I) - Chairman: S. Csorna (Vanderbilt)

Future Accelerator Plans at DESY - B. Wiik (DESY)

Experiments (PETRA, continued) - Chairman: B. Wiik (DESY)

Results from MARK J - Min Chen (MIT)

Results on QCD, QED and $\gamma\gamma$ Reactions from PLUTO - G. Knies (DESY)*

Recent Results from the TASSO Experiments at PETRA - R. J. Barlow (Oxford)

Theory (II) - Chairman: E. L. Berger (Argonne National Laboratory)

Jets, Gluons, QCD - T. Walsh (DESY)

Friday, May 2

Theory (III) - Chairman: E. L. Berger (ANL)

Low Energy Signals for Hypercolor - E. Eichten (Harvard)

Experiments (SPEAR) - Chairman: K. Berkelman (Cornell)

Selected Results from MARK II at SPEAR - D. Scharre (SLAC)

Physics Near Charm Threshold via the Crystal Ball - D. Coyne (Princeton)

Theory (IV) - Chairman: E. L. Berger (ANL)

Decays of D and B Mesons - M. Chanowitz (LBL)

New Accelerator Plans (II) - Chairman: D. Coyne (Princeton)

SLAC Single Pass Collider - R. Steining (SLAC)*

Experiments (DORIS)

Results from DASP II Experiment - H. Schröder (DESY)

Results from DHII Experiment - C. Rippich (Carnegie-Mellon)

New Accelerator Plans (III)

Plans for LEP - E. Keil (CERN)

Saturday, May 3

Instrumentation - Chairman: P. Stein (Cornell)

Overview of Highlights of e^+e^- Instrumentation - D. Hitlin (Cal Tech)

Experiments (CESR) - Chairman: P. Stein (Cornell)

Results from Columbia/Stony Brook Experiment - R. Schamberger (Stony Brook)

Results from CLEO Experiment - D. Kreinick (Cornell)

CONFERENCE PROGRAM (Cont'd.)

Theory (V) - Chairman: J. Rosner (Minnesota)
Photon-Photon Interactions - T. DeGrand (U.C.-Santa Barbara)
Quarkonia - K. Gottfried (Cornell)

*Paper not prepared for publication in these proceedings.

ADVISORS

E. L. Berger (Argonne)
K. Berkelman (Cornell)
J. D. Bjorken (SLAC/Fermilab)
D. Coyne (Princeton)
E. Lohrmann (DESY)
A. Odian (SLAC)

Participants

K. Abe, University of Pennsylvania
M. S. Alam, Vanderbilt University
A. Barbaro-Galtieri, LBL
R. Barlow, Oxford University
V. Barnes, Purdue University
B. A. Barnett, Johns Hopkins University
D. Bechis, Rutgers University
E. L. Berger, ANL
K. Berkelman, Cornell University
B. Brabson, Indiana University
W. Carithers, LBL
R. Cassell, University of Illinois
M. Cavalli-Sforza, Princeton University
M. Chanowitz, LBL
M. Chen, MIT
D. Coyne, Princeton University
H. Crater, University of Tennessee Space Institute
S. E. Csorna, Vanderbilt University
H. Cui, University of Illinois
C. W. Darden, University of South Carolina
T. DeGrand, University of California at Santa Barbara
E. Eichten, Harvard University
T. Ferguson, Cornell University
J. Ficenec, Virginia Polytechnic Institute
G. Finocchiaro, SUNY at Stony Brook
K. Foley, BNL
M. Franklin, SLAC
D. Fryberger, SLAC
M. Gettner, Northeastern University
B. Gittleman, Cornell University
A. T. Goshaw, Duke University
K. Gottfried, Cornell University
D. E. Groom, University of Utah
R. Hamilton, LBL
R. D. Heuer, University of Heidelberg
D. Hitlin, Caltech
D. Hood, David Lipscomb College
R. L. Imlay, Louisiana State University
K. Jaeger, ANL
J. A. Kadyk, LBL
G. Kalbfleisch, University of Oklahoma
J. Kandaswamy, Syracuse University
E. Keil, CERN
G. Knies, DESY
D. Kreinick, Cornell University
J. Koerner, DESY
N. Kwak, University of Kansas
K. E. Lasilla, Iowa State University
E. Loh, University of Utah

J. Ludwig, Caltech
A. Marini, Frascati
S. Meshkov, National Bureau of Standards
R. Mickens, Fisk University
R. Mozley, SLAC
J. Nieves, University of Pennsylvania
H. Ogren, Indiana University
S. Ozaki, BNL
R. S. Panvini, Vanderbilt University
R. Poling, Rochester University
F. Porter, Caltech
J. S. Poucher, Vanderbilt University
C. Rippich, Carnegie-Mellon University
J. Rosner, University of Minnesota
H. Sadrusinsky, University of California at Santa Cruz
J. J. Sakurai, UCLA
T. Schalk, University of California at Santa Cruz
R. Schamberger, SUNY at Stony Brook
D. Scharre, SLAC
H. Schneider, University of Karlsruhe
H. Schroder, DESY
V. Schroder, DESY
A. Seiden, University of California at Santa Cruz
E. Shibata, Purdue University
D. Sivers, ANL
P. Stein, Cornell University
R. Steining, SLAC
L. Trasatti, Frascati
P. Van Alstine, Vanderbilt University
H. Vogel, Max Planck Institute
K. Wacker, Harvard University
T. Walsh, DESY
B. H. Wiik DESY
E. Williams, Vanderbilt University
W. Wisniewski, University of Illinois
S. L. Wu, University of Wisconsin

AIP Conference Proceedings

		L.C. Number	ISBN
No.1	Feedback and Dynamic Control of Plasmas	70-141596	0-88318-100-2
No.2	Particles and Fields - 1971 (Rochester)	71-184662	0-88318-101-0
No.3	Thermal Expansion - 1971 (Corning)	72-76970	0-88318-102-9
No.4	Superconductivity in d-and f-Band Metals (Rochester, 1971)	74-18879	0-88318-103-7
No.5	Magnetism and Magnetic Materials - 1971 (2 parts) (Chicago)	59-2468	0-88318-104-5
No.6	Particle Physics (Irvine, 1971)	72-81239	0-88318-105-3
No.7	Exploring the History of Nuclear Physics	72-81883	0-88318-106-1
No.8	Experimental Meson Spectroscopy - 1972	72-88226	0-88318-107-X
No.9	Cyclotrons - 1972 (Vancouver)	72-92798	0-88318-108-8
No.10	Magnetism and Magnetic Materials - 1972	72-623469	0-88318-109-6
No.11	Transport Phenomena - 1973 (Brown University Conference)	73-80682	0-88318-110-X
No.12	Experiments on High Energy Particle Collisions - 1973 (Vanderbilt Conference)	73-81705	0-88318-111-8
No.13	π-π Scattering - 1973 (Tallahassee Conference)	73-81704	0-88318-112-6
No.14	Particles and Fields - 1973 (APS/DPF Berkeley)	73-91923	0-88318-113-4
No.15	High Energy Collisions - 1973 (Stony Brook)	73-92324	0-88318-114-2
No.16	Causality and Physical Theories (Wayne State University, 1973)	73-93420	0-88318-115-0
No.17	Thermal Expansion - 1973 (lake of the Ozarks)	73-94415	0-88318-116-9
No.18	Magnetism and Magnetic Materials - 1973 (2 parts) (Boston)	59-2468	0-88318-117-7
No.19	Physics and the Energy Problem - 1974 (APS Chicago)	73-94416	0-88318-118-5
No.20	Tetrahedrally Bonded Amorphous Semiconductors (Yorktown Heights, 1974)	74-80145	0-88318-119-3
No.21	Experimental Meson Spectroscopy - 1974 (Boston)	74-82628	0-88318-120-7
No.22	Neutrinos - 1974 (Philadelphia)	74-82413	0-88318-121-5
No.23	Particles and Fields - 1974 (APS/DPF Williamsburg)	74-27575	0-88318-122-3

No. 24	Magnetism and Magnetic Materials - 1974 (20th Annual Conference, San Francisco)	75-2647	0-88318-123-1
No. 25	Efficient Use of Energy (The APS Studies on the Technical Aspects of the More Efficient Use of Energy)	75-18227	0-88318-124-X
No. 26	High-Energy Physics and Nuclear Structure - 1975 (Santa Fe and Los Alamos)	75-26411	0-88318-125-8
No. 27	Topics in Statistical Mechanics and Biophysics: A Memorial to Julius L. Jackson (Wayne State University, 1975)	75-36309	0-88318-126-6
No. 28	Physics and Our World: A Symposium in Honor of Victor F. Weisskopf (M.I.T., 1974)	76-7207	0-88318-127-4
No. 29	Magnetism and Magnetic Materials - 1975 (21st Annual Conference, Philadelphia)	76-10931	0-88318-128-2
No. 30	Particle Searches and Discoveries - 1976 (Vanderbilt Conference)	76-19949	0-88318-129-0
No. 31	Structure and Excitations of Amorphous Solids (Williamsburg, VA., 1976)	76-22279	0-88318-130-4
No. 32	Materials Technology - 1975 (APS New York Meeting)	76-27967	0-88318-131-2
No. 33	Meson-Nuclear Physics - 1976 (Carnegie-Mellon Conference)	76-26811	0-88318-132-0
No. 34	Magnetism and Magnetic Materials - 1976 (Joint MMM-Intermag Conference, Pittsburgh)	76-47106	0-88318-133-9
No. 35	High Energy Physics with Polarized Beams and Targets (Argonne, 1976)	76-50181	0-88318-134-7
No. 36	Momentum Wave Functions - 1976 (Indiana University)	77-82145	0-88318-135-5
No. 37	Weak Interaction Physics - 1977 (Indiana University)	77-83344	0-88318-136-3
No. 38	Workshop on New Directions in Mossbauer Spectroscopy (Argonne, 1977)	77-90635	0-88318-137-1
No. 39	Physics Careers, Employment and Education (Penn State, 1977)	77-94053	0-88318-138-X
No. 40	Electrical Transport and Optical Properties of Inhomogeneous Media (Ohio State University, 1977)	78-54319	0-88318-139-8
No. 41	Nucleon-Nucleon Interactions - 1977 (Vancouver)	78-54249	0-88318-140-1
No. 42	Higher Energy Polarized Proton Beams (Ann Arbor, 1977)	78-55682	0-88318-141-X
No. 43	Particles and Fields - 1977 (APS/DPF, Argonne)	78-55683	0-88318-142-8
No. 44	Future Trends in Superconductive Electronics (Charlottesville, 1978)	77-9240	0-88318-143-6

No.45	New Results in High Energy Physics - 1978 (Vanderbilt Conference)	78-67196	0-88318-144-4
No.46	Topics in Nonlinear Dynamics (La Jolla Institute)	78-057870	0-88318-145-2
No.47	Clustering Aspects of Nuclear Structure and Nuclear Reactions (Winnepeg, 1978)	78-64942	0-88318-146-0
No.48	Current Trends in the Theory of Fields (Tallahassee, 1978)	78-72948	0-88318-147-9
No.49	Cosmic Rays and Particle Physics - 1978 (Bartol Conference)	79-50489	0-88318-148-7
No.50	Laser-Solid Interactions and Laser Processing - 1978 (Boston)	79-51564	0-88318-149-5
No.51	High Energy Physics with Polarized Beams and Polarized Targets (Argonne, 1978)	79-64565	0-88318-150-9
No.52	Long-Distance Neutrino Detection - 1978 (C.L. Cowan Memorial Symposium)	79-52078	0-88318-151-7
No.53	Modulated Structures - 1979 (Kailua Kona, Hawaii)	79-53846	0-88318-152-5
No.54	Meson-Nuclear Physics - 1979 (Houston)	79-53978	0-88318-153-3
No.55	Quantum Chromodynamics (La Jolla, 1978)	79-54969	0-88318-154-1
No.56	Particle Acceleration Mechanisms in Astrophysics (La Jolla, 1979)	79-55844	0-88318-155-X
No. 57	Nonlinear Dynamics and the Beam-Beam Interaction (Brookhaven, 1979)	79-57341	0-88318-156-8
No. 58	Inhomogeneous Superconductors - 1979 (Berkeley Springs, W.V.)	79-57620	0-88318-157-6
No. 59	Particles and Fields - 1979 (APS/DPF Montreal)	80-66631	0-88318-158-4
No. 60	History of the ZGS (Argonne, 1979)	80-67694	0-88318-159-2
No. 61	Aspects of the Kinetics and Dynamics of Surface Reactions (La Jolla Institute, 1979)	80-68004	0-88318-160-6
No. 62	High Energy e^+e^- Interactions (Vanderbilt , 1980)	80-53377	0-88318-161-4